AF389521

Advances in Primary Progressive Aphasia

Advances in Primary Progressive Aphasia

Editors

Jordi A. Matias-Guiu
Robert Jr Laforce
Rene L. Utianski

MDPI • Basel • Beijing • Wuhan • Barcelona • Belgrade • Manchester • Tokyo • Cluj • Tianjin

Editors

Jordi A. Matias-Guiu
Department of Neurology
Universidad Complutense
de Madrid
Madrid
Spain

Robert Jr Laforce
Clinique Interdisciplinaire
de Mémoire (CIME) du CHU
de Québec
Département des Sciences
Neurologiques
Québec
Canada

Rene L. Utianski
Department of Neurology
Mayo Clinic
Rochester
United States

Editorial Office
MDPI
St. Alban-Anlage 66
4052 Basel, Switzerland

This is a reprint of articles from the Special Issue published online in the open access journal *Brain Sciences* (ISSN 2076-3425) (available at: www.mdpi.com/journal/brainsci/special_issues/ Primary_Progressive_Aphasia).

For citation purposes, cite each article independently as indicated on the article page online and as indicated below:

LastName, A.A.; LastName, B.B.; LastName, C.C. Article Title. *Journal Name* **Year**, *Volume Number, Page Range.*

ISBN 978-3-0365-4408-3 (Hbk)
ISBN 978-3-0365-4407-6 (PDF)

Contents

MDPI

Editorial

Advances in Primary Progressive Aphasia

Jordi A. Matias-Guiu [1,*], Robert Laforce, Jr. [2], Monica Lavoie [2] and Rene L. Utianski [3]

[1] Department of Neurology, Hospital Clínico San Carlos, Health Research Institute "San Carlos" (IdISCC), Universidad Complutense de Madrid, 28040 Madrid, Spain
[2] Clinique Interdisciplinaire de Mémoire, Département des Sciences Neurologiques du CHU de Québec, Faculté de Médecine, Université Laval, Quebec City, QC G1V 0A6, Canada; robert.laforce@fmed.ulaval.ca (R.L.J.); monica.lavoie.1@ulaval.ca (M.L.)
[3] Department of Neurology, Mayo Clinic, Rochester, MN 55902, USA; utianski.rene@mayo.edu
* Correspondence: jordi.matias-guiu@salud.madrid.org

Citation: Matias-Guiu, J.A.; Laforce, R., Jr.; Lavoie, M.; Utianski, R.L. Advances in Primary Progressive Aphasia. *Brain Sci.* **2022**, *12*, 636. https://doi.org/10.3390/brainsci12050636

Received: 30 April 2022
Accepted: 9 May 2022
Published: 12 May 2022

Publisher's Note: MDPI stays neutral with regard to jurisdictional claims in published maps and institutional affiliations.

Primary progressive aphasia (PPA) is a neurodegenerative syndrome characterized by progressive and predominant language impairment. Our knowledge of this disorder has evolved significantly in recent years. Notably, correlations between clinical findings and pathology have improved, and main clinical, neuroimaging, and genetic features have been described. However, as in other neurodegenerative syndromes, diagnosis is often challenging, a better understanding of natural history is needed, and successful therapies are lacking.

In this context, this Special Issue of *Brain Sciences* is focused on "Advances in Primary Progressive Aphasia (PPA)" and includes articles addressing advances in the diagnosis, expected progression, and treatment of PPA, each of which is elucidated further in what follows.

1. Diagnosis

In "Contribution of the Cognitive Approach to Language Assessment to the Differential Diagnosis of Primary Progressive Aphasia" [1], the authors reviewed the contribution of the assessment of specific language abilities in the differential diagnosis of PPA, the main cognitive processes involved in each task, and the findings supportive of each variant.

In "Breakdowns in Informativeness of Naturalistic Speech Production in Primary Progressive Aphasia" [2], the authors examined 101 participants, including 70 patients with a diagnosis of PPA (19 svPPA patients, 26 lvPPA patients, and 25 nfvPPA) and 31 age-matched controls, using the "Picnic Scene" from the Western Aphasia Battery-Revised. The informativeness of speech was quantified. Relative informativeness, or efficiency, of speech was preserved in non-fluent variant PPA patients and impaired in logopenic and semantic variants. These findings support the value of assessing and quantifying functional communication in PPA, which could be useful in the diagnosis and complement other parameters of spontaneous speech analysis.

In "Verbal Short-Term Memory Disturbance in the Primary Progressive Aphasias: Challenges and Distinctions in a Clinical Setting" [3], the authors examined short-term memory profiles in four well-characterized patients with lvPPA, nfvPPA, svPPA, and Alzheimer's disease. The authors also discussed the adequate tasks to evaluate short-term memory, the influence of other cognitive functions, and the relevance of examining visuospatial short-term and working memory.

In "Primary Progressive Aphasia: Use of Graphical Markers for an Early and Differential Diagnosis" [4], the authors analyzed writing pressures during linguistic and non-linguistic tasks in patients with PPA, Alzheimer's disease, and healthy controls. Several differences were found between groups depending on the type of task.

Two studies examined the application of electroencephalography (EEG) in the diagnosis of PPA and its variants. In "Application of Machine Learning to Electroencephalography for the Diagnosis of Primary Progressive Aphasia: A Pilot Study" [5], a cross-sectional study with 40 PPA patients and 20 controls was performed. Patients underwent resting-state EEG. Several data extraction procedures were performed, and several machine learning algorithms were evaluated. Diagnostic capacity was relatively high for the discrimination between PPA and controls, while the classification between variants was lower. The most important features in the classification were those derived from network analysis based on graph theory. In this regard, in the study "A Preliminary Report of Network Electroencephalographic Measures in Primary Progressive Apraxia of Speech and Aphasia" [6], the authors evaluated EEG changes in the agrammatic PPA and primary progressive apraxia of speech variants using graph theory analysis. Several network alterations were found, and interestingly, there were correlations between EEG graph theory measures and certain clinical measures. Overall, both studies suggest the potential for further application of EEG in PPA.

2. Longitudinal Change

In "Longitudinal Changes in Cognition, Behaviours, and Functional Abilities in the Three Main Variants of Primary Progressive Aphasia: A Literature Review" [7], the authors provided a literature review on the natural history of the three main variants of PPA. The review focused on cognitive, behavioral, and functional changes associated with the syndromes. Findings from 15 studies included in the review showed that the svPPA was associated with more behavioral disturbances both at baseline and over the course of the disease, whereas lvPPA experienced worse cognitive decline and faster progression to dementia. The most significant decline in language production and functional abilities was found in individuals with nfvPPA. This review highlighted the need for more data on lvPPA, surprisingly, given it is the most frequent subtype of PPA. The authors also reported a lack of data regarding the prodromal and last stages of PPA, which could be very helpful for patients and families.

In "Survival in the Three Common Variants of Primary Progressive Aphasia: A Retrospective Study in a Tertiary Memory Clinic" [8], the authors analyzed survival data in a cohort of 83 deceased patients with a diagnosis of PPA. They reported a significantly longer survival from symptom onset and diagnosis in svPPA than in the two other variants. Indeed, the mean survival from symptom onset was 7.6 years for lvPPA, 7.1 years for nfvPPA, and 12 years for svPPA. The most common causes of death were natural cardio-pulmonary arrest and pneumonia. These findings provide invaluable data to healthcare professionals, as well as patients and families, about the progression of the disease and the end stages of life.

3. Treatment

In "Treatment for Anomia in Bilingual Speakers with Progressive Aphasia" [9], the authors explored the impact of the lexical retrieval cascade treatment approach on a group of bilingual patients with heterogeneous clinical presentations, including right-sided temporal frontotemporal dementia and semantic and logopenic PPA. Overall, participants demonstrated a significant treatment effect in each of the targeted languages and showed a cross-linguistic transfer for trained cognates in both languages that were maintained up to one-year post-treatment. While there was a decline in clinical measures of language and cognition, patient and care partner reported outcomes indicated communication was "somewhat better." The findings of the study support the important conclusions that (1) monolingual clinicians may be able to select cross-linguistic cognates as a means to support gains across languages, even for words trained in a single language, and (2) the importance of including patient-reported outcome measures in intervention studies.

In "Cognitive Intervention Strategies Directed to Speech and Language Deficits in Primary Progressive Aphasia: Practice-Based Evidence from 18 Cases" [10], the authors demonstrated the importance of symptom-targeted intervention in a group of patients with PPA, again with heterogeneous clinical presentations. While there was no control cohort or within-group comparison to determine the impact of alternative treatment approaches, the patients showed improved performance in trained items at post-test, with an individualized focus on either naming, sentence production, motor speech functioning, or phonological functioning. While 18 patients completed their personalized treatment, this was 56% of those recruited; it is also important to note that 25% of patients recruited did not complete the intervention because of frustration, anxiety, motivation, or other practical–logistical barriers. Many of the participants who completed the program had an objective decline in function at follow-up but reported subjective improvement that was not otherwise quantified, again highlighting the importance of patient-reported outcomes. Overall, the study supports the potential for symptom-targeted intervention and suggests that completing this in the early stages of the disease may improve adherence and the subsequent possibility of positive treatment outcomes.

In "Semantic Variant Primary Progressive Aphasia: Practical Recommendations for Treatment from 20 Years of Behavioural Research" [11], the authors highlighted the different sources of word-finding difficulties in PPA, including impairment in semantic knowledge in semantic, lexical access, and phonological impairment in logopenic, and post-lexical execution in non-fluent/agrammatic. With the focus on semantic PPA, they discussed the important implications of left and right atrophy, where patients with right-sided temporal atrophy may have greater behavioral changes, loss of insight, and altered pragmatics. It is crucial to recognize that these non-aphasic cognitive-communication challenges are well within the scope of speech pathology to address in intervention. In this review, the authors discussed the outcomes of different naming treatments and the benefit of capitalizing upon preserved long-term memory systems. They noted that in maintenance or compensatory approaches, the severity of impairment should inform the nature of the intervention. For instance, patients may not be able to learn how to augmentative or alternative communication devices later in the disease, so it is important to incorporate this before the skills to acquire its use are lost. They also mentioned the benefit of interdisciplinary treatment, with collaboration between speech and occupation therapy, and focusing on activities of daily living. Finally, they discussed the benefit of education and support group programs as a safe forum for discussing experiences and sharing resources and strategies, further highlighting the needs of both patient and care partners should both be addressed, simultaneously.

4. Conclusions

Taken together, the papers in this Special Issue, addressing the diagnosis, treatment, and expected progression of PPA, contribute to the literature and our understanding of this heterogeneous patient population. We strongly believe that speech–language pathologists, neurologists, and neuroscientists alike will benefit from the original research studies and reviews amassed in this collection, and as the Guest Editors of this Special Issue, we thank the authors for their contributions.

Author Contributions: Writing—review and editing, J.A.M.-G., R.L.J., M.L., R.L.U.; visualization, J.A.M.-G., R.L.J., M.L., R.L.U.; supervision, J.A.M.-G., R.L.J., M.L., R.L.U. All authors have contributed equally to writing the manuscript. All authors have read and agreed to the published version of the manuscript.

Funding: This research received no external funding.

Institutional Review Board Statement: Not applicable.

Informed Consent Statement: Not applicable.

Data Availability Statement: Not applicable.

Conflicts of Interest: The authors declare no conflict of interest.

References

1. Macoir, J.; Legaré, A.; Lavoie, M. Contribution of the cognitive approach to language assessment to the differential diagnosis of primary progressive aphasia. *Brain Sci.* **2021**, *11*, 815. [CrossRef] [PubMed]
2. Gallé, J.; Cordella, C.; Fedorenko, E.; Hochberg, D.; Touroutogiou, A.; Quimby, M.; Dickerson, B.C. Breakdowns in informativeness of naturalistic speech production in primary progressive aphasia. *Brain Sci.* **2021**, *11*, 130. [CrossRef] [PubMed]
3. Foxe, D.; Cheung, S.C.; Cordato, N.J.; Burrell, J.R.; Ahmed, R.M.; Taylor-Rubin, C.; Irish, M.; Piguet, O. Verbal short-term memory disturbance in the primary progressive aphasias: Challenges and distinctions in a clinical setting. *Brain Sci.* **2021**, *11*, 1060. [CrossRef] [PubMed]
4. Plonka, A.; Mouton, A.; Macoir, J.; Train, T.M.; Derremaux, A.; Robert, P.; Manera, V.; Gros, A. Primary progressive aphasia: Use of graphical markers for an early and differential diagnosis. *Brain Sci.* **2021**, *11*, 1198. [CrossRef] [PubMed]
5. Moral-Rubio, C.; Balugo, P.; Fraile-Pereda, A.; Pytel, V.; Fernández-Romero, L.; Delgado-Alonso, C.; Delgado-Álvarez, A.; Matías-Guiu, J.; Matias-Guiu, J.A.; Ayala, J.L. Application of machine learning to electroencephalography for the diagnosis of primary progressive aphasia: A pilot study. *Brain Sci.* **2021**, *11*, 1262. [CrossRef] [PubMed]
6. Utianski, R.L.; Botha, H.; Caviness, J.N.; Worrell, G.A.; Duffy, J.R.; Clark, H.M.; Whitwell, J.L.; Josephs, K.A. A preliminary report of network electroencephalographic measures in primary progressive apraxia of speech and aphasia. *Brain Sci.* **2022**, *12*, 378. [CrossRef] [PubMed]
7. De la Sabionnière, J.; Tastevin, M.; Lavoie, M.; Laforce, R., Jr. Longitudinal changes in cognition, behaviours, and functional abilities in the three main variants of primary progressive aphasia: A literature review. *Brain Sci.* **2021**, *11*, 1209. [CrossRef] [PubMed]
8. Tastevin, M.; Lavoie, M.; de la Sabionnière, J.; Carrier-Auclair, J.; Laforce, R., Jr. Survival in the three common variants of primary progressive aphasia: A retrospective study in a tertiary memory clinic. *Brain Sci.* **2021**, *11*, 1113. [CrossRef] [PubMed]
9. Grasso, S.M.; Peña, E.D.; Kazemi, N.; Mirzapour, H.; Neupane, R.; Bonakdarpour, B.; Gorno-Tempini, M.L.; Henry, M.L. Treatment for anomia in bilingual speakers with progressive aphasia. *Brain Sci.* **2021**, *11*, 1371. [CrossRef] [PubMed]
10. Machado, T.H.; Carthery-Goulart, M.T.; Campanha, A.C.; Caramelli, P. Cognitive intervention strategies directed to speech and language deficits in primary progressive aphasia: Practice-based evidence from 18 cases. *Brain Sci.* **2021**, *11*, 1268. [CrossRef] [PubMed]
11. Suárez-González, A.; Savage, S.A.; Bier, N.; Henry, M.L.; Jokel, R.; Nickels, L.; Taylor-Rubin, C. Semantic variant primary progressive aphasia: Practical recommendations for treatment from 20 years of behavioral research. *Brain Sci.* **2021**, *11*, 1552. [CrossRef] [PubMed]

Article

A Preliminary Report of Network Electroencephalographic Measures in Primary Progressive Apraxia of Speech and Aphasia

Rene L. Utianski [1,*], Hugo Botha [1], John N. Caviness [2], Gregory A. Worrell [1], Joseph R. Duffy [1], Heather M. Clark [1], Jennifer L. Whitwell [3] and Keith A. Josephs [1]

[1] Department of Neurology, Mayo Clinic, Rochester, MN 55905, USA; botha.hugo@mayo.edu (H.B.); worrell.gregory@mayo.edu (G.A.W.); jduffy@mayo.edu (J.R.D.); clark.heather1@mayo.edu (H.M.C.); josephs.keith@mayo.edu (K.A.J.)
[2] Department of Neurology, Mayo Clinic, Scottsdale, AZ 85259, USA; jcaviness@mayo.edu
[3] Department of Radiology, Mayo Clinic, Rochester, MN 55905, USA; whitwell.jennifer@mayo.edu
* Correspondence: utianski.rene@mayo.edu

Abstract: The objective of this study was to characterize network-level changes in nonfluent/agrammatic Primary Progressive Aphasia (agPPA) and Primary Progressive Apraxia of Speech (PPAOS) with graph theory (GT) measures derived from scalp electroencephalography (EEG) recordings. EEGs of 15 agPPA and 7 PPAOS patients were collected during relaxed wakefulness with eyes closed (21 electrodes, 10–20 positions, 256 Hz sampling rate, 1–200 Hz bandpass filter). Eight artifact-free, non-overlapping 1024-point epochs were selected. Via Brainwave software, GT weighted connectivity and minimum spanning tree (MST) measures were calculated for theta and upper and lower alpha frequency bands. Differences in GT and MST measures between agPPA and PPAOS were assessed with Wilcoxon rank-sum tests. Of greatest interest, Spearman correlations were computed between behavioral and network measures in all frequency bands across all patients. There were no statistically significant differences in GT or MST measures between agPPA and PPAOS. There were significant correlations between several network and behavioral variables. The correlations demonstrate a relationship between reduced global efficiency and clinical symptom severity (e.g., parkinsonism, AOS). This preliminary, exploratory study demonstrates potential for EEG GT measures to quantify network changes associated with degenerative speech–language disorders.

Keywords: electroencephalography (EEG); network analysis; graph theory; primary progressive aphasia; progressive apraxia of speech

Citation: Utianski, R.L.; Botha, H.; Caviness, J.N.; Worrell, G.A.; Duffy, J.R.; Clark, H.M.; Whitwell, J.L.; Josephs, K.A. A Preliminary Report of Network Electroencephalographic Measures in Primary Progressive Apraxia of Speech and Aphasia. *Brain Sci.* **2022**, *12*, 378. https://doi.org/10.3390/brainsci12030378

Academic Editor: Yang Zhang

Received: 27 January 2022
Accepted: 10 March 2022
Published: 12 March 2022

Publisher's Note: MDPI stays neutral with regard to jurisdictional claims in published maps and institutional affiliations.

1. Introduction

1.1. EEG Graph Theory Measures

The use of electroencephalography (EEG) has expanded from identifying and characterizing seizure disorders to differentiating many different cerebral functions. Past research has demonstrated that clinical EEG is sensitive to dementia associated with Alzheimer's (AD) [1] and Parkinson's diseases (PD) [2], and nonfluent/agrammatic Primary Progressive Aphasia (agPPA) [3], but not Primary Progressive Apraxia of Speech (PPAOS; patients who present with isolated apraxia of speech (AOS)) [4]. However, clinical EEG studies describe overall brain health and do not quantify interactions among multiple brain areas, or network activity.

Graph theory is a branch of mathematics that is central to much of the modern "network neuroscience." It is premised on representing a system or network as a collection of nodes, with the interaction among them represented by edges. Node, edge, subgraph, and global metrics can then be calculated and compared between groups or to a behavioral measure. For example, degree centrality is a node-level metric calculated as the number of edges, or the total weight of edges, to a given node. Nodes can be grouped in modules, representing nodes that tend to connect to each other more than other nodes and potentially reflect specialized processing. Some high-degree nodes connect many modules and are referred to as hubs. At a global scale, most real-world networks balance integration, or a high level of connectivity between nodes, and segregation, reflecting distinct modules in a network. The extent to which this balance is optimized is captured in the small world-ness of the network. In EEG studies, the nodes are represented by the electrodes and the edges by a measure of coherence within a selected frequency band [5].

1.2. EEG Graph Theory in Neurodegenerative Disease

Studies have shown changes in EEG graph theory measures in dementia associated with PD [6], AD [7–10], and frontotemporal dementia (FTD) [11]. More specifically, EEGs of cognitively unimpaired patients with PD showed increased local integration across frequency bands when compared to cognitively unimpaired controls; those with dementia associated with PD had decreased integration in the lower alpha band relative to the cognitively unimpaired PD patients [6], suggesting the latter change was related to cognitive changes, not simply the presence of the disease. Analysis of brain networks of patients with AD-related dementia have shown decreased connectivity (or increased randomness), with loss of hubs compared to cognitively unimpaired controls [9,10].

Different types of network change have been shown in FTD. There were no differences in clustering coefficient or path length measures; however, the lower alpha band degree correlation increased in FTD relative to cognitively unimpaired controls, suggesting reduced segregation [11]. Overall, while AD patients showed less order, FTD patients showed a more ordered structure, possibly reflecting the differing underlying pathophysiology. However, in that study, the behavioral variant and semantic dementia were the only clinical phenotypes represented. Overall, it seems that patterns of network breakdown may be evident in neurodegenerative cognitive disorders and may be specific to the clinical syndromes and/ or causative pathology. To date, EEG graph theory measures have not been described in PPAOS and only one study has addressed this in agPPA [12], two other clinical syndromes associated with FTD pathology.

1.3. Primary Progressive Aphasia and Apraxia of Speech

Briefly, PPA encompasses a group of neurodegenerative syndromes characterized by progressive and predominant language impairment [13]. The agPPA subtype is characterized by grammatical errors in speech and writing and, not infrequently, accompanied by AOS, a motor speech disorder characterized by disruption in sensorimotor planning and/or programming [14]. When AOS, and not aphasia, is the initial manifestation of neurodegenerative disorders it is referred to as PPAOS [15,16]. In the context of PPAOS, some patients eventually develop aphasia that remains milder in severity than the AOS [17]. Research has suggested the initial or combination of speech (i.e., AOS) and language (i.e., aphasia) features may have implications for imaging findings, underlying pathology and the anticipated progression of the neurodegenerative disorder [18–21]. Given that more cortical imaging findings have been associated with the presence of aphasia, we opted to group those with aphasia, with or without AOS and regardless of predominance, into a single group referred to as agPPA. Many patients with PPAOS have normal MRIs, with FDG-PET considered the most sensitive imaging biomarker [22]. Unfortunately, FDG PET scans are not ubiquitously available and are sometimes cost-prohibited.

1.4. Present Study

The primary goal of this study was to provide foundational information on which to build our understanding of the network breakdowns in patients with progressive AOS and/or aphasia. Ultimately, this might inform our theoretical understanding of the neuropathophysiology underlying these clinical presentations, and clinically, inform a more widely available and cost-effective method to support differential diagnosis. Toward that end, we describe graph theory network measures and correlate them with indices of speech and language deficits to better understand their relationship.

2. Materials and Methods

2.1. Participants

The study was approved by Mayo Clinic's Institutional Review Board (#17-002468 on 19 July 2017); all patients were native English speakers and gave written consent according to the Declaration of Helsinki. Between October 2016 and December 2019, a total of 22 patients with agPPA (n = 15) or PPAOS (n = 7) completed a clinical EEG recording as part of a larger study conducted by the Neurodegenerative Research Group (NRG).

2.2. Clinical Measures

A comprehensive speech–language evaluation was conducted by an experienced speech–language pathologist (SLP). Clinical judgments regarding the presence, nature (i.e., type), and severity of AOS and aphasia were made by the examining clinician and subsequently confirmed by consensus agreement with at least one other non-examining SLP. The SLPs were experienced in differential diagnosis of neurodegenerative speech and language disorders.

Severity ratings reflected gestalt clinical judgment on a 5-point scale (0 = absent, 1 = mild, 2 = moderate, 3 = marked, 4 = severe). Other formal measures were administered and used to inform the overall judgments. The Western Aphasia Battery-Revised (WAB-R) Aphasia Quotient (WAB-AQ) [23], as a composite measure of global language ability, and the Northwestern Anagram Test (NAT) [24], a non-speech sentence-production task, were administered. A conversational speech sample, including narrative picture description, was collected as a part of the WAB-R. Additionally, supplementary speech and speech-like tasks (alternating and sequential motion rates) were elicited. The speech samples were used to reach consensus about the predominance of phonetic or prosodic speech characteristics by the same SLPs, as previously described [25]. The speech samples were also used to score the Apraxia of Speech Rating Scale—version 3 (ASRS-3) [25,26], an index of abnormal speech features and severity of AOS.

As part of the neurological evaluation, the Montreal Cognitive Assessment (MoCA) [27], a screening test of general cognition, was completed. The Movement Disorder Society-Sponsored Revision of the Unified Parkinson's Disease Rating Scale, Motor section (MDS-UPDRS III) [28], an index of motor functioning, was scored.

2.3. Electroencephalographic (EEG) Recording

Scalp EEG recordings were collected with XLTEK utilizing 21 electrodes placed with standard 10–20 positions, recording reference electrode of CPZ, a sampling rate of 256 Hz, 1 Hz low-frequency filter, and 70 Hz high-frequency filter, during relaxed wakefulness, wherein patients sat quietly with their eyes closed for 90% of the 45 to 55 min recording. A Natus EMU40EX Wireless LTM Amplifier (Natus Medical Incorporated, CA, USA) was utilized. A time base of 30 mm/sec with patient-individualized sensitivity was utilized for ongoing monitoring of artifacts. Clinical protocols for "awake" EEG were followed; no request was made for sleep deprivation. Recording intervals that included mental activation were not included for analysis.

2.4. EEG Processing

The continuous EEG data were divided into non-overlapping 1024-point (1023 ms) epochs, dictated by the sampling rate (256 Hz). Each epoch was visually inspected for artifacts, though rejection of artifacts was uncommon due to the vigilant monitoring of the online acquisition. For detecting blinking and other eye-movement artifacts, comparison was made to the vertical and horizontal eye movement channels. Epochs with muscle artifacts were rejected if such artifact signals were present grossly. No specific criteria were applied, but rather gestalt judgment. Consistent with prior research [6], 8 artifact-free epochs were chosen for analysis.

2.5. Graph Theory Analysis

Graph theory network analysis was performed with Brainwave software (http://home.kpn.nl/stam7883/brainwave.html, accessed on 27 January 2022). Briefly, functional connectivity was assessed with phase lag index (PLI), as research has shown it is less affected by volume conduction than other measures [29]. Complementary traditional graph theory weighted connectivity and minimum spanning tree (MST) measures [30,31] were selected. Selected measures are shown in Table 1. All graph theory and MST measures were calculated for the following frequency bands (Hz): theta (4–8), alpha1 (8–10), and alpha2 (10–13), selected given prior demonstration of slowing and alterations in these ranges [4].

Table 1. Definition of utilized network measures (adapted from Van Steen [5]).

Measure	Definition
PLI, Phase lag index	Measure of functional connectivity between nodes
Gamma, Normalized weighted clustering coefficient	Measure of connectivity between nodes or the extent to which neighboring nodes are also neighbors with one another, calculated per node and averaged over the entire network.
Lambda, Normalized characteristic path length	Measure of the average number of connections in the shortest path between two nodes of the network
$Kappa_W$, Weighted degree divergence	Measure of the broadness of the weighted degree distribution, where weighted degree is the summed weights of all edges connected to a node
Modularity	Measure of the degree to which nodes are more connected to each other than to nodes outside a given cluster (i.e., module)
MST BC_{max}, Maximum MST betweenness centrality	Maximum number of paths between any two MST nodes running through a single node
MST Diameter	Maximum number of connections (distance) between two MST nodes
MST Eccentricity	Average maximum distance between any two MST nodes
MST Leaf, MST leaf fraction	Measure of the number of MST nodes with only one link relative to the maximum possible number of leaves

Utilizing Brainwave, the weighted network map of connections and minimum spanning tree were visualized for a given frequency band; this was performed for the whole cohort and separately for each subgroup (PPAOS and agPPA) based on an average of all individual epochs. In the weighed network map, the lines represent connections with PLI synchronization above the noted connectivity threshold.

2.6. Statistical Analysis

Differences in clinical characteristics between subgroups were assessed with Wilcoxon rank-sum tests. Differences between agPPA and PPAOS patients' graph theory and MST measures were assessed separately with Wilcoxon rank-sum tests, each collapsed across frequency bands. Spearman correlations were computed between behavioral and network measures in all frequency bands across all patients. Statistical analyses were performed utilizing the JMP computer software (JMP Software, version Pro 14; SAS Institute Inc., Cary, NC, USA) with significance set at $p < 0.05$. Multiple comparison corrections were not imposed due to the small sample size. Given the exploratory nature of the study, we prioritized avoiding type II error inflation which unfortunately results from all common multiple comparison corrections (see Figure 5 in the following reference) [32].

3. Results

Demographic information and clinical data for the cohort and each subgroup are detailed in Table 2. Overall, agPPA patients were slightly younger, with slightly longer disease durations, compared to PPAOS patients. Sex representation was equivalent (approximately 60% female in each group). Consistent with the diagnoses, indices of language functioning (e.g., NAT and WAB-AQ) were lower in agPPA compared to PPAOS. Scores on the index of general cognition (the MoCA) were lower and ratings of parkinsonism (on the MDS-UPDRS III) were slightly higher in agPPA compared to PPAOS. There was no difference in AOS severity or ASRS-3, a quantitative index of AOS, between subgroups. For all patients, objective testing aligned with the SLP's gestalt clinical judgment (i.e., normal language testing for those diagnosed PPAOS).

Table 2. Median clinical and demographic information for this cohort and subgroups.

	agPPA (*n* = 15)	PPAOS (*n* = 7)	All (*n* = 22)
Age at EEG *	69	74	73
Disease Duration at EEG *	4.1	2	3.95
Sex	9 F (60%)	4 F (57%)	13 F (59%)
MoCA* (/30)	21	27	25
MDS-UPDRS III (/81) *	15	12	15
ASRS-3 (/52)	21	16	21
NAT (/10) *	5	9	7
WAB-AQ (/100) *	88.775	97.9	96.4
Aphasia Severity (/4) *	1.5	0	1
AOS Severity (/4)	2	2	2

Note: Age and disease duration (years); MoCA = Montreal Cognitive Assessment; MDS-UPDRS III = Movement Disorder Society-sponsored version of the Unified Parkinson's Disease Rating Scale, Motor section; ASRS-3 = Apraxia of Speech Rating Scale-3; NAT = Northwestern Anagram Test; WAB-AQ = Western Aphasia Battery Revised Aphasia Quotient. Maximum score noted in row header, when applicable. Asterisk in row header indicates significant non-parametric test of differences between agPPA and PPAOS groups ($p < 0.05$).

Median network measures are reported in Table 3. Omnibus tests of differences did not support significant differences in either graph theory or MST measures between agPPA and PPAOS. The data are visualized in power maps and minimum spanning trees; results for the whole cohort are presented in Figures 1 and 2, respectively. Data were additionally visualized relative to the subgroups of agPPA and PPAOS, shown in Figures 3 and 4. The power maps show differences in the distribution of connectivity for agPPA compared to PPAOS. The MSTs for the agPPA in the alpha frequency bands show a relatively more "star-like" quality, with a more central node connecting to the majority of other nodes. The star-like quality typically relates to a more integrated network, with a smaller diameter and shorter path length; this MST configuration typically reflects efficient information transfer, although not always. One possible downfall is information overload at the central node with subsequent inefficiency.

To better understand the relationship between graph theory measures and clinical presentations, non-parametric correlations between network and behavioral variables were calculated across all patients; these are reported in Table 4. Statistically significant relationships were identified between: age and alpha2 gamma ($\rho = -0.60$), kappa$_w$ ($\rho = -0.42$), and MST leaf ($\rho = -0.48$); disease duration and theta modularity ($\rho = -0.58$); disease duration and alpha1 lambda ($\rho = 0.78$); MDS-UPDRS III and alpha1 PLI ($\rho = -0.55$) and kappa$_w$ ($\rho = -0.56$); MDS-UPDRS III and alpha2 MST leaf ($\rho = -0.47$); ASRS-3 and alpha1 gamma ($\rho = 0.54$) and lambda ($\rho = 0.82$); ASRS-3 and alpha2 lambda ($\rho = 0.059$). No significant relationships identified between graph theory or MST measures and the MoCA or WAB-AQ. Correlation scatter plots, with individual data points indicating group membership, are provided in Supplementary Materials.

Table 3. Median (interquartile range) for group-level network measures.

Measure	agPPA (*n* = 15)	PPAOS (*n* = 7)	All (*n* = 22)
Theta			
PLI	0.189 (0.180, 0.211)	0.187 (0.175, 0.209)	0.188 (0.178, 0.210)
Gamma	1.020 (1.007, 1.038)	1.030 (1.008, 1.040)	1.025 (1.007, 1.039)
Lambda	0.934 (0.932, 0.945)	0.934 (0.922, 0.947)	0.934 (0.930, 0.946)
$Kappa_W$	4.001 (3.787, 4.439)	3.978 (3.694, 4.440)	3.998 (3.723, 4.439)
Modularity	0.077 (0.068, 0.083)	0.081 (0.070, 0.082)	0.078 (0.070, 0.082)
MST BC_{max}	0.723 (0.692, 0.742)	0.711 (0.680, 0.722)	0.711 (0.691, 0.734)
MST Diameter	0.425 (0.413, 0.444)	0.406 (0.394, 0.438)	0.422 (0.405, 0.439)
MST Eccentricity	0.340 (0.324, 0.352)	0.330 (0.313, 0.347)	0.339 (0.323, 0.348)
MST Leaf	0.550 (0.519, 0.569)	0.544 (0.531, 0.575)	0.547 (0.530, 0.570)
Alpha1			
PLI	0.242 (0.242, 0.272)	0.257 (0.234, 0.274)	0.248 (0.234, 0.273)
Gamma	1.029 (1.022, 1.043)	1.033 (1.022, 1.039)	1.030 (1.022, 1.040)
Lambda	0.938 (0.937, 0.946)	0.935 (0.932, 0.941)	0.938 (0.935, 0.946)
$Kappa_W$	5.143 (4.992, 5.753)	5.434 (4.940, 5.833)	5.296 (4.980, 5.773)
Modularity	0.079 (0.073, 0.084)	0.081 (0.070, 0.084)	0.080 (0.072, 0.084)
MST BC_{max}	0.721 (0.707, 0.734)	0.733 (0.696, 0.757)	0.723 (0.705, 0.739)
MST Diameter	0.431 (0.388, 0.444)	0.394 (0.388, 0.431)	0.419 (0.388, 0.444)
MST Eccentricity	0.339 (0.306, 0.350)	0.314 (0.310, 0.348)	0.335 (0.309, 0.349)
MST Leaf	0.550 (0.531, 0.588)	0.581 (0.531, 0.600)	0.553 (0.531, 0.595)
Alpha2			
PLI	0.215 (0.193, 0.241)	0.207 (0.198, 0.219)	0.210 (0.196, 0.230)
Gamma	1.041 (1.012, 1.057)	1.029 (1.005, 1.044)	1.033 (1.012, 1.046)
Lambda	0.943 (0.932, 0.950)	0.933 (0.925, 0.938)	0.936 (0.928, 0.946)
$Kappa_W$	4.553 (4.054, 5.157)	4.327 (4.242, 4.620)	4.440 (4.156, 4.920)
Modularity	0.075 (0.071, 0.086)	0.080 (0.071, 0.080)	0.077 (0.071, 0.084)
MST BC_{max}	0.714 (0.700, 0.749)	0.734 (0.684, 0.742)	0.719 (0.700, 0.742)
MST Diameter	0.419 (0.388, 0.438)	0.406 (0.400, 0.419)	0.413 (0.398, 0.433)
MST Eccentricity	0.336 (0.309, 0.347)	0.320 (0.314, 0.332)	0.325 (0.314, 0.341)
MST Leaf	0.556 (0.538, 0.594)	0.563 (0.519, 0.581)	0.559 (0.536, 0.583)

Table 4. Non-parametric Spearman correlations between graph theory network and behavioral variables of interest.

	Age	Disease Duration	MoCA	MDS-UPDRS III	ASRS-3	WAB-AQ
Theta						
PLI	−0.1404	0.2893	−0.1254	0.0023	0.0788	−0.0589
Gamma	0.0583	0.1509	0.1904	−0.0736	0.0037	0.1218
Lambda	−0.1130	0.1524	−0.0325	−0.1586	−0.0283	−0.0558
$Kappa_W$	−0.1512	0.2995	−0.1005	−0.0068	0.0640	−0.0392
Modularity	−0.0261	**−0.5790** *	0.1266	−0.0739	−0.0382	0.1539
MST BC_{max}	−0.0798	0.0590	−0.2501	0.1687	−0.1280	−0.3221
MST Diameter	0.2293	0.2937	−0.0906	0.0923	0.3820	0.0083
MST Eccentricity	0. 3000	0.2640	−0.0495	0.0977	0.3879	−0.0021
MST Leaf	−0.3106	−0.0500	0.1542	−0.1328	−0.0315	0.0990
Alpha1						
PLI	−0.3447	−0.0483	0.1778	**−0.5537** *	−0.2069	0.1552
Gamma	0.0476	0.1421	0.1538	−0.1954	**0.5357** *	0.2598
Lambda	0.0340	**0.7833** *	−0.2614	0.4002	**0.8246** *	−0.0485
$Kappa_W$	−0.3505	−0.0596	0.1466	**−0.5593** *	−0.1625	0.1869
Modularity	0.0986	0.0301	0.2071	0.2310	0.0197	0.1260
MST BC_{max}	−0.0541	−0.2585	0.1364	−0.2593	−0.1016	0.0015
MST Diameter	0.0390	0.0542	−0.2057	0.2736	0.2427	0.0156
MST Eccentricity	−0.0456	0.1078	−0.1194	0.2591	0.1932	0.1147
MST Leaf	−0.0011	0.1291	0.1326	−0.3163	−0.0167	−0.0109

Table 4. *Cont.*

	Age	Disease Duration	MoCA	MDS-UPDRS III	ASRS-3	WAB-AQ
Alpha2						
PLI	−0.3771	0.0556	−0.1609	−0.0668	0.1822	0.0743
Gamma	**−0.5991 ***	0.0562	0.0650	−0.4110	0.1994	0.3263
Lambda	0.0102	0.3142	−0.2803	0.0221	**0.5871 ***	−0.1249
Kappa$_W$	**−0.4241 ***	0.0153	−0.0688	−0.1545	0.1883	0.1735
Modularity	0.2749	−0.1153	0.1483	0.1116	−0.4380	−0.1314
MST BC$_{max}$	−0.2936	0.3081	−0.0643	−0.0856	0.4436	0.3390
MST Diameter	0.3636	−0.0558	−0.0065	0.1252	−0.1891	−0.3696
MST Eccentricity	0.3539	−0.1016	−0.0295	0.1432	−0.2905	−0.4027
MST Leaf	**−0.4824 ***	0.0546	0.0546	**−0.4686 ***	0.0722	0.2962

Note: Age and disease duration (years); MoCA = Montreal Cognitive Assessment; MDS-UPDRS III = Movement Disorder Society-sponsored version of the Unified Parkinson's Disease Rating Scale, Motor section; ASRS-3 = Apraxia of Speech Rating Scale-3; WAB-AQ = Western Aphasia Battery Revised Aphasia Quotient. Significant correlations ($p < 0.05$) are indicated by bold font and *; correction for multiple comparisons was not applied.

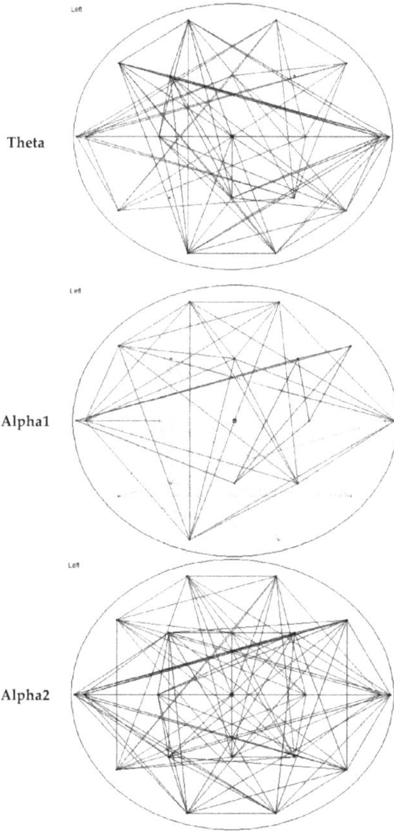

Figure 1. Average network maps for all patients for the theta, alpha1, and alpha2 frequency bands. The maps demonstrate the presence of correlations between pairs of channels, with threshold of 0.1 (PLI value) or correlations above that threshold.

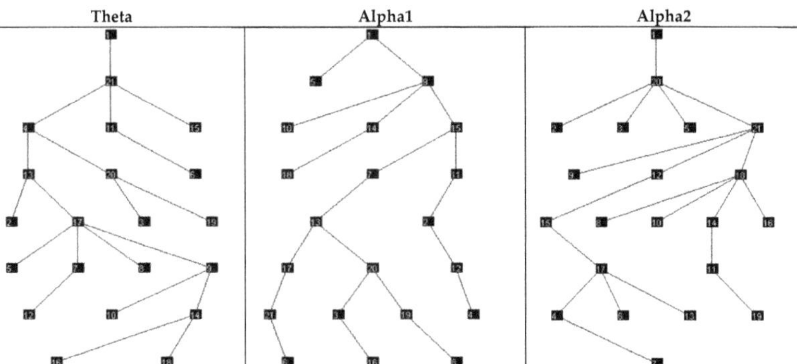

Figure 2. Minimum Spanning Trees for the cohort of all patients for the theta, alpha1, and alpha2 frequency bands. This visualization connects all nodes, maximizing synchronization. The numbers reflect electrode numbers; consistent with assessing mean connectivity, the relationships between specific electrodes were not explored in this study.

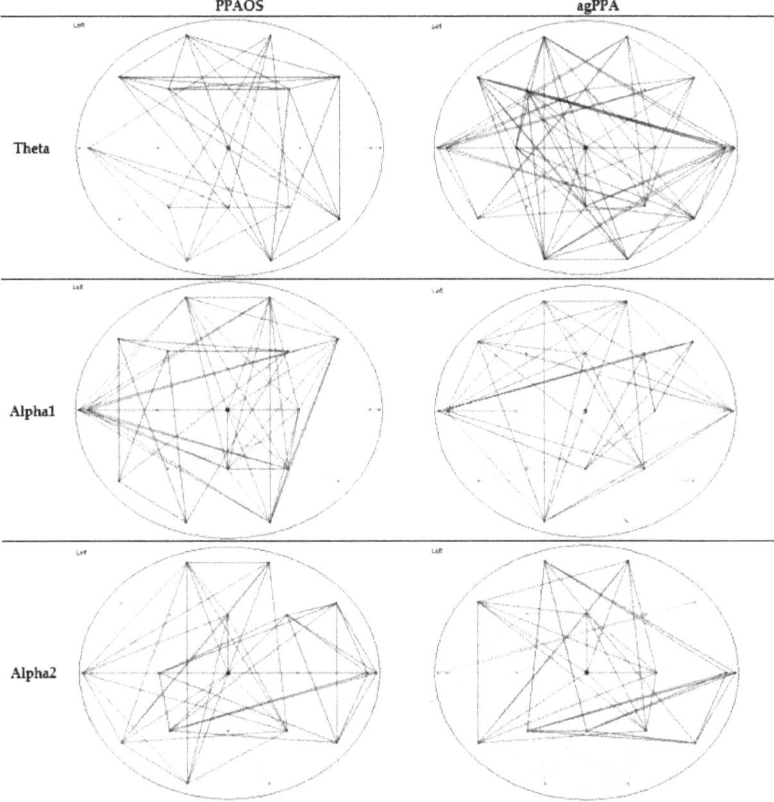

Figure 3. Average network maps separating the agPPA patients and PPAOS, for the theta, alpha1, and alpha2 frequency bands. The maps demonstrate the presence of correlations between pairs of channels, with threshold of 0.1 (PLI value) or correlations above that threshold.

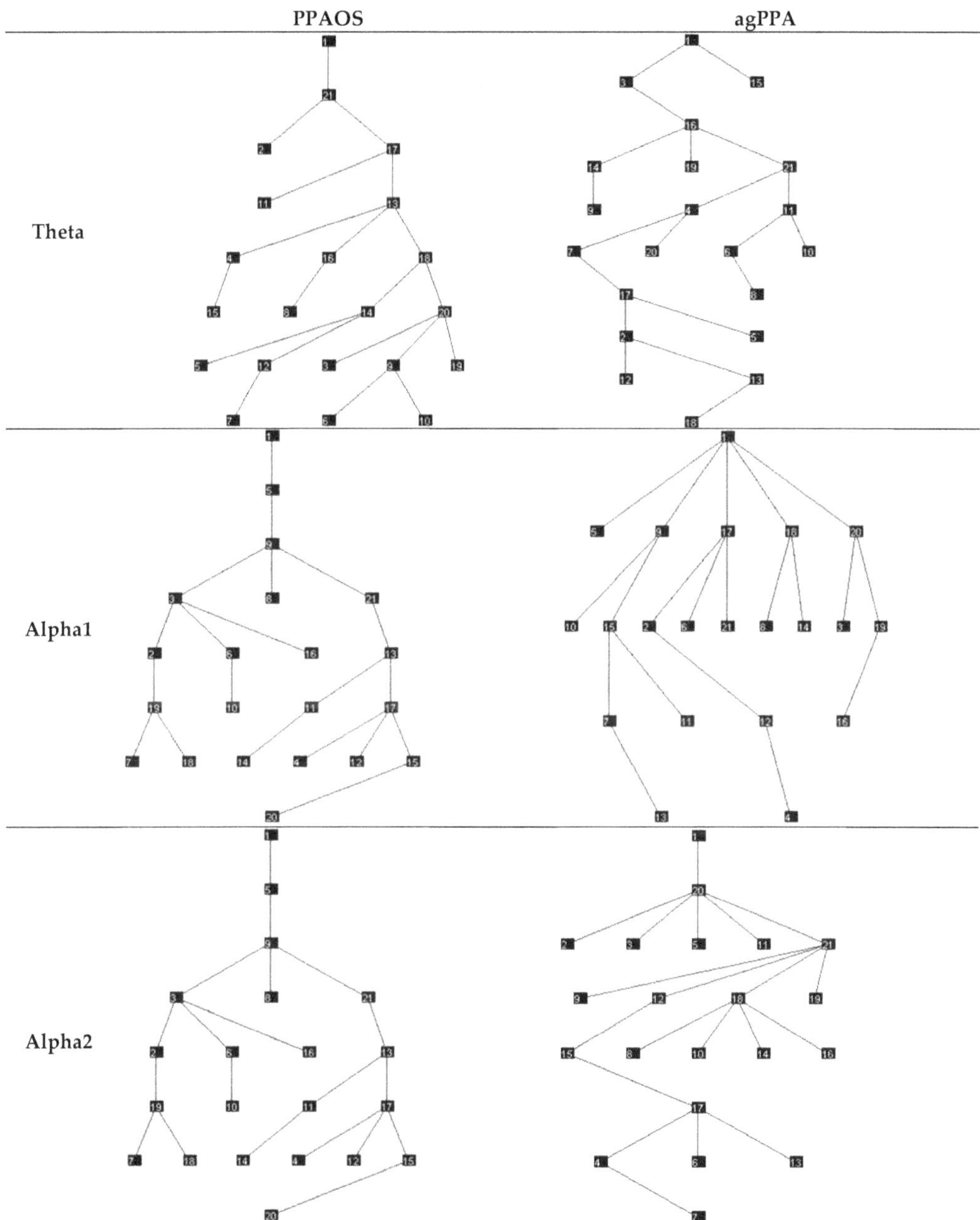

Figure 4. Minimum Spanning Trees (MSTs) separating the agPPA patients and PPAOS, for the theta, alpha1, and alpha2 frequency bands. This visualization connects all nodes, maximizing synchronization. The numbers reflect electrode numbers; consistent with assessing mean connectivity, the relationships between specific electrodes were not explored in this study.

4. Discussion

4.1. General Discussion

The results provide EEG evidence of network alteration in patients with agPPA and PPAOS. While it is difficult to fully describe or dismiss significant differences between groups due to the sample sizes, this exploratory study demonstrates potential for EEG graph theory measures to quantify network changes associated with degenerative speech and language disorders. The novelty of this study is the patient population and the correlation between EEG graph theory measures and certain clinical measures.

The results broadly suggest that increased global integration, or reduced network specificity, occurs in degenerative speech and language disorders. These network changes exist even in the absence of strong evidence for structural changes on magnetic resonance imaging [4] and it is therefore considered unlikely these are artifacts of atrophy. The visualization of the data supports the presence of network alterations, with correlation analyses offering insight into their clinical manifestations. This study explored global connectivity, rather than that of smaller cortical regions, which should be the focus of future studies. Further, it is not yet clear if the network changes represent direct disease effects or a compensatory response. For example, additional regional graph theory measures and correlational analyses might clarify whether connectivity in the region of suspected disease (e.g., precentral gyrus or supplementary motor area) is reduced and/ or whether there are downstream effects of hyperconnectivity in other areas working to compensate for that loss; alternatively, if hyperconnectivity is seen in the region of disease, it might reflect system stress. A more complete understanding of network disruption in neurodegenerative speech and language disorders, perhaps in the context of the cascading network failure model [33], might better elucidate the relationship between the underlying pathophysiology and clinical presentation. Toward that end, future studies will explore the graph theory measures and relationships with clinical measures longitudinally.

4.2. Tests of Differences and Correlations

In this study, the agPPA patients were, on average, slightly younger with slightly longer disease durations compared to PPAOS patients. These differences warrant caution when comparing the graph theory measures between the two groups. Scores on the index of general cognition (the MoCA) were lower and ratings of parkinsonism (on the MDS-UPDRS III) were also slightly higher in agPPA compared to PPAOS. However, it is important that there was no difference in AOS severity or ASRS-3, a quantitative index of AOS, between the subgroups.

Tests of differences did not support significant differences in either graph theory or MST measures between agPPA and PPAOS patients. Interestingly, differences in clinical EEGs were seen between the groups (i.e., relative to the presence of aphasia) [4] in a smaller subset of those patients included in this study, which is more consistent with the visualization of the data. In that study, patients with PPAOS (n = 5) had normal EEGS while two of three those with aphasia had theta slowing. The power maps and minimum spanning trees for the whole cohort (Figures 1 and 2, respectively) do not equally reflect the visualization of the agPPA and PPAOS subgroups (Figures 3 and 4). The MSTs for the agPPA in the alpha1 and alpha2 frequency bands show a more "star-like" quality, although given the unequal sample sizes, this should be interpreted cautiously. Future studies should systematically explore other possible sources of differences, including the subtype of AOS (i.e., phonetic or prosodic predominant speech disturbance [27]).

The correlation analysis offers insight into the relationship between graph theory measures and clinical presentations (see Table 4) and provides complementary support for reduced global efficiency and increased integration in patients with agPPA and PPAOS. There were negative relationships between the MDS-UPDRS III, a measure of motor impairment, and synchronicity, kappa, and MST leaf in the alpha band, likely reflecting severity (reduced synchronicity with increased motor dysfunction). The strongest correlation was noted between the ASRS-3, a measure of AOS severity, and lambda in the alpha1 frequency band, suggesting a relationship between reduced distance between nodes (measured by lambda) and more prominent AOS (indexed by the ASRS). This relationship supports the notion that network measures may better reflect more abstract process breakdowns (such as that of sensorimotor planning/programming in AOS) that have a less clear structural correlate, particularly compared to clinical EEG reads which were reportedly normal in these patients [4]. Interestingly, there were no significant relationships identified between graph theory or MST measures and the MoCA or WAB-AQ. This work lays the foundation to better understand whether these relationships (and lack thereof) represent the loss of ordered correlations (or anti-correlations) resulting from the disease. Frequency band differences require further exploration.

4.3. Relationship with Functional Connectivity Literature

While this is the first study of EEG graph theory measures in PPAOS, the broader literature on neurodegenerative disease provides helpful context for these findings. A recent study showed promising utility of EEG graph theory measures, in conjunction with machine learning, in distinguishing patients with PPA from controls [12]; however, the focus of the study was the machine-learning algorithms rather than the graph theory measures themselves. EEGs from patients with dementia associated with Lewy bodies had reduced connectivity strength in the alpha frequency band relative to cognitively unimpaired controls and patients with dementia from Alzheimer's disease, with additional evidence of reduced network efficiency. There were associations with clinical measures, including between leaf fraction and the Mini-Mental State Examination, a test of general cognition [34]. Another study showed increased connectivity in the theta band in patients with Alzheimer's disease dementia and mild cognitive impairment, relative to cognitively unimpaired controls; the connectivity measures were also correlated with neuropsychological test scores [35]. Finally, assessment of functional connectivity in multiple sclerosis via magnetoencephalography showed a less integrated network related to more severe cognitive impairment [36]. Together, these and other recent studies support the practical implications of EEG graph theory for accurate diagnosis, early detection, and disease monitoring [37]. It may be that a relative combination of graph theory metrics and their clinical correlates are most sensitive for diagnostic precision.

While a different modality, there have been at least four studies of functional connectivity in PPA and PPAOS via fMRI [38–41]. These studies have broadly demonstrated reduced connectivity in these populations. An fMRI study of functional connectivity in patients with PPAOS demonstrated reduced connectivity, specifically in the supplementary motor areas (SMA); reduced connectivity in the right SMA negatively correlated another measure of AOS, an articulatory error score, while connectivity in the left working memory network correlated with the ASRS [38]. These can serve as a foundation from which to formulate hypotheses for future regional analyses; for instance, it is hypothesized that there may be loss of ordered synchronization between frontal regions, supplementary motor areas, and, overall, regions in the left hemisphere compared to others.

Other fMRI studies of agPPA patients [40], patients with semantic variant PPA [39], and PPA patients more broadly [41] showed lower global integration and alteration in hub distribution in speech-predominant regions compared to cognitively unimpaired controls that were not entirely explained by structural changes. Taken together, there is support for looking at more functional, rather than structural, measures of disease burden in understanding clinical symptoms.

4.4. Limitations and Future Directions

There are limitations to the current study. While this is the largest documented EEG study of patients with PPAOS, the sample size was relatively small, which limited our ability to examine smaller subgroup influences (e.g., AOS type, phonetic or prosodic) on the findings. Further, given the results of the power analysis (which suggests the need for a much larger sample size; details not reported for brevity), we are unable to assess robust effects from this sample size. We are lacking an ideally age- and sex-matched cognitively unimpaired control cohort to expand the impact beyond patient group description and to assess the diagnostic power between impaired and unimpaired groups. The patient group comparisons offer important insight on which to base future hypotheses, but the groups are imbalanced in size, age, and disease duration. To explore the complex relationship between EEG network measures, clinical symptoms, and other explanatory variables (such as age and disease duration), regression models should be considered with relevant covariates.

The novelty of the current study lies in the relationship of network measures and clinical parameters. Stronger relationships are expected between regional, rather than mean, network measures, which should be explored in future studies. Additional limitations are methodological, including the use of 21 electrodes and a 256 Hz sampling rate, as well as PLI in favor of synchronization likelihood, another connectivity measure; different parameters, including exploring frequency bands beyond alpha and theta and frequency band measure ratios, could yield different results. Another modifiable parameter is sample length; here, the epoch length was limited by the sampling rate. While "clean" epochs were selected, no specific criteria were applied, which could impact replicability. Finally, differences in number of epochs and use of other connectivity measures could have influenced results [42], as could have the reference electrode [43]. While the recording parameters make it difficult to compare the results to those of published controls or other patient populations, methodological decisions were made to expedite transfer of these findings to clinical practice, which is considered a relative strength. Longitudinal assessments in a larger cohort, across the clinical severity spectrum and with different clinical phenotypes will also strengthen the interpretability and utility of these findings.

5. Conclusions

This study provides EEG evidence of network alteration and breakdown associated with primary progressive aphasia and apraxia of speech, although quantifiable differences between the groups are not yet clear. Nonetheless, this study demonstrates potential for EEG graph theory measures to quantify network changes that may reflect degenerative speech and language disturbances, given correlations with clinical measures. It remains important to compare these patterns to a healthy cognitively unimpaired control group. Describing network pathophysiology may have utility for understanding these diseases in a way not previously available, and, importantly, via a widely available and cost-effective method. This method may parlay into diagnostic EEG biomarkers, and ultimately, biomarkers for predicting disease progression and monitoring treatment-mediated improvements.

Supplementary Materials: The following supporting information can be downloaded at https://www.mdpi.com/article/10.3390/brainsci12030378/s1. Figure S1: Correlation scatter plots between graph theory and behavioral measures, with individual data points indicating frequency band and group membership.

Author Contributions: Conceptualization, R.L.U.; methodology, R.L.U.; software, J.N.C.; formal analysis, R.L.U.; investigation, R.L.U.; H.M.C.; J.R.D.; H.B. and K.A.J.; resources, K.A.J.; J.L.W. and G.A.W.; data curation, R.L.U.; writing—original draft preparation, R.L.U.; writing—review, editing, and approval, all authors; visualization, R.L.U.; funding acquisition, R.L.U.; K.A.J. and J.L.W. All authors have read and agreed to the published version of the manuscript.

Funding: This study was funded by the National Institutes of Health (NIH), National Institute on Deafness and Other Communication Disorders grants R01 DC014942 (PI: KAJ) and R01 DC012519 (PI: JLW), NIH National Institute of Neurological Disorders and Stroke grant R21 NS094684 (PI: KAJ), and the American Speech-Language-Hearing Foundation's New Investigator Research Grant (PI: RLU).

Institutional Review Board Statement: The study was conducted according to the guidelines of the Declaration of Helsinki, and approved by the Institutional Review Board at Mayo Clinic.

Informed Consent Statement: Informed consent was obtained from all study participants.

Data Availability Statement: Requests for the data presented in this study should be sent to the corresponding author for consideration.

Acknowledgments: We extend our gratitude to these patients and their families for their time and dedication to our research program.

Conflicts of Interest: R.L.U.; H.B.; J.R.D.; H.M.C.; J.L.W. and K.A.J. receive research support from the National Institutes of Health. There are no other financial or non-financial disclosures to report. The funders had no role in the design of the study; in the collection, analyses, or interpretation of data; in the writing of the manuscript, or in the decision to publish the results.

References

1. de Waal, H.; Stam, C.J.; Blankenstein, M.A.; Pijnenburg, Y.A.; Scheltens, P.; van der Flier, W.M. EEG abnormalities in early and late onset Alzheimer's disease: Understanding heterogeneity. *J. Neurol. Neurosurg. Psychiatry* **2011**, *82*, 67–71. [CrossRef] [PubMed]
2. Caviness, J.N.; Hentz, J.G.; Evidente, V.G.; Driver-Dunckley, E.; Samanta, J.; Mahant, P.; Connor, D.J.; Sabbagh, M.N.; Shill, H.A.; Adler, C.H. Both early and late cognitive dysfunction affects the electroencephalogram in Parkinson's disease. *Parkinsonism Relat. Disord.* **2007**, *13*, 348–354. [CrossRef] [PubMed]
3. Mesulam, M.M. Slowly progressive aphasia without generalized dementia. *Ann. Neurol.* **1982**, *11*, 592–598. [CrossRef] [PubMed]
4. Utianski, R.L.; Caviness, J.N.; Worrell, G.A.; Duffy, J.R.; Clark, H.M.; Machulda, M.M.; Whitwell, J.L.; Josephs, K.A. Electroencephalography in primary progressive aphasia and apraxia of speech. *Aphasiology* **2019**, *33*, 1410–1417. [CrossRef] [PubMed]
5. Van Steen, M. Graph theory and complex networks. An introduction. *Meterial* **2010**, *144*, 152.
6. Utianski, R.L.; Caviness, J.N.; van Straaten, E.C.; Beach, T.G.; Dugger, B.N.; Shill, H.A.; Driver-Dunckley, E.D.; Sabbagh, M.N.; Mehta, S.; Adler, C.H.; et al. Graph theory network function in Parkinson's disease assessed with electroencephalography. *Clin. Neurophysiol.* **2016**, *127*, 2228–2236. [CrossRef] [PubMed]
7. Miraglia, F.; Vecchio, F.; Rossini, P.M. Searching for signs of aging and dementia in EEG through network analysis. *Behav. Brain Res.* **2017**, *317*, 292–300. [CrossRef] [PubMed]
8. Vecchio, F.; Miraglia, F.; Iberite, F.; Lacidogna, G.; Guglielmi, V.; Marra, C.; Pasqualetti, P.; Tiziano, F.D.; Rossini, P.M. Sustainable method for Alzheimer's prediction in Mild Cognitive Impairment: EEG connectivity and graph theory combined with ApoE. *Ann. Neurol.* **2018**, *84*, 302–314. [CrossRef] [PubMed]
9. Stam, C.J.; De Haan, W.; Daffertshofer, A.; Jones, B.; Manshanden, I.; van Cappellen van Walsum, A.-M.; Montez, T.; Verbunt, J.; De Munck, J.; Van Dijk, B. Graph theoretical analysis of magnetoencephalographic functional connectivity in Alzheimer's disease. *Brain* **2009**, *132*, 213–224. [CrossRef] [PubMed]
10. Stam, C.J.; Jones, B.F.; Nolte, G.; Breakspear, M.; Scheltens, P. Small-world networks and functional connectivity in Alzheimer's disease. *Cereb. Cortex* **2007**, *17*, 92–99. [CrossRef] [PubMed]
11. de Haan, W.; Pijnenburg, Y.A.; Strijers, R.L.; van der Made, Y.; van der Flier, W.M.; Scheltens, P.; Stam, C.J. Functional neural network analysis in frontotemporal dementia and Alzheimer's disease using EEG and graph theory. *BMC Neurosci.* **2009**, *10*, 101. [CrossRef] [PubMed]
12. Moral-Rubio, C.; Balugo, P.; Fraile-Pereda, A.; Pytel, V.; Fernández-Romero, L.; Delgado-Alonso, C.; Delgado-Álvarez, A.; Matias-Guiu, J.; Matias-Guiu, J.A.; Ayala, J.L. Application of Machine Learning to Electroencephalography for the Diagnosis of Primary Progressive Aphasia: A Pilot Study. *Brain Sci.* **2021**, *11*, 1262. [CrossRef] [PubMed]
13. Mesulam, M.M. Primary progressive aphasia. *Ann. Neurol.* **2001**, *49*, 425–432. [CrossRef] [PubMed]
14. Gorno-Tempini, M.L.; Hillis, A.E.; Weintraub, S.; Kertesz, A.; Mendez, M.; Cappa, S.F.; Ogar, J.M.; Rohrer, J.D.; Black, S.; Boeve, B.F.; et al. Classification of primary progressive aphasia and its variants. *Neurology* **2011**, *76*, 1006–1014. [CrossRef] [PubMed]
15. Josephs, K.A.; Duffy, J.R.; Strand, E.A.; Machulda, M.M.; Senjem, M.L.; Master, A.V.; Lowe, V.J.; Jack, C.R.; Whitwell, J.L. Characterizing a neurodegenerative syndrome: Primary progressive apraxia of speech. *Brain* **2012**, *135*, 1522–1536. [CrossRef] [PubMed]
16. Duffy, J.R. Apraxia of Speech in degenerative neurologic disease. *Aphasiology* **2006**, *20*, 511–527. [CrossRef]
17. Utianski, R.L.; Duffy, J.R.; Clark, H.M.; Strand, E.A.; Boland, S.M.; Machulda, M.M.; Whitwell, J.L.; Josephs, K.A. Clinical Progression in Four Cases of Primary Progressive Apraxia of Speech. *Am. J. Speech-Lang. Pathol.* **2018**, *27*, 1303–1318. [CrossRef] [PubMed]
18. Josephs, K.A.; Duffy, J.R.; Clark, H.M.; Utianski, R.L.; Strand, E.A.; Machulda, M.M.; Botha, H.; Martin, P.R.; Pham, N.T.T.; Stierwalt, J.; et al. A molecular pathology, neurobiology, biochemical, genetic and neuroimaging study of progressive apraxia of speech. *Nat. Commun.* **2021**, *12*, 3452. [CrossRef] [PubMed]
19. Whitwell, J.L.; Duffy, J.R.; Machulda, M.M.; Clark, H.M.; Strand, E.A.; Senjem, M.L.; Gunter, J.L.; Spychalla, A.J.; Petersen, R.C.; Jack, C.R.; et al. Tracking the development of agrammatic aphasia: A tensor-based morphometry study. *Cortex* **2017**, *90*, 138–148. [CrossRef] [PubMed]
20. Whitwell, J.L.; Martin, P.; Duffy, J.R.; Clark, H.M.; Utianski, R.L.; Botha, H.; Machulda, M.M.; Strand, E.A.; Josephs, K.A. Survival analysis in primary progressive apraxia of speech and agrammatic aphasia. *Neurol. Clin. Pract.* **2021**, *11*, 249–255. [CrossRef] [PubMed]

21. Tetzloff, K.A.; Duffy, J.R.; Clark, H.M.; Utianski, R.L.; Strand, E.A.; Machulda, M.M.; Botha, H.; Martin, P.R.; Schwarz, C.G.; Senjem, M.L.; et al. Progressive agrammatic aphasia without apraxia of speech as a distinct syndrome. *Brain* **2019**, *142*, 2466–2482. [CrossRef]

22. Duffy, J.R.; Utianski, R.L.; Josephs, K.A. Primary progressive apraxia of speech: From recognition to diagnosis and care. *Aphasiology* **2021**, *35*, 560–591. [CrossRef] [PubMed]

23. Kertesz, A. *Western Aphasia Battery (Revised)*; PsychCorp: San Antonio, TX, USA, 2006.

24. Weintraub, S.; Mesulam, M.M.; Wieneke, C.; Rademaker, A.; Rogalski, E.J.; Thompson, C.K. The northwestern anagram test: Measuring sentence production in primary progressive aphasia. *Am. J. Alzheimers Dis. Other Demen.* **2009**, *24*, 408–416. [CrossRef] [PubMed]

25. Utianski, R.L.; Duffy, J.R.; Clark, H.M.; Strand, E.; Botha, H.; Schwarz, C.G.; Machulda, M.M.; Senjem, M.; Spychalla, A.J.; Jack, C.; et al. Prosodic and phonetic subtypes of primary progressive apraxia of speech. *Brain Lang.* **2018**, *184*, 54–65. [CrossRef] [PubMed]

26. Strand, E.A.; Duffy, J.R.; Clark, H.M.; Josephs, K.A. The Apraxia of Speech Rating Scale: A new tool for diagnosis and description of AOS. *J. Commun. Disord.* **2014**, *51*, 43–50. [CrossRef] [PubMed]

27. Nasreddine, Z.; Phillips, N.; Bedirian, V.; Charbonneau, S.; Whitehead, V.; Collin, I.; Cummings, J.; Chertkow, H. The Montreal Cognitive Assessment, MoCA: A brief screening tool for mild cognitive impairment. *J. Am. Geriatr. Soc.* **2005**, *53*, 669–695. [CrossRef] [PubMed]

28. Goetz, C.; Tilley, B.; Shaftman, S.; Stebbins, G.; Fahn, S.; Martinex-Martin, P.; Poewe, W.; Sampaio, C.; Stern, M.B.; Dodel, R.; et al. Movement Disorder Society-sponsored revision of the Unified Parkinson's Disease Rating Scale (MDS-UPDRS): Scale presentation and clinimetric testing results. *Mov. Disord.* **2008**, *23*, 2129–2170. [CrossRef] [PubMed]

29. Stam, C.J.; Nolte, G.; Daffertshofer, A. Phase lag index: Assessment of functional connectivity from multi channel EEG and MEG with diminished bias from common sources. *Hum. Brain Mapp.* **2007**, *28*, 1178–1193. [CrossRef] [PubMed]

30. Stam, C.J.; Tewarie, P.; Van Dellen, E.; van Straaten, E.C.W.; Hillebrand, A.; Van Mieghem, P. The trees and the forest: Characterization of complex brain networks with minimum spanning trees. *Int. J. Psychophysiol.* **2014**, *92*, 129–138. [CrossRef] [PubMed]

31. Tewarie, P.; van Dellen, E.; Hillebrand, A.; Stam, C.J. The minimum spanning tree: An unbiased method for brain network analysis. *NeuroImage* **2015**, *104*, 177–188. [CrossRef] [PubMed]

32. Midway, S.; Robertson, M.; Flinn, S.; Kaller, M. Comparing multiple comparisons: Practical guidance for choosing the best multiple comparisons test. *Test. PeerJ* **2020**, *8*, e10387. [CrossRef] [PubMed]

33. Jones, D.T.; Knopman, D.S.; Gunter, J.L.; Graff-Radford, J.; Vemuri, P.; Boeve, B.F.; Petersen, R.C.; Weiner, M.W.; Jack, C.R., Jr. Cascading network failure across the Alzheimer's disease spectrum. *Brain* **2016**, *139*, 547–562. [CrossRef] [PubMed]

34. van Dellen, E.; de Waal, H.; van der Flier, W.M.; Lemstra, A.W.; Slooter, A.J.; Smits, L.L.; van Straaten, E.C.; Stam, C.J.; Scheltens, P. Loss of EEG Network Efficiency Is Related to Cognitive Impairment in Dementia with Lewy Bodies. *Mov. Disord.* **2015**, *30*, 1785–1793. [CrossRef] [PubMed]

35. Musaeus, C.S.; Engedal, K.; Hogh, P.; Jelic, V.; Morup, M.; Naik, M.; Oeksengaard, A.R.; Snaedal, J.; Wahlund, L.O.; Waldemar, G.; et al. EEG Theta Power Is an Early Marker of Cognitive Decline in Dementia due to Alzheimer's Disease. *J. Alzheimers Dis.* **2018**, *64*, 1359–1371. [CrossRef] [PubMed]

36. Nauta, I.M.; Kulik, S.D.; Breedt, L.C.; Eijlers, A.J.; Strijbis, E.M.; Bertens, D.; Tewarie, P.; Hillebrand, A.; Stam, C.J.; Uitdehaag, B.M.; et al. Functional brain network organization measured with magnetoencephalography predicts cognitive decline in multiple sclerosis. *Mult. Scler. J.* **2021**, *27*, 1727–1737. [CrossRef] [PubMed]

37. Pievani, M.; de Haan, W.; Wu, T.; Seeley, W.W.; Frisoni, G.B. Functional network disruption in the degenerative dementias. *Lancet Neurol.* **2011**, *10*, 829–843. [CrossRef]

38. Botha, H.; Utianski, R.L.; Whitwell, J.L.; Duffy, J.R.; Clark, H.M.; Strand, E.A.; Machulda, M.M.; Tosakulwong, N.; Knopman, D.S.; Petersen, R.C.; et al. Disrupted functional connectivity in primary progressive apraxia of speech. *Neuroimage Clin.* **2018**, *18*, 617–629. [CrossRef] [PubMed]

39. Agosta, F.; Galantucci, S.; Valsasina, P.; Canu, E.; Meani, A.; Marcone, A.; Magnani, G.; Falini, A.; Comi, G.; Filippi, M. Disrupted brain connectome in semantic variant of primary progressive aphasia. *Neurobiol. Aging* **2014**, *35*, 2646–2655. [CrossRef]

40. Mandelli, M.L.; Welch, A.E.; Vilaplana, E.; Watson, C.; Battistella, G.; Brown, J.A.; Possin, K.L.; Hubbard, H.I.; Miller, Z.A.; Henry, M.L. Altered topology of the functional speech production network in non-fluent/agrammatic variant of PPA. *Cortex* **2018**, *108*, 252–264. [CrossRef]

41. Tao, Y.; Ficek, B.; Rapp, B.; Tsapkini, K. Different patterns of functional network reorganization across the variants of primary progressive aphasia: A graph-theoretic analysis. *Neurobiol. Aging* **2020**, *96*, 184–196. [CrossRef]

42. Baselice, F.; Sorriso, A.; Rucco, R.; Sorrentino, P. Phase Linearity Measurement: A Novel Index for Brain Functional Connectivity. *IEEE Trans. Med. Imaging* **2019**, *38*, 873–882. [CrossRef] [PubMed]

43. Anastasiadou, M.N.; Christodoulakis, M.; Papathanasiou, E.S.; Papacostas, S.S.; Hadjipapas, A.; Mitsis, G.D. Graph theoretical characteristics of EEG based functional brain networks in patients with epilepsy: The effect of reference choice and volume conduction. *Front. Neurosci.* **2019**, *13*, 221. [CrossRef] [PubMed]

Review

Semantic Variant Primary Progressive Aphasia: Practical Recommendations for Treatment from 20 Years of Behavioural Research

Aida Suárez-González [1,*,†], Sharon A. Savage [2], Nathalie Bier [3,4], Maya L. Henry [5,6], Regina Jokel [7,8], Lyndsey Nickels [9] and Cathleen Taylor-Rubin [9,10]

1   Dementia Research Centre, UCL Queen Square Institute of Neurology, University College London, London WC1N 3BG, UK
2   School of Psychological Sciences, College of Engineering, Science and Environment, The University of Newcastle, Callaghan, NSW 2308, Australia; sharon.savage@newcastle.edu.au
3   Centre de Recherche de l'Institut Universitaire de Gériatrie de Montréal, Montreal, QC H3W 1W4, Canada; nathalie.bier@umontreal.ca
4   School of Rehabilitation, Faculty of Medicine, Université de Montréal, Montreal, QC HEC 3J7, Canada
5   Department of Speech, Language, and Hearing Sciences, The University of Texas at Austin, Austin, TX 2504A, USA; maya.henry@austin.utexas.edu
6   Dell Medical School, The University of Texas at Austin, Austin, TX 78712, USA
7   Rotman Research Institute, Baycrest Health Sciences, Toronto, ON M6A 2E1, Canada; rjokel@research.baycrest.org
8   Department of Speech-Language Pathology, University of Toronto, Toronto, ON M5G 1V7, Canada
9   School of Psychological Sciences, Macquarie University, Sydney, NSW 2109, Australia; lyndsey.nickels@mq.edu.au (L.N.); Cathleen.Taylor@health.nsw.gov.au (C.T.-R.)
10 Department of Speech Pathology, Uniting War Memorial Hospital, South Eastern Sydney Local Health District War Memorial Hospital, Waverley, NSW 2024, Australia
*   Correspondence: aida.gonzalez@ucl.ac.uk
†   Current Address: UCL Queen Square Institute of Neurology, Dementia Research Centre, Box 16, 8-11 Queen Square, London WC1N 3BG, UK.

Citation: Suárez-González, A.; Savage, S.A.; Bier, N.; Henry, M.L.; Jokel, R.; Nickels, L.; Taylor-Rubin, C. Semantic Variant Primary Progressive Aphasia: Practical Recommendations for Treatment from 20 Years of Behavioural Research. *Brain Sci.* **2021**, *11*, 1552. https://doi.org/10.3390/brainsci11121552

Academic Editors: Jordi A. Matias-Guiu, Robert Jr Laforce and Rene L. Utianski

Received: 20 September 2021
Accepted: 16 November 2021
Published: 23 November 2021

Publisher's Note: MDPI stays neutral with regard to jurisdictional claims in published maps and institutional affiliations.

Abstract: People with semantic variant primary progressive aphasia (svPPA) present with a characteristic progressive breakdown of semantic knowledge. There are currently no pharmacological interventions to cure or slow svPPA, but promising behavioural approaches are increasingly reported. This article offers an overview of the last two decades of research into interventions to support language in people with svPPA including recommendations for clinical practice and future research based on the best available evidence. We offer a lay summary in English, Spanish and French for education and dissemination purposes. This paper discusses the implications of right- versus left-predominant atrophy in svPPA, which naming therapies offer the best outcomes and how to capitalise on preserved long-term memory systems. Current knowledge regarding the maintenance and generalisation of language therapy gains is described in detail along with the development of compensatory approaches and educational and support group programmes. It is concluded that there is evidence to support an integrative framework of treatment and care as best practice for svPPA. Such an approach should combine rehabilitation interventions addressing the language impairment, compensatory approaches to support activities of daily living and provision of education and support within the context of dementia.

Keywords: semantic dementia; semantic variant primary progressive aphasia; word finding; frontotemporal dementia; language therapy; behavioural therapy

1. Introduction

In the 1970s, Warrington's description of three individuals with a selective and profound inability to name and recognise objects [1] laid the foundation for what years later, in 1989, would be coined "semantic dementia" [2]. Semantic dementia, now widely referred to as semantic variant primary progressive aphasia (svPPA), is a neurodegenerative syndrome characterised by progressive loss of semantic knowledge in the context of otherwise well-preserved language and cognitive abilities [3,4]. Current consensus criteria require language impairment to be the most salient clinical symptom and the main cause of impairment in daily living activities [3,5]. Clinically, individuals with svPPA present with fluent speech (preserved repetition and speech production) and loss of semantic knowledge across all modalities of testing (e.g., picture naming, single-word comprehension and visual association tasks). As the disease progresses, behavioural features emerge, and speech becomes increasingly empty, culminating with mutism in the final stages [6]. An illustrative example is provided by the response of one woman with svPPA who, when asked about her symptoms, pointed to the trees in the hospital's courtyard and said, "I don't know what those green things are anymore".

SvPPA is estimated to account for one-third of all cases of frontotemporal dementia [7] with an average age at symptom onset of 60 years (64 years for diagnosis to be established). The prognosis for length of survival following diagnosis is highly variable, with a median of 12 years [8]. MRI brain scans typically reveal bilateral and asymmetric temporal pole atrophy (greater on the left) and asymmetric anterior hippocampal atrophy [9]. Furthermore, the anterior portion of the fusiform gyrus and adjacent regions are also critical areas systematically affected in svPPA and appear to play a pivotal role in semantic degradation [10–13]. Between 75% and 100% of all svPPA cases are associated with underlying TDP-43-C pathology, with the remainder mostly involving FTD tau [8,14–16] and a small proportion of cases showing concomitant Alzheimer's disease pathology [8,17].

There is no curative or disease-modifying treatment for svPPA. However, a growing body of research on non-pharmacological interventions has shown that people with svPPA may relearn lost vocabulary and benefit from other behavioural therapies. The first rehabilitation reports emerged in the literature in the late 1990s, inspired by patients who spontaneously engaged in self-practice as an attempted remedy for their anomia [18,19]. The proliferation of single case studies and small group studies over the next decades have demonstrated that people with svPPA who receive naming therapy can improve their recall of object labels in the short term, that the gains might be retained over time and that at least partial restoration of semantic knowledge may be possible (see reviews by Carthery-Gouland et al. [20], Jokel et al., [21], Cotelli et al., [22] and Pagnoni et al. [23] for an overview). Furthermore, the breadth of research into non-pharmacological interventions has by no means remained restricted to word retrieval. Therapeutic approaches targeting conversation [24], tasks and activities of daily living [25–27], psychoeducation programmes [28,29] and peer support groups [30] have made headway and are on the increase. Altogether they have set the stage for an integrative framework of clinical treatment and care in svPPA that combines rehabilitation interventions, compensatory approaches and provision of education and support, addressing the language impairment in svPPA within the context of dementia [31]. This article aimed to synthesise the learnings from 20 years of research in the non-pharmacological treatment and management of svPPA and lay out evidence-based recommendations for clinical practice and future research. For the purposes of education and dissemination beyond an academic audience, this article includes a lay summary available in English, Spanish and French (Supplementary Materials S1–S3).

2. Anomia in svPPA as a Sign of Semantic Breakdown

There is evidence that the anomia seen in svPPA stems from impairment in semantic knowledge [32]. This is different from the word retrieval impairments shown in the other PPA variants that arise at the lexical/phonological (logopenic variant PPA) or post lexical (non-fluent/agrammatic PPA) [33] stages. A basic understanding of how semantic memory architecture works is therefore required to develop effective treatments. A common theory is that semantic knowledge is organised in a hierarchy of specificity [1,34], ranging from very specific attributes at the bottom (e.g., the hummingbird is a small bird that can hover) to very general knowledge at the top (e.g., a hummingbird is an animal) (see Figure 1).

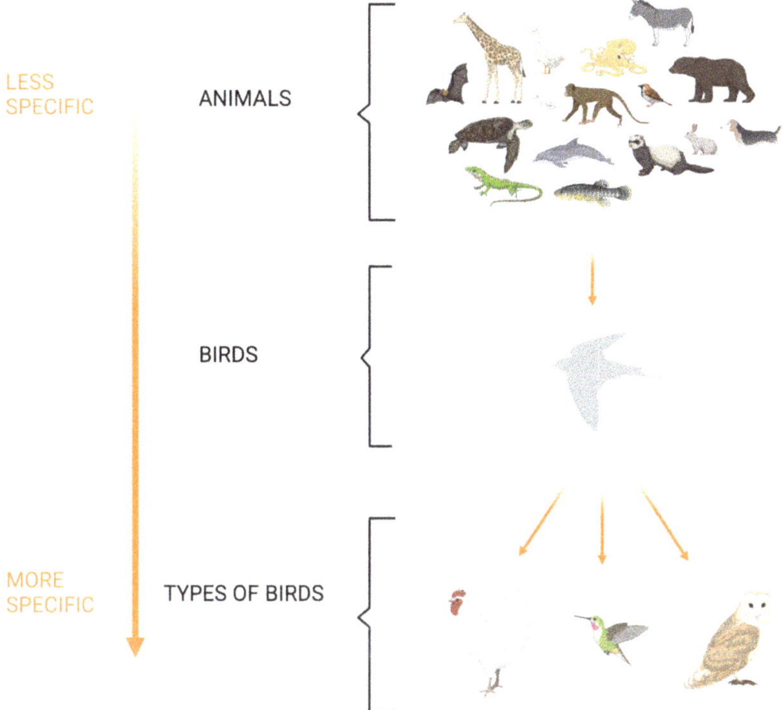

Figure 1. Organisation of the semantic memory category system and its implications for semantic breakdown in svPPA. The characteristic pattern of semantic organisation for the concept "birds" is illustrated in the picture above. Superordinate categories (e.g., animals) sit at the top of the semantic hierarchy. They display a high degree of generality and low specificity among the features shared by their members. Subordinate categories are a more specific level of categorisation, e.g.," birds" is a subordinate category of "animals" and "hummingbird" is a subordinate category of "birds". At the bottom of the hierarchy sit the most specific attributes, which are also those to degrade first in svPPA, e.g., "a hummingbird is a very small bird, feeds on flower nectar and can hover". A typical patient with svPPA may initially name the picture of a hummingbird correctly, but as the disease progresses, errors and superordinate responses would emerge in the following pattern: Assessment 1: Hummingbird → "hummingbird" (named correctly); Assessment 2: Hummingbird → "sparrow" (named as a semantically similar category coordinate); Assessment 3: Hummingbird → "bird" (named as a higher-familiarity typical member of the category); Assessment 4: Hummingbird → "animal" (named as the superordinate category); Assessment 5: Hummingbird → "I don't know".

Specific attributes are hypothesised to degrade first in a continuum of progressive degeneration that continues with the loss of general attributes and culminates in the disappearance of the concept. For instance, a person may identify a hummingbird as a living thing without being able to identify its specific properties (e.g., that it can fly and feeds on flower's nectar). This means that, during cognitive and language assessment, partial provision of information should not be interpreted as unequivocal proof of complete semantic preservation. Further investigation of semantic integrity should always be pursued in people with svPPA in preparation for therapy.

Leveraging Episodic Memory

Episodic memory (e.g., the ability to remember where you parked your car, what you did yesterday evening, or the plumage of a bird that is new to you) is a main entry point of semantic information into the memory system. This new information is integrated into existing bodies of knowledge by a dual system supported by the hippocampus (allowing quick capture of episodes) and neocortical structures (allowing a slower but effective integration into a long-term database) [35]. More specifically, in this second neocortical stage, information is consolidated in integrated, generalisable representations across a network distributed along the neocortex, tapping into the sensory, motor and linguistic systems [36]. Cross-modal interaction of these areas has been hypothesised to be anatomically supported by the anterior temporal lobes (ATLs) that operate as a hub where different forms of semantic information converge and connect [37,38]. This ATL region is affected at an early stage by the bilateral pathological aggregation of proteins associated with svPPA. However, the brain structures supporting episodic memory, such as the posterior area of the hippocampus and posterior cingulate cortex [39], are usually reasonably preserved. This suggests that, in principle, the episodic memory gateway to inputs that will eventually transform into re-learned concepts may remain functional. Consequently, this mechanism may be used, in conjunction with partially degraded neocortical structures, to the advantage of rehabilitation goals [40].

3. Differences between Left and Right Variants: Implications for Practice

The usual pattern of brain atrophy in svPPA (left greater than right) is reversed in approximately 30% of cases (i.e., right greater than left), giving rise to left and right-sided variants (left-svPPA and right-svPPA respectively) [41–43] (see Figure 2). Left-svPPA is characterised by poorer performance on verbal tasks compared to right-svPPA [13,43,44]. At the time of presentation, the prevalence of word-finding difficulties in left-svPPA is reported to be 94%, compared to 36% in right-svPPA, while impairments in single-word comprehension are reported in 67% of left-svPPA and 18% of right [43]. In contrast, individuals with right-svPPA show greater impairment of non-verbal semantics [38,42,43]. In up to 91% of cases with right-svPPA, the clinical picture is characterised by prosopagnosia (a difficulty in recognising faces) that for these individuals is associated with person-specific semantic knowledge breakdown [42,45–49]. Behavioural changes, although reported in both variants, seem to be more pronounced and appear earlier in right-svPPA, with social awkwardness and loss of insight are commonly reported (present in 64% and 55% of individuals respectively) [43] along with loss of empathy, disinhibition, apathy and compulsiveness [42,45,48].

Analysis of the types of naming errors produced by each group suggests that individuals with right-svPPA might have more difficulty accessing semantic knowledge through visual than verbal modalities (e.g., more difficulty recognising a famous face by looking at a photograph than by listening to a description of the person) (see Table 1). Individuals with left-svPPA show a larger proportion of circumlocutions in response to naming difficulties (e.g., "when it rains" for umbrella) and omissions [44,50] compared to right-svPPA, while those with right-svPPA make more coordinate and superordinate semantic errors (e.g., coordinate: "cat" for "dog" and superordinate: "animal" for "dog") [44,50]. The reduced ability of these individuals to access knowledge through visual features has been proposed as a possible mechanism that contributes to their greater difficulty in producing semantic associations, predisposing them to production of more taxonomic (coordinated and superordinate) semantic errors [44].

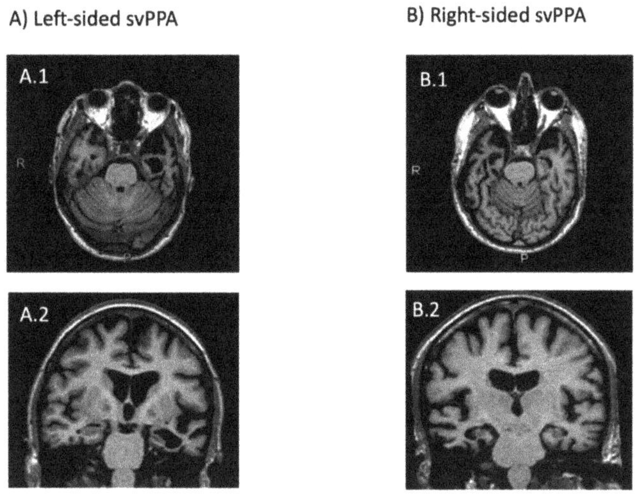

A) Left-sided svPPA

B) Right-sided svPPA

Figure 2. (**A.1**) Axial T1 MR image: anterior temporal lobe displaying marked atrophy on the left pole. (**A.2**) Coronal T1 MR image: marked left temporal atrophy with dilation of the temporal horn and left hippocampal shrinkage. (**B.1**) Axial T1 MR image: anterior temporal lobe displaying bilateral atrophy more marked on the right. (**B.2**) Coronal T1 MR image: marked right temporal atrophy with dilation of the temporal horn and right hippocampal shrinkage.

Table 1. Differences between right and left variant: implications for clinical practice.

	Left-Sided svPPA	Right-Sided svPPA
Verbal tasks		
Verbal tasks	Poorer	Better
Single word comprehension	+ impaired	- impaired
Naming	+ impaired	- impaired
Type of naming errors		
Circumlocutions	+ frequent	- frequent
omissions	+ frequent	- frequent
Semantic errors	- frequent	+ frequent
Visual/non-verbal tasks		
Non-language semantics	Better	Poorer
Prosopagnosia *	- frequent	+ frequent
Behaviour		
Social awkwardness	- frequent	+ frequent
Loss of insight	- frequent	+ frequent
Loss of empathy	- frequent	+ frequent
Disinhibition	- frequent	+ frequent
Compulsiveness	- frequent	+ frequent
Apathy	- frequent	+ frequent

(+) means more; (-) means less; * "prosopagnosia" is a term that refers to impaired ability to recognise faces. It was used by previous authors in the clinical description of the syndrome. It is however worth noting that the recognition deficit seen in right-svPPA is not restricted to faces but encompasses multimodal person knowledge as well. Grey background indicates features more severely impaired or more frequent symptoms in one variant compared to the other.

In light of this evidence, clinicians should pay particular attention to a few factors. First, whether verbal material (e.g., audio recordings, verbal descriptions and sounds) may be preferable to visual (e.g., photographs and real objects) should be considered when treating individuals with right-svPPA. Second, individuals with left-svPPA seem better able to access residual associated semantic knowledge and use this to describe the target when attempting to name. This can be used as a therapeutic opportunity, for instance, by encouraging the individual to retrieve this residual knowledge and relink it with the label.

4. The Current Evidence Informing Treatment and Management of Anomia and Word Comprehension Deficits

4.1. How Should Therapies for Anomia Be Designed and Administered?

Typically, lexical training therapies have consisted of a set of items given to individuals to practice. Therapy in svPPA should focus on maintaining or improving access to both names and semantic representations. Below, we present a summary of how these therapies should be planned and administered in svPPA based on a synthesis of current evidence.

4.2. Who Benefits from Anomia Therapy?

Benefits of therapy have been shown across a range of severities of anomia, provided some level of spoken language is preserved (i.e., there are no studies of individuals who are mute). This suggests that, in principle, the level of severity should not prevent any individual with svPPA from being considered for treatment, although the nature of the intervention would differ based on the level of severity. People in the early stages may have the advantage of retaining more semantic knowledge on which to build the therapy. They are also more likely to be free of other cognitive or behavioural symptoms that may impact successful engagement with therapy and, in fact, circumscribed semantic impairments longer than 6 years post-onset have been reported in some individuals [51–54].

4.3. How Many Sessions, of What Length and How Many Items per Session?

Current evidence suggests that 20–60 min of daily (or almost daily) practice is effective to produce short-term benefits [51,52,55–57], although some individuals have also shown benefits from less. Significant improvements should be expected within the first month of consistent practice [40,52,53,58–64] but may be evident sooner. Most studies to date have combined face-to-face sessions with the therapist with self-administered home programmes. Usual set size is between 15 and 30 items per session [51–54,58,60,63,65].

4.4. What Kind of Items and Naming Therapy?

Two kinds of words have been targeted in therapy: those that still are associated with some residual semantic knowledge and those that are not. A word is considered to have residual knowledge when the person can produce or comprehend at least partial information about it (e.g., "it's food" for an egg, without being able to connect the association between an egg and a hen). These words are by far the most investigated in the lexical retrieval literature. Words where meaning is completely lost have, on the contrary, been less investigated and the few studies looking at the use of conceptual enrichment therapies to treat words destitute of semantic knowledge have produced mixed results [66,67]. A list of the techniques used in the svPPA rehabilitation literature is shown in Table 2.

Table 2. List of lexical retrieval techniques used in the svPPA rehabilitation literature.

Technique	Example
Reading and repetition in the presence of a picture	Picture presented + corresponding printed word [18,40,52,53,55,58,60,62,63,66,68–71]
	Picture presented + corresponding printed word + audio recording of the object name (some authors have also included audio recorded descriptions of the treated item and some others also require a written response) [51,54,57,65,67,72]
Semantic treatment	Picture presented + corresponding spoken + written name + specific attributes [59]
	Semantic feature analysis—this technique requires patients to describe each feature of a word in a systematic way by answering a set of questions about group, use, action, properties, location and association [73,74]
	Conceptual enrichment therapy—this technique manipulates the encoding of new learning to promote flexible learning by placing the trained item in a personally meaningful temporal and spatial context [66,67,70]
	Feature generation from a list of sentence cues for personally relevant episodic or semantic information [75]
	Elaboration of items within subcategories, sorting pictures and words by subcategory, identifying semantic attributes of exemplars, usage of a picture dictionary organised by categories [76]
Sentence generation	Picture presented + name of the item + example sentence using the word + blank line for the participant to write their own [65]
Semantic, phonological, orthographic and/or autobiographical cueing/treatment	Sequence of tasks to engage semantic, phonemic, and orthographic self-cues and/or autobiographic memories, e.g., prompt semantic description by asking "what do you use it for?" [56,62,64,77–83]

Note: This is not intended to be a systematic review of naming therapy techniques. It rather aims to offer a practical overview of commonly administered training strategies. See [66] for a review of methods used in svPPA studies up until 2014 and [23] for methods used in PPA studies in general.

One of the most common approaches to improving naming is the "Look, listen and repeat" (LLR) or "Repetition (and reading) in the presence of the picture (RRIPP)". A picture of the target concept is presented, along with the name as a spoken and/or written word for the individual to repeat/read aloud, sometimes preceded by an attempt at naming, with or without (semantic or phonological) cues. Multiple variations of this approach have proven effective for improving production of vocabulary that the person with svPPA can still comprehend (see Table 2). However, this technique can lead to rote-learning (rigid and context-specific) and poor generalisation when semantic knowledge of the trained item is very impaired (e.g., the person can no longer comprehend either the lexical label or a picture of the object). Restitutive training of words/concepts that the person can no longer comprehend has been less explored in the literature. The suitability of a semantic approach to treating these items (e.g., working on characteristics of an object's usage and location and linking it with other related memories) is supported by two types of studies. The first consists of studies looking at the direct restoration of semantic knowledge [66,67,70] and the second capitalising on residual semantic information to boost word retrieval [18,40,52,53,55,58,60,62,63,66,68–71]. Both contribute to understanding the importance of the semantic system in the rehabilitation of svPPA. For instance, the naming of items with residual semantic knowledge appears to be easier to rehabilitate than that of items completely devoid of meaning [53,58]. Likewise, greater success is achieved with familiar items—familiar concepts degrade slower due to the larger and stronger network of semantic connections that are regularly reinforced with use (e.g., the concept of a toaster, used daily for breakfast, will be retained for longer than a hammer that is borrowed from a neighbour and used occasionally) [53,58,60,84]. In this same vein, some authors have introduced photos of individuals' own items within their therapy material (rather than generic exemplars), to harness familiarity and personal significance [51,54,65,82]. Others have identified semantic attributes of exemplars [76] or sorted items within semantic categories [18,51,54,65] to further reinforce the semantic concept (but randomise the order of items with each presentation to avoid rote learning).

4.5. Are These Therapies Well Accepted by People with svPPA?

Most studies of word retrieval therapy in svPPA have shown good adherence of people to practice. In many cases, participants completed home programmes consistently for many months. The first lexical retrieval therapy studies were prompted by individuals who started self-practice on their own initiative, evidencing their keenness to play an active role in their treatment [18]. Inevitably, individuals reported in the literature are those who volunteered for research and are probably particularly motivated to pursue therapy, which may not be the case when extrapolating to the broader clinical population. It has been reported that, in clinical settings, individuals with PPA who receive lexical retrieval therapy show a rate of adherence of 60% [85]. The authors of that study found that adherence was more likely when the treatment commenced in the year after diagnosis and when the patient was motivated, and mood was stable. Clearly, there will be people with svPPA who may prefer not to engage in lexical therapy for various reasons. In these cases, there is still a wide range of therapeutic options that can be offered (e.g., use of compensatory techniques, environmental adaptations, partner training and psychological support).

4.6. Are People with svPPA Aware of Their Deficits?

People with svPPA typically recognise that their language performance has weakened. However, some individuals appear to have difficulties evaluating their past knowledge of words (even in realising that certain words ever existed) and the extent of the impoverishment of their language content. For instance, Savage et al. [86] reported that people with svPPA who have mild to moderate semantic impairments showed no awareness of obvious mislabelling errors when naming components of objects. The authors of the study warn about the implications that this may have regarding patients' role and input into rehabilitation planning and recommend that rehabilitation programmes should not be based on patients' judgment alone and instead also involve family members and friends.

4.7. How Long Does the Effect of Therapy Last?

Many studies have demonstrated that the significant improvements in naming are often very well maintained over the first month after ceasing practice [40,52,54,56,59,63,81]. Outcomes beyond this, however, are variable. For some people with svPPA, a high proportion (73–82%) of the words named at the end of treatment can still be successfully named 3 to 6 months later [54,60,63,82]. In others, levels of retention in that time window are modest (e.g., around 65% of trained words) [53,62,65] or low (e.g., only 10–40% of words are maintained) [58,68]. Encouragingly, the majority of studies report performance that continues to be above baseline levels for up to 6 months after completing treatment [87].

These benefits have also been observed 12 months post-treatment in a small number of studies [82].

The extent of retention may be influenced by the degree of semantic knowledge still retained for an item (i.e., meaningful items persist longer [58]) and the opportunity to continue rehearsing items in everyday life [54,60,68]. This is consistent with observations that autobiographical experience and subsequent conversations regarding such experiences, may enhance semantic knowledge and preserve these words over time [81,82,84]. While this integration of the use of words in everyday life plays an important role in retaining vocabulary, many words (e.g., stove, plate) may not be used often enough in everyday conversation to allow regular practice, requiring alternative strategies for ongoing reinforcement. One feasible alternative is maintaining regular revision of the re-learned words. While daily practice may be needed in the early phases of an intervention, successful maintenance revisions (to maintain at least 80% of therapy items) require less practice [54]. For instance, when monitored over a 6 month period, people with svPPA with a moderate level of impairment needed less than 10 revision sessions over 6 months to maintain their naming. For those with more severe semantic impairment, Savage et al. [51,54] found that regular, weekly practice was needed to restore the benefits of the initial intense training. In particular, performance at around 2 months post-intervention appears to be a useful indicator of the frequency of revision that could be required for sustained maintenance—implying that this is a useful time point for clinicians to monitor and then formulate the revision programme for those people with mild to moderate svPPA.

A practical consideration for people with svPPA and their families then becomes how long to continue with interventions. In some cases [54,68], the practice simply becomes part of the usual routine or there may be enjoyment gained from it. Consistent with this, some studies have reported ongoing practice persisting for 1–2 years [55,88]. For some individuals, however, where declines in performance may become upsetting or practice becomes stressful, it may not be desirable to continue. In these circumstances, individuals with PPA and their families should be prepared for declines to emerge over the months that follow.

4.8. Does This Learning Generalise?

An important aspect of any rehabilitation programme is the degree to which improvements extend from the intervention to assist the person in their daily living. The generalisation of benefits in svPPA has been usually evaluated in two ways: (1) whether naming improvements extend from trained to untrained words and (2) whether words can be used by the person with svPPA in contexts that differ from the training format. Generalisation of naming improvements, extending from trained to untrained words, have been observed in some individuals with non-progressive aphasia, but usually only when the impairment is one of phonological encoding, in the absence of significant semantic or lexical deficits [89]. A consistent finding across most svPPA treatment studies is that untrained words do not improve [21,25] with very few exceptions showing the opposite result [81,82].

An alternative way of considering the generalisation of naming therapies is to evaluate the extent to which trained words can be used by the person with svPPA in contexts that differ from the training format. Broadly, this may be divided into "near transfer"—wherein the demonstration of knowledge is highly similar to the original training context (e.g., asking the person to produce the word in response to a different exemplar of the stimulus—see Figure 2 in Heredia et al. [60]) or "far transfer"—where knowledge must be applied more flexibly (e.g., by completing a different kind of language task such as verbal comprehension) [90]. Successful naming has been observed when people with svPPA are tested on alternative versions of trained items [52,63] or photographs of target items taken from different views [60,66] but much less when they are required to name visually dissimilar versions of the trained item [40,60,72,91]. Encouragingly, evidence of producing trained words in other contexts after word training, such as fluency tasks (in one individual [63], naming to description [66], describing short videos of everyday scenes [51] or in production of a simple sentence construction, have also been observed [91].

To increase the chances of people with svPPA being able to correctly use the trained words in their everyday lives, it is helpful to tailor training stimuli to visually match the objects found within a person's home (likewise in actual object use, which was found to depend on personal familiarity with object exemplars [92]).

4.9. What Evidence Do We Have about Prophylactic Treatment in svPPA?

Prophylactic/preventative treatment aims to help retain current abilities by practising intact skills or items. There is some evidence suggesting that such preventative interventions may hold value in svPPA [73,78]. Several studies have found that treatment of items that could already be successfully named may slow the progression of semantic loss and anomia for those items [52–54,56].

4.10. Can We Deliver These Therapies Remotely? What Evidence Do We Have?

Digital technologies in treatment programmes provide opportunities to increase access for those with svPPA and their families who struggle to access expert care because of geographic location. Delivery of treatment via telehealth is highly relevant given the limited access to services for many individuals with PPA [78,83,93,94]. Significant improvements in word retrieval have been achieved after completing home-based programmes using either hardcopy or computer-mediated materials [18,40,53,54,58,60,65,66,68,95]. Rogalski and colleagues examined the feasibility of teletherapy for 28 individuals with PPA [96] showing that treatment delivered via video conferencing has the potential to improve access to care for people with PPA. Two studies conducted on people with svPPA show that lexical retrieval therapy can be delivered in-person or by teletherapy with similar results [78,83].

4.11. What Are the Barriers and Facilitators of Online Therapy?

Recognition of the barriers to, and facilitators of, successful implementation of remote digital therapy, however, is extremely important in both the research and clinical setting. Disease severity has been noted by several studies to be a contraindication for remote therapy and there is a recommendation that individuals participating in remote therapy should preferably be in the early to mid-stages of disease progression [83,91,96]. The inherent requirements of a technology can also be a barrier with the quality of audio and the stability of the internet connection being a prerequisite to successful participation online. In addition, the individual must possess adequate computer skills or a suitable support person to facilitate participation, particularly when carrying out intervention independently at home rather than supervised over the internet.

An example of these barriers, acting in concert, is provided by Taylor-Rubin et al. [91], reporting a series of single-case design treatment studies where lexical retrieval treatment was delivered via a computer-mediated home programme. Two of three svPPA participants had significant improvement in verb and noun production, following lexical retrieval treatment. However, a third participant, Nsv, showed only marginal gains over two blocks of lexical retrieval treatment. The authors hypothesised that as Nsv was five years post-onset, the severity of impairment may have contributed to less positive treatment results. Practice logs indicated poor adherence with computer operating difficulties preventing completion of all treatment schedule sessions in the second block of treatment [91].

A further barrier can be the lack of contact with the therapist. Caregivers in Rogalski's study reported that less than optimum opportunities for face-to-face support for the person with PPA, in times of distress, was a limitation of participation in the web-based treatment programme [96]. Similarly, caregivers of people with PPA, including svPPA, reported, in a study of treatment adherence, that home treatment programmes can be lonely and socially isolating and this would be anticipated to reduce adherence, "It is easier to fall off the wagon with a programme at home" [85]. Finally, the barrier of social isolation could be minimised by innovative networking; pairing peers with svPPA in small online groups, thus incorporating support, increased social participation and positive experiences [30].

5. Compensatory Approaches to Support Communication in svPPA

Aside from the direct treatment of language, a number of single-case studies have explored the benefit of using a compensatory approach to support language difficulties, particularly naming, in svPPA [26,97–99]. Compensatory approaches include the use of external devices to support communication, such as compensatory augmentative and alternative communication (AAC) systems [100]. These can be based on low (e.g., paper communication board or notebooks) or high technology (e.g., smartphones or tablets and computers); people with svPPA may use them in conjunction with verbal communication in a multi-modal way, multiplying the communication options available to them [100].

In two case studies, Bier and collaborators [26,97] explored the potential of using smartphone applications to help two people with svPPA learn how to search for information related to lost concepts through Internet search engines or a visual dictionary application named ARCUS©. This application aimed to support the retrieval of people's names from a virtual name directory using clues or information chosen by the person with svPPA. ARCUS© was successfully used by ND, a recreation therapist in a senior living facility with early svPPA [97]. In his work, ND had to identify a large number of people by name each day. At the start of the study, he used a paper notebook to do this, organised into several columns, each linked to a different piece of information (e.g., resident's room number or employee's job type). The authors converted ND's notebook into a smartphone application to ease its use and reduce the stigma associated with it. ND phased out his paper-based compensatory system in favour of this new, more flexible name retrieval system. Four years later, ND had extended the use of ARCUS© by adapting it to record information about grocery stores and food items to buy before he went shopping.

Another recent study has combined the classical use of mobile technology to develop CoChat, an app constructed on natural language processing (NLP) features, social media use, and just-in-time principles that was tested in two people with svPPA [98]. In this app the user takes a photograph with the tablet's built-in camera, shares the pictures with the person's simulated social network (e.g., family and friends) and sees comments to the images in real time. Results suggest that CoChat may improve word retrieval in a natural conversational context making conversations easier when using the app. As AAC devices and systems are becoming common practice in aphasia, further studies will have to deepen our understanding of how these types of tools can be optimised in svPPA.

Semantic deficits may sometimes prevent people with svPPA from understanding task requirements and limit their ability to learn certain functions of assistive technological devices [26,97] (e.g., being able to remember the series of actions required to obtain an Internet connection, but not understanding why). Nevertheless, taken together, these case studies suggest that it is possible to teach the use of practical, portable solutions to compensate for semantic memory deficits. Considering the degenerative nature of svPPA, it is important to integrate AAC with other treatment approaches as early as possible in the disease process so that they are well practised before the skills to acquire their use are lost [98,100]. Finally, although strategies of functional communication have been explored in individuals with PPA in general, there is a lack of studies examining non-AAC compensatory strategies targeting svPPA in particular.

6. Interventions to Support Activities of Daily Living

Complementary approaches that support engagement or re-engagement in meaningful activities of daily living are also important. Participation in meaningful activity is the primary focus of these kinds of interventions, without special consideration for language skills—although these may also benefit. They are oriented toward two objectives: (1) capitalising upon preserved episodic (e.g., what you had for lunch) and procedural memory functions (e.g., how to perform different skills, such as tying your shoes); (2)

focusing on significant and meaningful everyday activities that will have immediate results and a potential impact on well-being.

To our knowledge, only two studies have explored the engagement, or re-engagement, in meaningful daily living activities in svPPA [27,101]. In the first study, Bier et al. [27] did so by combining the repeated practice of an activity that the person had stopped doing (e.g., meal preparation) with a step-by-step cognitive assistive technology (SemAssist©). The objective was to support EC, a woman with left svPPA, to relearn how to prepare a specific recipe of her choice. This study showed that EC mainly used SemAssist© to follow the current steps during the activity. While she made many mistakes before the therapy sessions began, she was able to complete the recipe without error by the end of the process. Interestingly, EC also resumed spontaneous preparation of other recipes, showing that she had acquired new "knowledge" about the ingredients from the recipes she practised (e.g., "goes in the shrimp recipe") and did not overgeneralise. In the second study, O'Connor et al. [101] applied the Tailored Activity Program (TAP) with a person with svPPA who had highly repetitive routine behaviours. The TAP intervention resulted in this person engaging well in prescribed activities, with scores reflecting reduced carer distress regarding challenging behaviours and improved caregiver vigilance.

It therefore seems appropriate and promising to combine traditional language-based approaches with an interdisciplinary intervention that also incorporates a participatory approach such as occupational therapy or other meaningful activity interventions in svPPA.

7. Support Groups and Educational Programmes

One of the most recent developments in therapy for primary progressive aphasia is in group-based programmes offering education and support. While none of the published reports are specific to a particular PPA variant, they do include individuals with svPPA. In 2017, Jokel and colleagues published the first report of a group intervention programme that included both individuals with PPA and their caregivers [102]. The group members not only shared the intervention focus but, importantly, actively participated in defining it. Half of each session was spent on education, counselling and/or training communication strategies in dyads. The other half was separated into language activities for people with PPA and networking activities for caregivers. All participants reported valuing learning about the nature, progression and types of PPA, becoming familiar with current research in PPA, and several other aspects of the intervention. Components that were reported to be beneficial included receiving information on nutrition and lifestyle to support brain health, learning strategies for managing stress and depression, feeling understood by others in the group when experiencing difficulties during verbal communication, and getting support from multiple disciplines.

Although not specific to svPPA, to date, three more group interventions for PPA have also been reported [29,30,98]. Mooney and colleagues [98] developed a PPA group treatment model that incorporated elements of three methodologies used in language rehabilitation: communication strategies from augmentative and alternative communication, communication partner training from aphasia rehabilitation, and systematic instruction from dementia management. Morhardt et al. [29] describe the development of a programme that offered education, communication strategies, strategies to "live well" with PPA and non-language-based activities (e.g., watercolour painting and horticultural therapy). Finally, Taylor-Rubin et al. [30] delivered PPA education and support for a group of people with PPA and their caregivers in the early post-diagnostic period. In the post-intervention interview, participants highlighted the reduced feelings of isolation, increased feelings of support, increased knowledge of coping strategies and improved understanding of PPA as a result of participating in the programme soon after the receipt of the PPA diagnosis.

Based on the outcomes of these group interventions in PPA, several factors have emerged that may be critical to PPA care. First and foremost, the needs of both patients and caregivers should be addressed, preferably simultaneously [28,30]. A successful intervention programme for PPA should provide not just language activities and education but also a safe forum for discussing important and difficult issues, for sharing successes and failures, and for peer education. Such a programme is likely to ultimately result in improvements in confidence and well-being for both individuals with PPA and their caregivers. Published studies underline that PPA-specific education and ecologically valid context (i.e., group format) are positive elements highlighted by all participants. In addition, having consistent peer support helps to "normalise" daily challenges. It has been suggested that self-help groups may be beneficial in maintaining the group intervention benefits and they are recommended even in the absence of professional input [103].

As more and more services are being offered online, the support for individuals dealing with svPPA may also migrate to virtual space. A review of virtual support groups for dementia caregivers [104] suggests that weekly or monthly sessions can provide participants with knowledge about dementia, caregiving skills, coping strategies and access to resources. While occasional barriers, such as technology and access, were identified, there are also numerous economic and geographical advantages to online group sessions. Extrapolating from the broad dementia field, we may predict that the trend towards virtual care in svPPA will continue.

8. Future Directions in Behavioural Therapies in svPPA

We have shown in this synthesis of evidence that there has been relatively little research on intervention for words and concepts that the person with svPPA can no longer understand, and that therapy gains for such words show limited generalisation. Far from indicating that conceptual restoration or generalisation is not possible, we argue that optimal treatments may not yet have been found, and that this should motivate future research. On the other hand, the use of compensatory approaches to supporting communication and activities of daily living (e.g., assistive technologies) is promising and has the potential to make a difference to the lives of people with svPPA. The next steps should therefore be directed towards: (1) the development of more precise naming therapies, tailored to the level of semantic degradation of the words and concepts treated; (2) finding ways to guarantee transfer and generalisation of therapy gains to daily life; (3) expanding research into the use of assistive technologies, compensatory strategies, programmes to support daily living and how and when to combine these components.

9. Conclusions

The last two decades have witnessed rapid advances in the understanding and treatment of svPPA. The current body of research suggests that people with svPPA who have access to non-pharmacological therapies show favourable outcomes and long-lasting effects that can have benefits for health outcomes. Moreover, these treatments are generally well accepted. Although there is a lack of empirical research examining what the optimal combination and timing for treatments are, there are general guidelines for delivering language therapy at different stages of PPA that offer pragmatic advice about how to combine different therapy approaches in a meaningful way [105]. Current ongoing research around the staging of PPA (including svPPA) will make it easier to match therapies to impairments in the future. We therefore advocate for the svPPA care pathway to include a wide range of therapeutic options including both restorative and compensatory strategies and educational and support groups for people with svPPA as well as their care partners. These therapeutic options have the potential to become more accessible due to the advent of telemedicine, which has overcome geographical barriers and can provide care of similar efficacy to face-to-face therapy. Finally, to facilitate dissemination beyond an academic audience we have included a lay summary in English, Spanish and French (Supplementary Materials S1–S3).

Supplementary Materials: The following are available online at https://www.mdpi.com/article/10.3390/brainsci11121552/s1. Supplementary 1 (English). Semantic Variant Primary Progressive Aphasia (svPPA): evidence-based recommendations for therapy and management; Supplementary 2 (French). Variante sémantique de l'aphasie progressive primaire (vsAPP): indications fondées sur les données probantes pour le traitement et la prise en charge; Supplementary 3 (Spanish). Variante semántica de la afasia progresiva primaria (APP-s): recomendaciones basadas en evidencia para el manejo terapéutico.

Author Contributions: A.S.-G. and C.T.-R. contributed to the conceptualisation, original draft preparation, critical review and editing. S.A.S., N.B. and R.J. contributed to the original draft preparation, critical review and editing. M.L.H. and L.N. contributed to the critical review and editing. All authors have read and agreed to the published version of the manuscript.

Funding: A.S.-G. was funded by a UKRI Healthy Ageing Challenge Catalyst Award (ES/W006405/1) and a grant jointly funded by the Economic and Social Research Council (UK) and the National Institute for Health Research (UK) (ES/S010467/1). N.B. was supported by a research scholarship from the Fonds de la recherche du Québec-Santé. M.H. was funded by the National Institutes of Health (NIDCD R01DC016291) and by the Darrell K Royal Research Fund for Alzheimer's Disease.

Institutional Review Board Statement: Not applicable.

Informed Consent Statement: Not applicable.

Data Availability Statement: Not applicable.

Conflicts of Interest: The authors declare no conflict of interest.

References

1. Warrington, E.K. The Selective Impairment of Semantic Memory. *Q. J. Exp. Psychol.* **1975**, *27*, 635–657. [CrossRef] [PubMed]
2. Julie, S.; Goulding, P.J.; Neary, D. Semantic dementia: A form of circumscribed cerebral atrophy. *Behav. Neurol.* **1989**, *2*, 167–182. [CrossRef]
3. Gorno-Tempini, M.L.; Hillis, A.E.; Weintraub, S.; Kertesz, A.; Mendez, M.; Cappa, S.F.; Ogar, J.M.; Rohrer, J.D.; Black, S.; Boeve, B.F.; et al. Classification of primary progressive aphasia and its variants. *Neurology* **2011**, *76*, 1006–1014. [CrossRef]
4. Neary, D.; Snowden, J.; Gustafson, L.; Passant, U.; Stuss, D.; Black, S.; Freedman, M.; Kertesz, A.; Robert, P.H.; Albert, M.; et al. Frontotemporal lobar degeneration: A consensus on clinical diagnostic criteria. *Neurology* **1998**, *51*, 1546–1554. [CrossRef] [PubMed]
5. Mesulam, M.M. Primary progressive aphasia. *Ann. Neurol.* **2001**, *49*, 425–432. [CrossRef] [PubMed]
6. Harciarek, M.; Sitek, E.J.; Kertesz, A. The patterns of progression in primary progressive aphasia—Implications for assessment and management. *Aphasiology* **2014**, *28*, 964–980. [CrossRef]
7. Coyle-Gilchrist, I.T.; Dick, K.M.; Patterson, K.; Rodríquez, P.V.; Wehmann, E.; Wilcox, A.; Lansdall, C.J.; Dawson, K.E.; Wiggins, J.; Mead, S.; et al. Prevalence, characteristics, and survival of frontotemporal lobar degeneration syndromes. *Neurology* **2016**, *86*, 1736–1743. [CrossRef] [PubMed]
8. Hodges, J.R.; Mitchell, J.; Dawson, K.; Spillantini, M.G.; Xuereb, J.H.; McMonagle, P.; Nestor, P.J.; Patterson, K. Semantic dementia: Demography, familial factors and survival in a consecutive series of 100 cases. *Brain* **2010**, *133*, 300–306. [CrossRef]
9. Chan, D.; Fox, N.C.; Scahill, R.I.; Crum, W.R.; Whitwell, J.L.; Leschziner, G.; Rossor, A.M.; Stevens, J.M.; Cipolotti, L.; Rossor, M.N. Patterns of temporal lobe atrophy in semantic dementia and Alzheimer's disease. *Ann. Neurol.* **2001**, *49*, 433–442. [CrossRef]
10. Chen, K.; Ding, J.; Lin, B.; Huang, L.; Tang, L.; Bi, Y.; Han, Z.; Lv, Y.; Guo, Q. The neuropsychological profiles and semantic-critical regions of right semantic dementia. *NeuroImage Clin.* **2018**, *19*, 767–774. [CrossRef]
11. Ding, J.; Chen, K.; Chen, Y.; Fang, Y.; Yang, Q.; Lv, Y.; Lin, N.; Bi, Y.; Guo, Q.; Han, Z. The Left Fusiform Gyrus is a Critical Region Contributing to the Core Behavioral Profile of Semantic Dementia. *Front. Hum. Neurosci.* **2016**, *10*, 215. [CrossRef]
12. Yang, J.; Pan, P.; Song, W.; Shang, H.-F. Quantitative Meta-Analysis of Gray Matter Abnormalities in Semantic Dementia. *J. Alzheimers Dis.* **2012**, *31*, 827–833. [CrossRef] [PubMed]
13. Mion, M.; Patterson, K.; Acosta-Cabronero, J.; Pengas, G.; Izquierdo-Garcia, D.; Hong, Y.T.; Fryer, T.D.; Williams, G.B.; Hodges, J.R.; Nestor, P. What the left and right anterior fusiform gyri tell us about semantic memory. *Brain* **2010**, *133*, 3256–3268. [CrossRef]
14. Hodges, J.R.; Davies, R.R.; Xuereb, J.H.; Casey, B.; Broe, M.; Bak, T.H.; Kril, J.J.; Halliday, G.M. Clinicopathological correlates in frontotemporal dementia. *Ann. Neurol.* **2004**, *56*, 399–406. [CrossRef] [PubMed]
15. Davies, R.R.; Hodges, J.R.; Kril, J.J.; Patterson, K.; Halliday, G.M.; Xuereb, J.H. The pathological basis of semantic dementia. *Brain* **2005**, *128*, 1984–1995. [CrossRef]
16. Grossman, M.; Wood, E.M.; Moore, P.; Neumann, M.; Kwong, L.; Forman, M.S.; Clark, C.M.; McCluskey, L.F.; Miller, B.L.; Lee, V.M.; et al. TDP-43 Pathologic Lesions and Clinical Phenotype in Frontotemporal Lobar Degeneration with Ubiquitin-Positive Inclusions. *Arch. Neurol.* **2007**, *64*, 1449. [CrossRef]
17. Santos-Santos, M.A.; Rabinovici, G.D.; Iaccarino, L.; Ayakta, N.; Tammewar, G.; Lobach, I.; Henry, M.L.; Hubbard, I.; Mandelli, M.L.; Spinelli, E.; et al. Rates of Amyloid Imaging Positivity in Patients with Primary Progressive Aphasia. *JAMA Neurol.* **2018**, *75*, 342. [CrossRef] [PubMed]
18. Graham, K.S.; Patterson, K.; Pratt, K.H.; Hodges, J.R. Relearning and subsequent forgetting of semantic category exemplars in a case of semantic dementia. *Neuropsychology* **1999**, *13*, 359–380. [CrossRef]

19. Funnell, E. A case of forgotten knowledge. In *Broken Memories: Case Studies in Memory Impairment*; Campbell, R., Conway, M.A., Eds.; Wiley-Blackwell: Oxford, UK, 1995; pp. 225–236.

20. Carthery-Goulart, M.T.; Silveira, A.D.; Machado, T.H.; Mansur, L.L.; Parente, M.A.; Senaha, M.L.; Brucki, S.M.; Nitrini, R. Nonpharmacological interventions for cognitive impairments following primary progressive aphasia: A systematic review of the literature. *Dement. Neuropsychol.* **2013**, *7*, 122–131. [CrossRef]

21. Jokel, R.; Graham, N.L.; Rochon, E.; Leonard, C. Word retrieval therapies in primary progressive aphasia. *Aphasiology* **2014**, *28*, 1038–1068. [CrossRef]

22. Cotelli, M.; Manenti, R.; Ferrari, C.; Gobbi, E.; Macis, A.; Cappa, S.F. Effectiveness of language training and non-invasive brain stimulation on oral and written naming performance in Primary Progressive Aphasia: A meta-analysis and systematic review. *Neurosci. Biobehav. Rev.* **2020**, *108*, 498–525. [CrossRef] [PubMed]

23. Pagnoni, I.; Gobbi, E.; Premi, E.; Borroni, B.; Binetti, G.; Cotelli, M.; Manenti, R. Language training for oral and written naming impairment in primary progressive aphasia: A review. *Transl. Neurodegener.* **2021**, *10*, 24. [CrossRef] [PubMed]

24. Taylor-Rubin, C.; Croot, K.; Power, E.; Savage, S.A.; Hodges, J.R.; Togher, L. Communication behaviors associated with successful conversation in semantic variant primary progressive aphasia. *Int. Psychogeriatr.* **2017**, *29*, 1619–1632. [CrossRef] [PubMed]

25. Croot, K.; Nickels, L.; Laurence, F.; Manning, M. Impairment- and activity/participation-directed interventions in progressive language impairment: Clinical and theoretical issues. *Aphasiology* **2009**, *23*, 125–160. [CrossRef]

26. Bier, N.; Brambati, S.; Macoir, J.; Paquette, G.; Schmitz, X.; Belleville, S.; Faucher, C.; Joubert, S. Relying on procedural memory to enhance independence in daily living activities: Smartphone use in a case of semantic dementia. *Neuropsychol. Rehabil.* **2015**, *25*, 913–915. [CrossRef]

27. Bier, N.; Macoir, J.; Joubert, S.; Bottari, C.; Chayer, C.; Pigot, H.; Giroux, S.; Team, S. Cooking "Shrimp à la Créole": A pilot study of an ecological rehabilitation in semantic dementia. *Neuropsychol. Rehabil.* **2011**, *21*, 455–483. [CrossRef] [PubMed]

28. Jokel, R.; Meltzer, J. Group intervention for individuals with primary progressive aphasia and their spouses: Who comes first? *J. Commun. Disord.* **2017**, *66*, 51–64. [CrossRef]

29. Morhardt, D.J.; O'Hara, M.C.; Zachrich, K.; Wieneke, C.; Rogalski, E.J. Development of a Psycho-Educational Support Program for Individuals with Primary Progressive Aphasia and their Care-Partners. *Dementia* **2019**, *18*, 1310–1327. [CrossRef]

30. Taylor-Rubin, C.; Azizi, L.; Croot, K.; Nickels, L. Primary Progressive Aphasia Education and Support Groups: A Clinical Evaluation. *Am. J. Alzheimers Dis. Demen.* **2020**, *35*, 1533317519895638. [CrossRef]

31. Taylor-Rubin, C.; Croot, K.; Nickels, L. Speech and language therapy in primary progressive aphasia: A critical review of current practice. *Expert Rev. Neurother.* **2021**, *21*, 419–430. [CrossRef]

32. Lambon Ralph, M.A.; McClelland, J.L.; Patterson, K.; Galton, C.J.; Hodges, J.R. No right to speak? The relationship between object naming and semantic impairment: Neuropsychological evidence and a computational model. *J. Cogn. Neurosci.* **2001**, *13*, 341–356. [CrossRef]

33. Rohrer, J.D.; Knight, W.D.; Warren, J.E.; Fox, N.C.; Rossor, M.N.; Warren, J.D. Word-finding difficulty: A clinical analysis of the progressive aphasias. *Brain J. Neurol.* **2008**, *131 Pt 1*, 8–38. [CrossRef]

34. Hodges, J.R.; Graham, N.; Patterson, K. Charting the progression in semantic dementia: Implications for the organisation of semantic memory. *Memory* **1995**, *3*, 463–495. [CrossRef] [PubMed]

35. O'Reilly, R.C.; Bhattacharyya, R.; Howard, M.D.; Ketz, N. Complementary learning systems. *Cogn. Sci.* **2014**, *38*, 1229–1248. [CrossRef] [PubMed]

36. Patterson, K.; Nestor, P.J.; Rogers, T.T. Where do you know what you know? The representation of semantic knowledge in the human brain. *Nat. Rev. Neurosci.* **2007**, *8*, 976–987. [CrossRef]

37. Patterson, K.; Lambon Ralph, M.A. The Hub-and-Spoke Hypothesis of Semantic Memory. In *Neurobiology of Language*; Elsevier: Amsterdam, The Netherlands, 2016; pp. 765–775.

38. Ralph, M.A.L.; Jefferies, E.; Patterson, K.; Rogers, T.T. The neural and computational bases of semantic cognition. *Nat. Rev. Neurosci.* **2017**, *18*, 42–55. [CrossRef]

39. Irish, M.; Bunk, S.; Tu, S.; Kamminga, J.; Hodges, J.R.; Hornberger, M.; Piguet, O. Preservation of episodic memory in semantic dementia: The importance of regions beyond the medial temporal lobes. *Neuropsychologia* **2016**, *81*, 50–60. [CrossRef] [PubMed]

40. Mayberry, E.J.; Sage, K.; Ehsan, S.; Ralph, M.A.L. Relearning in semantic dementia reflects contributions from both medial temporal lobe episodic and degraded neocortical semantic systems: Evidence in support of the complementary learning systems theory. *Neuropsychologia* **2011**, *49*, 3591–3598. [CrossRef]

41. Kumfor, F.; Landin-Romero, R.; Devenney, E.; Hutchings, R.; Grasso, R.; Hodges, J.R.; Piguet, O. On the right side? A longitudinal study of left- versus right-lateralized semantic dementia. *Brain* **2016**, *139*, 986–998. [CrossRef]

42. Thompson, S.A.; Patterson, K.; Hodges, J.R. Left/right asymmetry of atrophy in semantic dementia: Behavioral-cognitive implications. *Neurology* **2003**, *61*, 1196–1203. [CrossRef]

43. Hodges, J.R.; Patterson, K. Semantic dementia: A unique clinicopathological syndrome. *Lancet Neurol.* **2007**, *6*, 1004–1014. [CrossRef]

44. Woollams, A.M.; Patterson, K. Cognitive consequences of the left-right asymmetry of atrophy in semantic dementia. *Cortex.* **2018**, *107*, 64–77. [CrossRef]

45. Ulugut Erkoyun, H.; Groot, C.; Heilbron, R.; Nelissen, A.; van Rossum, J.; Jutten, R.; Koene, T.; van der Flier, W.M.; Wattjes, M.P.; Scheltens, P.; et al. A clinical-radiological framework of the right temporal variant of frontotemporal dementia. *Brain J. Neurol.* **2020**, *143*, 2831–2843. [CrossRef]

46. Josephs, K.A.; Whitwell, J.L.; Vemuri, P.; Senjem, M.L.; Boeve, B.F.; Knopman, D.S.; Smith, G.E.; Ivnik, R.J.; Petersen, R.C.; Jack, C.R. The anatomic correlate of prosopagnosia in semantic dementia. *Neurology* **2008**, *71*, 1628–1633. [CrossRef] [PubMed]

47. Nakachi, R.; Muramatsu, T.; Kato, M.; Akiyama, T.; Saito, F.; Yoshino, F.; Mimura, M.; Kashima, H. Progressive prosopagnosia at a very early stage of frontotemporal lobar degeneration. *Psychogeriatrics* **2007**, *7*, 155–162. [CrossRef]

48. Henry, M.L.; Wilson, S.M.; Ogar, J.M.; Sidhu, M.S.; Rankin, K.P.; Cattaruzza, T.; Miller, B.L.; Gorno-Tempini, M.L.; Seeley, W.W. Neuropsychological, behavioral, and anatomical evolution in right temporal variant frontotemporal dementia: A longitudinal and post-mortem single case analysis. *Neurocase* **2014**, *20*, 100–109. [CrossRef] [PubMed]

49. Suárez-González, A.; Crutch, S.J. Relearning knowledge for people in a case of right variant frontotemporal dementia. *Neurocase* **2016**, *22*, 130–134. [CrossRef] [PubMed]

50. Snowden, J.S.; Harris, J.M.; Thompson, J.; Kobylecki, C.; Jones, M.; Richardson, A.M.; Neary, D. Semantic dementia and the left and right temporal lobes. *Cortex* **2018**, *107*, 188–203. [CrossRef] [PubMed]

51. Savage, S.A.; Piguet, O.; Hodges, J.R. Giving words new life: Generalization of word retraining outcomes in Semantic Dementia. *J. Alzheimers Dis.* **2014**, *40*, 309–317. [CrossRef]

52. Jokel, R.; Anderson, N.D. Quest for the best: Effects of errorless and active encoding on word re-learning in semantic dementia. *Neuropsychol. Rehabil.* **2012**, *22*, 187–214. [CrossRef]

53. Jokel, R.; Rochon, E.; Leonard, C. Treating anomia in semantic dementia: Improvement, maintenance, or both? *Neuropsychol. Rehabil.* **2006**, *16*, 241–256. [CrossRef]

54. Savage, S.A.; Piguet, O.; Hodges, J.R. Cognitive Intervention in Semantic Dementia: Maintaining Words Over Time. *Alzheimers Dis. Assoc. Disord.* **2015**, *29*, 55–62. [CrossRef]

55. Senaha, M.L.H.; Brucki, S.M.D.; Nitrini, R. Rehabilitation in semantic dementia: Study of effectiveness of lexical reacquisition in three patients. *Dement. Neuropsychol.* **2010**, *4*, 306–312. [CrossRef]

56. Macoir, J.; Leroy, M.; Routhier, S.; Auclair-Ouellet, N.; Houde, M.; Laforce, R. Improving verb anomia in the semantic variant of primary progressive aphasia: The effectiveness of a semantic-phonological cueing treatment. *Neurocase* **2015**, *21*, 448–456. [CrossRef]

57. Jokel, R.; Kielar, A.; Anderson, N.; Black, S.E.; Rochon, E.; Graham, S.; Freedman, M.; Tang-Wai, D.F. Behavioural and neuroimaging changes after naming therapy for semantic variant primary progressive aphasia. *Neuropsychologia* **2016**, *89*, 191–216. [CrossRef] [PubMed]

58. Snowden, J.S.; Neary, D. Relearning of verbal labels in semantic dementia. *Neuropsychologia* **2002**, *40*, 1715–1728. [CrossRef]

59. Bier, N.; Macoir, J.; Gagnon, L.; Van der Linden, M.; Louveaux, S.; Desrosiers, J. Known, lost, and recovered: Efficacy of formal-semantic therapy and spaced retrieval method in a case of semantic dementia. *Aphasiology* **2009**, *23*, 210–235. [CrossRef]

60. Heredia, C.G.; Sage, K.; Ralph, M.A.L.; Berthier, M.L. Relearning and retention of verbal labels in a case of semantic dementia. *Aphasiology* **2009**, *23*, 192–209. [CrossRef]

61. Robinson, S.; Druks, J.; Hodges, J.; Garrard, P. The treatment of object naming, definition, and object use in semantic dementia: The effectiveness of errorless learning. *Aphasiology* **2009**, *23*, 749–775. [CrossRef]

62. Dressel, K.; Huber, W.; Frings, L.; Kümmerer, D.; Saur, D.; Mader, I.; Hüll, M.; Weiller, C.; Abel, S. Model-oriented naming therapy in semantic dementia: A single-case fMRI study. *Aphasiology* **2010**, *24*, 1537–1558. [CrossRef]

63. Jokel, R.; Rochon, E.; Anderson, N.D. Errorless learning of computer-generated words in a patient with semantic dementia. *Neuropsychol. Rehabil.* **2010**, *20*, 16–41. [CrossRef]

64. Beales, A.; Cartwright, J.; Whitworth, A.; Panegyres, P.K. Exploring generalisation processes following lexical retrieval intervention in primary progressive aphasia. *Int. J. Speech Lang. Pathol.* **2016**, *18*, 299–314. [CrossRef] [PubMed]

65. Savage, S.A.; Ballard, K.J.; Piguet, O.; Hodges, J.R. Bringing words back to mind—Improving word production in semantic dementia. *Cortex* **2013**, *49*, 1823–1832. [CrossRef] [PubMed]

66. Suárez-González, A.; Heredia, C.G.; Savage, S.A.; Gil-Néciga, E.; García-Casares, N.; Franco-Macías, E.; Berthier, M.L.; Caine, D. Restoration of conceptual knowledge in a case of semantic dementia. *Neurocase* **2014**, *21*, 309–321. [CrossRef]

67. Krajenbrink, T.; Croot, K.; Taylor-Rubin, C.; Nickels, L. Treatment for spoken and written word retrieval in the semantic variant of primary progressive aphasia. *Neuropsychol. Rehabil.* **2020**, *30*, 915–947. [CrossRef] [PubMed]

68. Graham, K.S.; Patterson, K.; Pratt, K.H.; Hodges, J.R. Can repeated exposure to 'forgotten' vocabulary help alleviate word-finding difficulties in semantic dementia? An illustrative case study. *Neuropsychol. Rehabil.* **2001**, *11*, 429–454. [CrossRef]

69. Hoffman, P.; Clarke, N.; Jones, R.W.; Noonan, K.A. Vocabulary relearning in semantic dementia: Positive and negative consequences of increasing variability in the learning experience. *Neuropsychologia* **2015**, *76*, 240–253. [CrossRef] [PubMed]

70. Suárez-González, A.; Savage, S.A.; Caine, D. Successful short-term re-learning and generalisation of concepts in semantic dementia. *Neuropsychol. Rehabil.* **2018**, *28*, 1095–1109. [CrossRef]

71. Montagut, N.; Borrego-Écija, S.; Castellví, M.; Rico, I.; Reñé, R.; Balasa, M.; Lladó, A.; Sánchez-Valle, R. Errorless Learning Therapy in Semantic Variant of Primary Progressive Aphasia. *J. Alzheimers Dis.* **2021**, *79*, 415–422. [CrossRef]

72. Croot, K.; Raiser, T.; Taylor-Rubin, C.; Ruggero, L.; Ackl, N.; Wlasich, E.; Danek, A.; Scharfenberg, A.; Foxe, D.; Hodges, J.R.; et al. Lexical retrieval treatment in primary progressive aphasia: An investigation of treatment duration in a heterogeneous case series. *Cortex* **2019**, *115*, 133–158. [CrossRef]

73. Flurie, M.; Ungrady, M.; Reilly, J. Evaluating a Maintenance-Based Treatment Approach to Preventing Lexical Dropout in Progressive Anomia. *J. Speech Lang. Hear Res.* **2020**, *63*, 4082–4095. [CrossRef]

74. Lavoie, M.; Bier, N.; Laforce, R.; Macoir, J. Improvement in functional vocabulary and generalization to conversation following a self-administered treatment using a smart tablet in primary progressive aphasia. *Neuropsychol. Rehabil.* **2020**, *30*, 1224–1254. [CrossRef] [PubMed]

75. Evans, W.S.; Quimby, M.; Dickey, M.W.; Dickerson, B.C. Relearning and Retaining Personally-Relevant Words using Computer-Based Flashcard Software in Primary Progressive Aphasia. *Front. Hum. Neurosci.* **2016**, *10*, 561. [CrossRef] [PubMed]

76. Henry, M.; Beeson, P.; Rapcsak, S. Treatment for Anomia in Semantic Dementia. *Semin Speech Lang.* **2008**, *29*, 060–070. [CrossRef] [PubMed]

77. Meyer, A.M.; Tippett, D.C.; Friedman, R.B. Prophylaxis and remediation of anomia in the semantic and logopenic variants of primary progressive aphasia. *Neuropsychol. Rehabil.* **2018**, *28*, 352–368. [CrossRef]

78. Meyer, A.M.; Getz, H.R.; Brennan, D.M.; Hu, T.M.; Friedman, R.B. Telerehabilitation of anomia in primary progressive aphasia. *Aphasiology* **2016**, *30*, 483–507. [CrossRef]

79. Meyer, A.M.; Tippett, D.C.; Turner, R.S.; Friedman, R.B. Long-Term maintenance of anomia treatment effects in primary progressive aphasia. *Neuropsychol. Rehabil.* **2019**, *29*, 1439–1463. [CrossRef]

80. Newhart, M.; Davis, C.; Kannan, V.; Heidler-Gary, J.; Cloutman, L.; Hillis, A.E. Therapy for naming deficits in two variants of primary progressive aphasia. *Aphasiology* **2009**, *23*, 823–834. [CrossRef]

81. Henry, M.L.; Rising, K.; DeMarco, A.T.; Miller, B.L.; Gorno-Tempini, M.L.; Beeson, P.M. Examining the value of lexical retrieval treatment in primary progressive aphasia: Two positive cases. *Brain Lang.* **2013**, *127*, 145–156. [CrossRef]

82. Henry, M.L.; Hubbard, H.I.; Grasso, S.M.; Dial, H.R.; Beeson, P.M.; Miller, B.L.; Gorno-Tempini, M.L. Treatment for Word Retrieval in Semantic and Logopenic Variants of Primary Progressive Aphasia: Immediate and Long-Term Outcomes. *J. Speech Lang. Hear Res.* **2019**, *62*, 2723–2749. [CrossRef]

83. Dial, H.R.; Hinshelwood, H.A.; Grasso, S.M.; Hubbard, H.I.; Gorno-Tempini, M.-L.; Henry, M.L. Investigating the utility of teletherapy in individuals with primary progressive aphasia. *Clin. Interv. Aging* **2019**, *14*, 453–471. [CrossRef]

84. Snowden, J.S.; Griffiths, H.; Neary, D. Semantic dementia: Autobiographical contribution to preservation of meaning. *Cogn. Neuropsychol.* **1994**, *11*, 265–288. [CrossRef]

85. Taylor-Rubin, C.; Croot, K.; Nickels, L. Adherence to lexical retrieval treatment in Primary Progressive Aphasia and implications for candidacy. *Aphasiology* **2019**, *33*, 1182–1201. [CrossRef]

86. Savage, S.A.; Piguet, O.; Hodges, J.R. 'Knowing What You Don't Know': Language Insight in Semantic Dementia. *J. Alzheimers Dis.* **2015**, *46*, 187–198. [CrossRef]

87. Cadório, I.; Lousada, M.; Martins, P.; Figueiredo, D. Generalization and maintenance of treatment gains in primary progressive aphasia (PPA): A systematic review. *Int. J. Lang. Commun. Disord.* **2017**, *52*, 543–560. [CrossRef] [PubMed]

88. Reilly, J. How to constrain and maintain a lexicon for the treatment of progressive semantic naming deficits: Principles of item selection for formal semantic therapy. *Neuropsychol. Rehabil.* **2016**, *26*, 126–156. [CrossRef]

89. Best, W.; Greenwood, A.; Grassly, J.; Herbert, R.; Hickin, J.; Howard, D. Aphasia rehabilitation: Does generalisation from anomia therapy occur and is it predictable? A case series study. *Cortex* **2013**, *49*, 2345–2357. [CrossRef] [PubMed]

90. Subedi, B.S. Emerging trends of research on transfer of learning. *Int. Educ. J.* **2004**, *5*, 591–599.

91. Taylor-Rubin, C.; Nickels, L.; Croot, K. Exploring the effects of verb and noun treatment on verb phrase production in primary progressive aphasia: A series of single case experimental design studies. *Neuropsychol. Rehabil.* **2021**, 1–43. [CrossRef]

92. Bozeat, S.; Ralph, M.A.L.; Patterson, K.; Hodges, J.R. The Influence of Personal Familiarity and Context on Object Use in Semantic Dementia. *Neurocase* **2002**, *8*, 127–134. [CrossRef]

93. Rutherford, S. Our journey with primary progressive aphasia. *Aphasiology* **2014**, *28*, 900–908. [CrossRef]

94. Taylor, C.; Kingma, R.M.; Croot, K.; Nickels, L. Speech pathology services for primary progressive aphasia: Exploring an emerging area of practice. *Aphasiology* **2009**, *23*, 161–174. [CrossRef]

95. Jokel, R.; Rochon, E.; Leonard, C. Therapy for anomia in semantic dementia. *Brain Cogn.* **2002**, *49*, 241–244. [PubMed]

96. Rogalski, E.J.; Saxon, M.; McKenna, H.; Wieneke, C.; Rademaker, A.; Corden, M.E.; Borio, K.; Mesulam, M.-M.; Khayum, B. Communication Bridge: A pilot feasibility study of Internet-based speech-language therapy for individuals with progressive aphasia. *Alzheimers Dement. Transl. Res. Clin. Interv.* **2016**, *2*, 213–221. [CrossRef] [PubMed]

97. Bier, N.; Paquette, G.; Macoir, J. Smartphone for smart living: Using new technologies to cope with everyday limitations in semantic dementia. *Neuropsychol. Rehabil.* **2018**, *28*, 734–754. [CrossRef] [PubMed]

98. Mooney, A.; Bedrick, S.; Noethe, G.; Spaulding, S.; Fried-Oken, M. Mobile technology to support lexical retrieval during activity retell in primary progressive aphasia. *Aphasiology* **2018**, *32*, 666–692. [CrossRef]

99. Routhier, S.; Macoir, J.; Imbeault, H.; Jacques, S.; Pigot, H.; Giroux, S.; Cau, A.; Bier, N. From smartphone to external semantic memory device: The use of new technologies to compensate for semantic deficits. *Non-Pharmacol. Ther. Dement.* **2011**, *2*, 81.

100. Fried-Oken, M.; Beukelman, D.R.; Hux, K. Current and future AAC research considerations for adults with acquired cognitive and communication impairments. *Assist. Technol.* **2011**, *24*, 56–66. [CrossRef]

101. O'Connor, C.M.; Clemson, L.; Brodaty, H.; Gitlin, L.N.; Piguet, O.; Mioshi, E. Enhancing caregivers' understanding of dementia and tailoring activities in frontotemporal dementia: Two case studies. *Disabil. Rehabil.* **2016**, *38*, 704–714. [CrossRef] [PubMed]

102. Lanyon, L.E.; Rose, M.L.; Worrall, L. The efficacy of outpatient and community-based aphasia group interventions: A systematic review. *Int. J. Speech Lang. Pathol.* **2013**, *15*, 359–374. [CrossRef]

103. Davies, K.; Howe, T. Experiences of Living with Primary Progressive Aphasia: A Scoping Review of Qualitative Studies. *Am. J. Alzheimers Dis. Dement.* **2020**, *35*, 1533317519886218. [CrossRef] [PubMed]

104. Armstrong, M.J.; Alliance, S. Virtual Support Groups for Informal Caregivers of Individuals with Dementia: A Scoping Review. *Alzheimers Dis. Assoc. Disord.* **2019**, *33*, 362–369. [CrossRef] [PubMed]

105. Jokel, R. Planning for patient deterioration. *ASHA Lead.* **2021**, *26*, 32–41.

Article

Treatment for Anomia in Bilingual Speakers with Progressive Aphasia

Stephanie M. Grasso [1,*], Elizabeth D. Peña [2], Nina Kazemi [1], Haideh Mirzapour [1], Rozen Neupane [1], Borna Bonakdarpour [3], Maria Luisa Gorno-Tempini [4] and Maya L. Henry [1,5]

[1] Department of Speech, Language and Hearing Sciences, University of Texas, Austin, TX 78705, USA; kazeminina@gmail.com (N.K.); haideh39@yahoo.com (H.M.); neupane.rozen@gmail.com (R.N.); maya.henry@austin.utexas.edu (M.L.H.)
[2] School of Education, University of California, Irvine, CA 92697, USA; edpena@uci.edu
[3] Feinberg School of Medicine, Northwestern University, Chicago, IL 60611, USA; bbk@northwestern.edu
[4] Memory and Aging Center, Department of Neurology, University of California San Francisco, San Francisco, CA 94143, USA; MariaLuisa.GornoTempini@ucsf.edu
[5] Department of Neurology, Dell Medical School, University of Texas, Austin, TX 78705, USA
* Correspondence: smgrasso@austin.utexas.edu

Abstract: Anomia is an early and prominent feature of primary progressive aphasia (PPA) and other neurodegenerative disorders. Research investigating treatment for lexical retrieval impairment in individuals with progressive anomia has focused primarily on monolingual speakers, and treatment in bilingual speakers is relatively unexplored. In this series of single-case experiments, 10 bilingual speakers with progressive anomia received lexical retrieval treatment designed to engage relatively spared cognitive-linguistic abilities and promote word retrieval. Treatment was administered in two phases, with one language targeted per phase. Cross-linguistic cognates (e.g., rose and *rosa*) were included as treatment targets to investigate their potential to facilitate cross-linguistic transfer. Performance on trained and untrained stimuli was evaluated before, during, and after each phase of treatment, and at 3, 6, and 12 months post-treatment. Participants demonstrated a significant treatment effect in each of their treated languages, with maintenance up to one year post-treatment for the majority of participants. Most participants showed a significant cross-linguistic transfer effect for trained cognates in both the dominant and nondominant language, with fewer than half of participants showing a significant translation effect for noncognates. A gradual diminution of translation and generalization effects was observed during the follow-up period. Findings support the implementation of dual-language intervention approaches for bilingual speakers with progressive anomia, irrespective of language dominance.

Keywords: bilingualism; primary progressive aphasia; treatment; intervention

Citation: Grasso, S.M.; Peña, E.D.; Kazemi, N.; Mirzapour, H.; Neupane, R.; Bonakdarpour, B.; Gorno-Tempini, M.L.; Henry, M.L. Treatment for Anomia in Bilingual Speakers with Progressive Aphasia. *Brain Sci.* **2021**, *11*, 1371. https://doi.org/10.3390/brainsci11111371

Academic Editors: Jordi A. Matias-Guiu, Robert Laforce, Jr. and Rene L. Utianski

Received: 25 August 2021
Accepted: 9 October 2021
Published: 20 October 2021

Publisher's Note: MDPI stays neutral with regard to jurisdictional claims in published maps and institutional affiliations.

1. Introduction

The majority of individuals worldwide speak two or more languages (e.g., [1,2]); nonetheless, most studies that have evaluated the benefits of speech-language intervention for individuals with aphasia have focused on monolingual speakers (e.g., [3–8]). This disparity is even more striking in aphasia caused by neurodegenerative disease (e.g., [9,10]). In the United States, bilingual speakers are more likely to belong to historically minoritized populations (e.g., [10]). Therefore, the lack of evidence regarding treatment for bilingual speakers with aphasia disproportionately impacts individuals from historically marginalized populations, which, in turn, contributes to health disparities in these groups. In an era of globalization, speech-language pathologists are increasingly called upon to provide services for individuals who speak more than one language [11,12]. This necessitates careful consideration of therapeutic manipulations that may be used to support multiple languages for bilingual speakers, especially given the shortage of bilingual speech-language pathologists in the United States [13].

In this study, we sought to investigate whether a lexical retrieval intervention that has largely been evaluated in monolingual speakers [14–16] would be efficacious for bilingual speakers with progressive anomia. The treatment was adapted to include distinct targets treated in each of the participants' languages. We also examined whether inclusion of targets with shared phonology (i.e., cross-linguistic cognates, such as dentist and its Spanish translation equivalent *dentista*) may promote naming accuracy across languages. In the following sections, we briefly review neurodegenerative syndromes that may present with progressive anomia, summarize the literature examining restitutive interventions in monolingual and bilingual speakers with progressive anomia, and present evidence for treatment-induced cross-linguistic transfer in bilingual aphasia.

1.1. Progressive Anomia

Anomia is a ubiquitous feature of aphasia syndromes and distinct etiologies can result in word-retrieval difficulty. This study includes patients with anomia in the context of a number of neurodegenerative disorders. Primary progressive aphasia (PPA) is a neurodegenerative syndrome characterized by gradual worsening of speech and language ability, with relative sparing of other cognitive domains [17]. International consensus criteria delineate three clinical variants of PPA [18]: the nonfluent/agrammatic variant, the semantic variant, and the logopenic variant. Each subtype presents with a distinct profile of speech and/or language impairments and pattern of brain atrophy (e.g., [19]). Anomia is a core feature of both the logopenic and semantic PPA variants but for different underlying reasons. The logopenic variant of PPA (lvPPA) presents with a core deficit in phonological processing, which manifests as impaired word retrieval in spontaneous speech and naming, and impaired repetition of phrases and sentences [20]. In this syndrome, cortical atrophy is typically observed in left temporoparietal regions implicated in phonological processing and phonological working memory [20,21]. LvPPA is most often associated with Alzheimer's pathology [22].

The semantic variant of PPA (svPPA) presents with left greater than right atrophy in the anterior temporal lobes [23,24]. Individuals with svPPA have impaired confrontation naming and single-word comprehension due to a gradual degradation of conceptual knowledge [18]. In cases where right anterior temporal atrophy is greater than that in the left hemisphere, individuals are characterized using different diagnostic terminology, either behavioral variant frontotemporal dementia or right temporal variant of FTD (e.g., [25,26]). These individuals are also anomic; however, their anomia is typically less pronounced than deficits in affect processing, and person and social semantic knowledge [25–28]. Both left and right temporal variants of FTD that present with primary deficits in semantic processing are associated with TDP-43 proteinopathy [22].

1.2. Treatment for Progressive Anomia in Primary Progressive Aphasia

At present, there are no pharmacological interventions proven to successfully treat the speech and language symptoms that accompany PPA or FTD. There is, however, a growing body of evidence documenting the utility of behavioral speech-language interventions to improve targeted communication skills in PPA. Most of this work has centered on treating anomia in the context of PPA, with the overwhelming majority of studies focusing on monolingual speakers (for reviews, see [29–36]).

Treatment for anomia has been shown to result in improved naming in all three PPA variants; however, given the scope of the current paper, we will focus on outcomes reported in sv and lvPPA. The treatment approaches used to treat anomia in sv and lvPPA range from rehearsal of spoken and/or written word forms [31,37–49] to more varied training tasks, some of which are designed to encourage self-cueing through the recruitment of residual semantic and word form knowledge [14,15,40,50–52]. Maintenance of treatment gains has been more frequently observed in lv versus svPPA, with gains observed up to 12 [15,53] and 15 months post-treatment [54], respectively. In both variants, generalization to untrained items has been reported. Generalization is more often reported in studies that have utilized approaches that incorporate more elaborated training tasks and/or that encourage self-cueing, e.g., [14–16,40,50–52,55].

Only one study [10] has examined the effects of naming intervention administered to a bilingual speaker (Norwegian-English) with PPA (i.e., lvPPA). The treatment, administered only in English, began with eight in-person sessions, which were then followed by 11 months of home practice. In general, the participant showed a decline in both languages from pre- to post-treatment, with the exception of written naming accuracy. More specifically, the participant demonstrated better written naming for trained versus untrained items in English. Despite the fact that treatment was only offered in English, evidence for cross-language transfer was observed in oral naming and naming-to-definition in Norwegian. The results of this study suggest that cross-language transfer is possible in bilingual PPA, despite progressive worsening.

In sum, research addressing speech-language treatment for monolingual speakers with PPA documents that intervention is efficacious and may have long-term benefits for some individuals. In bilingual speakers, additional research is needed in order to evaluate the effects of intervention within and between languages, and to investigate optimal treatment designs to promote cross-linguistic transfer.

1.3. Cross-Linguistic Transfer in Treatment for Anomia in Bilingual Aphasia and the Role of Cognates

Studies of linguistic processing in healthy bilingual speakers can inform predictions regarding treatment-induced cross-linguistic transfer in bilingual aphasia. Perhaps the closest analogue to cross-linguistic transfer in neurotypical bilingual speakers is that of translation. Evidence from studies of healthy bilingual speakers has shown asymmetry in translation directionality, such that backward translation (L2 to L1) is faster and more accurate than forward translation (L1 to L2, e.g., [56–59]), particularly for those who learn their L2 subsequent to their L1. This pattern is thought to reflect weakened links between the L2 lexicon and conceptual representations relative to the L1. This is in addition to stronger lexical links from the L2 to the L1 (bilingual speakers may access conceptual information via the L1, particularly at lower levels of L2 proficiency, as is described in the revised hierarchical model [58]). Interestingly, bilingual speakers who speak languages that share cross-linguistic cognates (i.e., words that share meaning and form across languages, such as telephone and *teléfono*) tend to demonstrate a cognate facilitation effect, wherein cognates are named faster and are translated more quickly and reliably relative to noncognates (e.g., [56,57,60–67]). This may be possible due to shared conceptual representations activating lexical items in both languages, with cognates benefiting from increased activation from shared phonological segments.

Studies examining cross-linguistic transfer effects following treatment for anomia in stroke-induced aphasia have reported different patterns of transfer (e.g., [3–5,68]). The majority of naming intervention studies report transfer or generalization from participants' trained L2 to their untrained L1 (e.g., [69–73]). Other studies have found transfer to the untrained L2 following L1 treatment (e.g., [74–76]). Taken together, these studies illustrate that bidirectional transfer is possible, but not uniform, in the context of aphasia treatment. In addition, a series of studies (e.g., [72,75,77]) has shown that the effects of an intervention targeting semantic bases of naming can result in within- and between-language transfer (to translation equivalents of trained items and to untrained items). Other work has investigated whether the inclusion of cognates in treatment may result in greater cross-linguistic transfer effects.

The effect of including cognates as treatment targets for bilingual speakers with aphasia has been examined primarily in the context of naming intervention. A handful of studies has reported positive transfer effects for cognate items [78,79] in individuals with nonfluent aphasia. Other studies have not observed such an effect [69,80,81]. The variability of cognate transfer effects reported from single cases in the literature may be attributed to a number of factors, including participant characteristics and differences in methodology (e.g., treatment approaches/tasks). Given that the majority of studies examining treatment for bilingual aphasia have focused on stroke-induced aphasia, a study examining the effect of treatment and potential for cross-language transfer in progressive aphasia is warranted.

1.4. The Present Study

The aim of this study was to investigate the effects of an established lexical retrieval training approach in a series of bilingual speakers with progressive anomia. Each individual underwent treatment using a single-subject multiple baseline design, with treatment administered in each of their languages in distinct phases. We assessed performance on items trained in each language, as well as cross-linguistic transfer effects (performance of untrained translation equivalents in one language that were trained items in the other language). Performance on trained and untrained items as well as standardized tests was assessed before, during, and after treatment, with follow-up testing at 3, 6, and 12 months post-treatment. Consistent with previous research [15], we predicted that treatment would result in improved naming for trained items, with maintenance of gains in the follow-up period, and with some participants demonstrating evidence of generalization to untrained items. We also hypothesized that treated cognates would show significant cross-linguistic transfer and that the magnitude of transfer would be significantly greater than that for noncognates.

2. Materials and Methods

2.1. Participants

Ten bilingual speakers with progressive anomia were recruited for this study. Participants included one individual with the right temporal variant of frontotemporal dementia, four participants with the semantic variant of PPA, and five participants with the logopenic variant PPA. With the exception of the participant with right temporal variant FTD, participants with PPA met current diagnostic criteria for PPA and subtype [17,18]. Inclusionary criteria required that individuals presented with progressive anomia, and attained a conceptual or composite score ([82–84]; where an appropriate response in either language is counted as correct) of 15 or higher on the Mini-Mental State Exam [85] at pre-treatment. In addition, we recruited only bilingual individuals who reported speaking both languages at the time of enrollment and who were in favor of undergoing treatment in both of their languages. Bilingual individuals who reported no longer using one of their languages were not enrolled in the current study but were enrolled in a separate study evaluating the effects of intervention provided in English only.

Six individuals were male and nine were right-handed, with one participant reporting ambidexterity. The mean age of participants was 67 years ($\pm$7) and, on average, individuals were 3.5 years ($\pm$2) post symptom onset. All participants spoke English and another language (n = 5 Spanish, n = 2 Farsi, n = 1 Portuguese, and n = 1 French; see Table 1). Participants gave written informed consent, and all procedures were approved by the institutional review board at The University of Texas at Austin. Structural magnetic resonance imaging was acquired for five participants prior to the commencement of treatment and voxel-based morphometry analysis was conducted, comparing each participant to 30 healthy age-matched controls (see Figure 1). The results from these analyses revealed the expected pattern of atrophy for each individual (left > right anterior temporal lobe atrophy in svPPA (right > left for right temporal variant) and left > right temporoparietal atrophy in lvPPA). All participants lived at a distance from the research site; therefore, assessment and treatment were conducted via HIPAA-compliant videoconferencing software (Fuze, Adobe Connect or Zoom). Previous work from our group has shown that treatment delivery modality (face-to-face versus telerehabilitation) does not impact treatment outcomes (i.e., performance on the primary outcome measure and maintenance and generalization effects) for the intervention used in this study [86].

Table 1. Individual Demographic and Language History Profiles.

Participant	rtFTD1		SV1		SV2		SV3		SV4		LV1		LV2		LV3		LV4		LV5	
Demographics																				
Sex	M		M		M		F		F		F		M		F		M		M	
Age (years)	67		72		64		60		63		78		80		59		64		62	
Education (years)	16		12		18		20		20		16		20		18		18		13	
Years Post Onset	3		5		3		3		9		2		4		2		2.5		1.5	
Handedness	Right		Right		Right		Ambidextrous		Right		Right		Right		Right		Right		Right	
Language History Variables																				
Language	Span	Eng	Span	Eng	Farsi	Eng	Span	Eng	Span	Eng	French	Eng	Port	Eng	Span	Eng	Farsi	Eng	Span	Eng
Age of acquisition (years)	Birth	6	Birth	5	Birth	18	11	Birth	9	Birth	17	Birth	Birth	Birth	Birth	16	Birth	14	Birth	17
Premorbid proficiency (5-point scale; with 5 indicating native-like proficiency)	3	5	4	5	4	5	3	5	5	5	5	5	4	5	5	5	5	4	5	5
Premorbid daily usage (out of 100%)	7%	93%	16%	85%	60%	40%	8%	93%	12%	88%	13%	88%	10%	90%	20%	80%	37%	63%	48%	52%
Weekday	13%	87%	18%	82%	38%	62%	15%	85%	12%	88%	13%	88%	13%	87%	20%	80%	53%	47%	46%	54%
Weekend	1%	99%	13%	87%	82%	18%	0%	100%	12%	88%	13%	88%	6%	94%	20%	80%	21%	79%	50%	50%
Postmorbid proficiency (5-point scale; with 5 indicating native-like proficiency)	3	5	3	5	4	5	2	4	2	3	5	5	2	4	2	2	5	4	3	4
Postmorbid daily usage (out of 100%)	5%	94%	7%	93%	60%	40%	8%	93%	97%	3%	13%	88%	6%	94%	10%	90%	80%	20%	85%	15%
Weekday	6%	91%	7%	93%	38%	62%	15%	85%	100%	0%	13%	88%	6%	94%	10%	90%	80%	20%	87%	13%
Weekend	3%	97%	7%	93%	82%	18%	0%	100%	94%	6%	13%	88%	6%	94%	10%	90%	80%	20%	83%	17%
Self-reported dominance	English		English		Farsi		English		English		English		English		English		Farsi		Spanish	
Dominance index (lower BNT score/ higher BNT score)	0.39		0.15		0.50		0.33		0.82		0.50		0.14		0.22		0.47		0.97	

Note: rtFTD1 = participant with right temporal variant frontotemporal dementia; sv = semantic variant PPA; lv = logopenic variant PPA. See Gollan et al., 2010; 2012 for details regarding dominance index. Span = Spanish, Eng = English, Fre = French, Port = Portuguese.

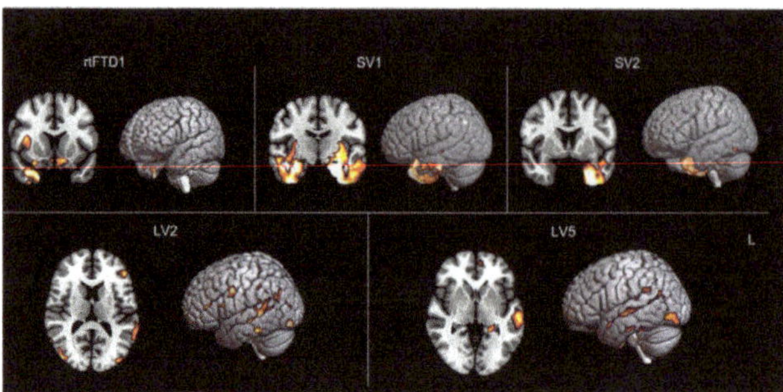

Figure 1. Results of whole brain voxel-based morphometry analysis showing atrophy patterns for each participant relative to controls ($n = 30$, FWE < 0.05, k = 100, total intracranial volume and age included as covariates). Note that scans were available for only five participants. rtFTD = right temporal variant of frontotemporal dementia; SV = semantic variant; LV = logopenic variant.

A language use history questionnaire (subset of items from Kiran et al. [87]) was used to gain information regarding individuals' use and exposure to each of their languages. A summary of each individual's language history is provided in Table 1. There was a range in age of second language acquisition (birth-18 years) and seven participants reported dominance in English. All participants received a comprehensive cognitive-linguistic evaluation prior to the initiation of treatment in order to confirm diagnosis and clinical subtype. Aphasia with prominent anomia and a history of progressive decline were confirmed in all participants. In general, participants demonstrated better performance in their dominant language. Pre-treatment assessment scores are presented in Table 2.

2.2. Treatment Design and Procedures

Treatment was administered following a single subject multiple-baseline design, with two intervention phases (one language per phase; see Figure 2 for the training schedule). An adapted form of Lexical Retrieval Cascade Treatment [14,15] was used to target individually tailored word sets for all participants. The treatment cascade targets naming via guided retrieval of residual semantic, phonological, and orthographic information, with the goal of retraining specific vocabulary as well as instilling strategies for word retrieval more broadly (see Table 3 for the sequence of training tasks). Treatment sessions occurred twice weekly. Daily homework consisted of Copy and Recall Treatment [88], involving repeated rehearsal and delayed recall of spoken and written target words.

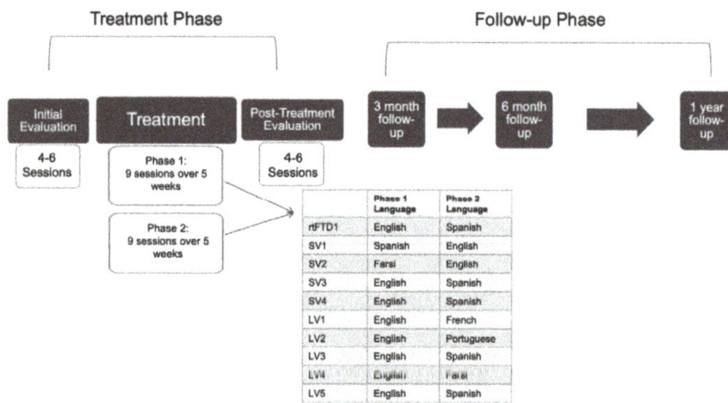

	Phase 1 Language	Phase 2 Language
rtFTD1	English	Spanish
SV1	Spanish	English
SV2	Farsi	English
SV3	English	Spanish
SV4	English	Spanish
LV1	English	French
LV2	English	Portuguese
LV3	English	Spanish
LV4	English	Farsi
LV5	English	Spanish

Figure 2. Schematic depicting chronology and duration of participation. rtFTD = right temporal variant of frontotemporal dementia; SV = semantic variant; LV = logopenic variant.

Table 2. Pre-Treatment Assessment Battery.

Participant ID	rtFTD1		SV1		SV2		SV3		SV4		LV1		LV2		LV3		LV4		LV5	
Language	Span	Eng	Span	Eng	Farsi	Eng	Span	Eng	Span	Eng	Fre	Eng	Port	Eng	Spa	Eng	Farsi	Eng	Span	Eng
Mini-Mental State Examination [1] (30)	23	22	15	25	30	27	17	23	14	17	23	26	6	29	9	14	29	27	26	27
CVLT Total (36) [2]	15	16	13	18	-	15	0	13	-	13	-	19	-	17	-	11	-	24	9	11
CVLT 10-min Recall [2]	1	3	0	0	-	0	0	0	-	1	-	5	-	2	-	3	-	3	0	0
Stroop Color naming [2]	26	38	12	45	-	38	-	35	11	48	-	42	-	52	-	7	-	69	38	38
Stroop interference [2]	14	24	7	30	-	12	-	21	9	31	-	31	-	20	-	4	-	49	23	22
Complex Figure Copy (17) [2]	-	14	-	14	-	15	-	17	-	17	-	13	-	15	-	7	-	16	16	-
Complex Figure Recall (17) [2]	-	6	-	3	-	13	-	15	-	11	-	10	-	6	-	4	-	17	5	-
Calculations (5) [2]	-	5	-	4	-	5	-	-	-	-	-	3	-	-	-	0	-	5	-	-
Digit Span Forward [2]	4	5	5	6	-	6	5	7	4	6	-	6	-	5	-	3	-	6	3	3
Digit Span Backward [2]	3	4	3	5	-	5	4	5	5	5	-	4	-	4	-	2	-	4	4	3
PPVT Short (16) [2]	-	14	-	10	-	8	-	1	-	4	-	-	-	12	-	9	-	8	-	13
Western Aphasia Battery (AQ; 100) [3]	78.2	92.6	69.2	87.5	90.2	81.3	42.9	75.9	51	74.4	77.3	88.7	38.4	86.8	39.3	61.3	92	82.1	84.4	82.8
Motor Speech Eval: AOS (0–7) [4]	0	0	0	0	-	0	N/A	N/A	-	0	-	0	-	0	-	0	-	0	0	0
Motor Speech Eval: Dysarthria (0–7) [4]	0	0	0	0	-	0	N/A	N/A	-	0	-	0	-	0	-	0	-	0	0	0
Pyramids and Palm Trees Test [5] (short; 14 [6]; * = /25, ^ = /20 [7])	-	14	-	14	12	14	-	7	14 ^	14 *	-	13	-	14	-	13	-	14	-	13
Boston Naming Test (60; * = /18) [8]	11	28	4	27	8	4	1	3	2 *	2	17	34	4	29	2	9	43	20	33	34
UCSF Syntax Comprehension Test (%) [9]	-	97	-	100	-	97	-	-	-	-	-	97	-	100	-	75	-	92	-	-
BAT Syntax Comprehension Subtest (%) [10]	92	100	79	98	93	92	69	95	84	84	74	91	76	91	51	53	94	92	8	85
Arizona Phonological Battery (%) [11]	-	50	-	80	-	53	-	97	-	94	-	58	-	56	-	8	-	69	-	50

Note: rtFTD1 = participant with right temporal variant frontotemporal dementia; sv = semantic variant PPA; lv = logopenic variant PPA, BAT = Bilingual Aphasia Test, Span = Spanish, Eng = English, Fre = French, Port = Portuguese. [1] Folstein, Folstein & McHugh, 1975; [2] Kramer et al., 2003; [3] Kertesz, 1982; [4] Wertz, LaPointe & Rosenbek, 1984; [5] Howard & Patterson, 1982; [6] Breining et al., 2015; [7] Martínez-Cuitiño & Barreyro, 2010; [8] Kaplan, Goodglass & Weintraub, 2001; [9] Wilson, Dronkers, et al., 2010; [10] Paradis & Libben, 1987; [11] Beeson et al., 2010.

Table 3. Lexical Retrieval Cascade Used During Treatment Sessions (Henry et al., 2013; 2019).

1. (Picture is presented) Semantic self-cue	Clinician prompts semantic description with, "Tell me about it." Additional prompting follows, as needed: "Where would you find this? What is it used for? Do you have any memories about this?" (If the item is named in this step, the clinician proceeds to step 5.)
2. Orthographic self-cue	Clinician requests written form of the word: "Can you write the word?" If unable to, the participant is encouraged to think of the first letter and/or sound of the word and any other characteristics about the word (i.e., "Is it a long or a short word?"). If the participant cannot come up with the first letter, the clinician writes the first grapheme.
3. Phonemic self-cue	Clinician asks the participant to make the sound associated with the letter. (If the item is named in this step, the clinician proceeds to step 5.)
4. Oral reading	If the item is not yet named, the clinician writes out the remainder of the word and the participant reads it aloud.
5. Written and Spoken Repetition	The participant writes and says the word three times.
6. Semantic Plausibility Judgments	Clinician asks three yes/no questions regarding semantic features of the item (e.g., "would you find this in a toolbox?")
7. Recall	Clinician asks the participant to provide the most salient semantic features and write and say the word one time.

Treatment targets consisted of six sets of words, each containing 4 or 8 nouns (participants had different numbers of words per set for pragmatic reasons related to severity of anomia and the number of viable cognates that existed across different language pairs); therefore, the total treatment set contained either 24 or 48 nouns. Untrained items for each participant comprised a minimum of two sets (again containing 4 or 8 nouns); therefore, the total untrained set contained 8 to 24 items. Participants and their care partners provided images of items for inclusion in treatment; when possible, these items were prioritized for inclusion and were distributed across trained and untrained sets. When an insufficient number of items from the personal set were provided, functional items were supplemented by the clinician. In general, items were eligible for inclusion in treatment if participants did not name the item on two out of three occasions in both languages. However, for the first two participants, we required that they not name the item on two out of three occasions in the target language only (i.e., the language the item was assigned to for training; rtFTD1 and SV1). This means that some items treated in Spanish were accurately named in English on two out of three attempts and vice versa. For these two individuals, only the consistently unnamed subset was included when examining cross-language translation effects. As a result, for SV1, an insufficient number of items was present to assess translation effects for noncognates from Spanish to English.

For each language of treatment, half of the treated and untreated items were cross-linguistic cognates. Sets were trained for three sessions each in their assigned language. All word sets (trained and untrained) were balanced for frequency, length in letters (English, French, Spanish, and Portuguese), or phonemes (English and Farsi) within and across languages. When possible (i.e., when corpora contained these variables), sets were also balanced within and across languages for familiarity, imageability, and concreteness (English, Spanish, and French). Psycholinguistic parameters were attained from the following sources in each language: English = Medical Research Council Psycholinguistic Database [89], Corpus of Contemporary American English [90], and the CLEARPOND database [91]; Spanish= Corpus del Español [92], the CLEARPOND database, and Es-Pal [93]; French= Lexique [94], and the CLEARPOND database; Portuguese – Corpus do Portugues [95], and Farsi= TalkBank Persian [96,97].

The lead author (S.G.) administered treatment in both phases for individuals who spoke English and Spanish, and in the English phase for individuals who spoke English and a different language. For those participants who spoke English and a different language (French, Portuguese, and Farsi), clinicians were recruited and trained to administer treatment in the non-English treatment phase; in one case, a doctoral student in French linguistics assisted with assessment and treatment after extensive training and observation.

After the formal treatment period ended, participants were allowed to retain their homework materials and to practice their trained items. Allowing practice to take place after the immediate treatment period likely mirrors what occurs in typical clinical care with speech-language pathologists, wherein individuals are allowed and encouraged to practice with their treatment materials. This was consistent with procedures from the original studies demonstrating efficacy for this treatment approach [14,15].

2.3. Treatment Fidelity

Undergraduate and graduate students in speech-language pathology or linguistics, who spoke the language of treatment administration, were trained to conduct treatment fidelity ratings. Raters were provided with a template that included each treatment step (in the prescribed order). While reviewing each video, the rater indicated whether the clinician performed each step. If the same clinician provided treatment to a participant in both phases of treatment, then 25% (5/18) of the total number of sessions (sampled across phases) were independently reviewed by one student. If different clinicians administered each phase of treatment, 33% (3/9) of the total number of sessions were reviewed from each *phase* of treatment, except for two participants. For these two participants, videos were only available for 11% (1/9) or 22% (2/9) of sessions from one phase of treatment; however, a full set of videos was available for the other phase of treatment. The percentage of correctly administered treatment steps was calculated for each reviewed session. Fidelity ratings, averaged across participants, revealed that clinicians adhered to the treatment steps with 99.21% accuracy.

2.4. Self- and Communication Partner-Assessment of Change Following Treatment

Participants and their primary communication partners were asked to complete a post-treatment survey [14,15] documenting their perceptions regarding changes in communication from pre- to post-treatment. The survey consisted of 20 questions and a qualitative rating scale was used to capture respondents' perceptions (7 point scale: 3 = "A lot better," 2 = "Better," 1 = "Somewhat better," 0 = "Unchanged," −1 = "Somewhat worse," −2 = "Worse," and −3 = "A lot worse").

2.5. Follow-Up Assessment

Follow-up assessments were conducted at 3, 6, and 12 months post-treatment. Only one participant (lv1) was unavailable for follow-up assessment at 3 and 6 months post-treatment, due to health-related issues. Additionally, one individual had yet to complete the follow-up period (sv5) at the time that this paper was written. All remaining participants were available at one year post-treatment. Performance on standardized assessments at each time point is reported in Appendix A.

2.6. Outcome Measures and Statistical Analysis

Data were analyzed at the single-subject level. The primary outcome measure was the proportion of items named correctly during probing for trained and untrained stimuli in the target language. Cross-linguistic transfer was assessed by examining participants' responses to treatment probes for trained stimuli in the non-targeted language (i.e., translation effects). Probes were collected in each language three times at pre-treatment, once or twice at mid-treatment, twice at post-treatment, and once at each follow-up visit (3, 6, and 12 months post-treatment). Additionally, approximately half of items were probed at the beginning of each treatment session in a given language, so that all sets were probed once per week in each language.

Significance testing was conducted using a simulation technique [98]. An individual's percent accuracy was attained from each condition and probabilities of correct responses were used to create simulated datasets with parameters that mirror the observed data. This procedure was completed 10,000 times to create 10,000 simulated distributions of accuracy scores from each condition, at each time point. The resulting simulated datasets from two conditions were then directly compared to one another to calculate a p-value (i.e., the likelihood that post-treatment performance was greater than pre-treatment performance). In addition, using the simulated data, difference scores were calculated between conditions to determine the 95% confidence intervals of the observed differences. For comparing differences in the magnitude of effects (e.g., translation effects between cognate and noncognate items), the same process was followed with one additional step. Specifically, simulations were conducted for each condition and time point, but p-values were calculated by comparing difference scores between time points and conditions (e.g., the difference scores for trained and untrained stimuli for simulated post-treatment minus simulated pre-treatment performance).

We predicted that each participant would demonstrate a significant treatment effect, with maintenance in the follow-up period. We also predicted that some participants would demonstrate evidence of generalization to untrained items. It was hypothesized that each participant would show a significant cross-language translation effect for cognate items and that the magnitude of this effect would be greater for cognates relative to noncognates.

In addition, we assessed performance over time (pre-treatment versus subsequent time points) on a subset of assessments administered in the dominant and nondominant language using paired permutation tests at the group level in order to identify overall trends with respect to stability and/or progression. Specifically, the stability of general cognitive and linguistic function (MMSE and WAB-R; [85,99]) and overall naming ability (Boston Naming Test; BNT [100]) were evaluated. We predicted that performance on the BNT would improve at post-treatment, consistent with previous literature demonstrating generalization on this measure in monolingual speakers with PPA [15]. Analyses comparing subsequent timepoints on the BNT and for all timepoints for the other assessments (MMSE and WAB-R) were assessed using two-tailed tests, as performance on these measures was less predictable over time.

3. Results

In the following sections, outcomes that are directly related to the aforementioned hypotheses will be reported. In order to contextualize our reporting of the number of participants demonstrating significant improvement at the individual level, we also provide the average change and range of performance for the entire group. For additional treatment outcomes (including outcomes following the first treatment phase and cross-linguistic generalization effects to untrained items), please see the Supplementary Materials.

3.1. Treatment and Maintenance Effects

Simulation analyses revealed that each participant demonstrated a significant treatment response in both their dominant (M change = 70.37%; range = 31–92%) and nondominant language (M change = 65.03%; range = 30–97%, see Figure 3) from pre- to post-treatment (after training in both languages was completed). Of the eight participants for whom follow-up data were collected at 3 and 6 months post-treatment, all participants had significantly better performance at the 3-month follow-up, and all but one individual had significantly better performance at the 6-month follow-up relative to pre-treatment. At 12 months post-treatment, seven of nine participants demonstrated significantly better performance relative to pre-treatment in their dominant language, with six of eight participants showing this pattern in their nondominant language.

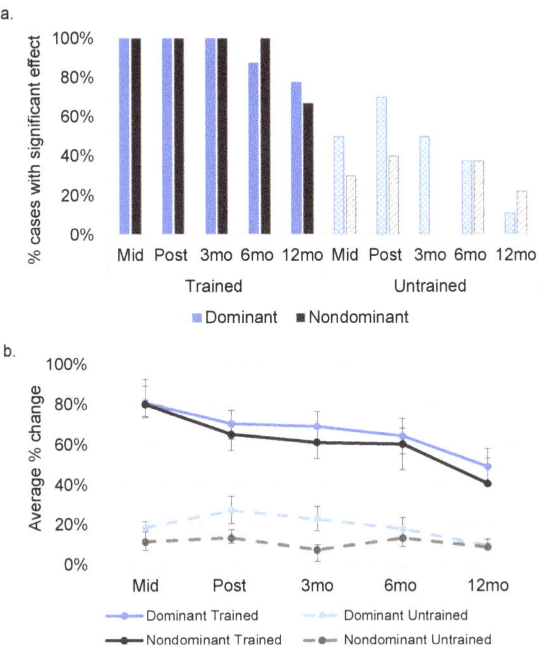

Figure 3. Within-language treatment and generalization effects at each time point relative to pre-treatment. (**a**). Depicts the percentage of cases demonstrating a significant effect at each time point relative to pre-treatment. (**b**). Depicts the average percent change at each time point relative to pre-treatment. At mid-treatment, seven of nine participants had received treatment in their dominant language and three had received treatment in their nondominant language; the figure shows performance for these subsets at mid-tx for trained items. See Supplemental Materials for data at the single-subject level. Mid = mid-treatment; mo = month.

3.2. Within-Language Generalization to Untrained Items

Seven of 10 individuals showed improvement on matched, untrained items in their dominant language (*M* change = 27.18%, range = −3–75%; see Figure 3), with four individuals showing this pattern in their nondominant language (*M* change = 13.50%, range = −4–38%), from pre- to post-treatment. A direct comparison of the magnitude of improvement on trained versus untrained items revealed a significant difference (with greater improvement for trained items) for six individuals from pre- to post-treatment in the dominant language (an additional three participants demonstrated a marginal or trending difference between trained and untrained items; *M* difference = 43.19%, range = –0.1–89%), and eight participants showing this pattern in the nondominant language (the remaining two participants demonstrated a marginal difference between trained and untrained items; *M* difference = 51.53%, range = 27–87%). Performance on untrained items showed gradual decline in the follow-up period.

3.3. Cross-Linguistic Translation Effects

Following both treatment phases, eight of 10 participants demonstrated a significant cross-linguistic translation effect relative to pre-treatment for cognates from the nondominant to the dominant language (*M* change = 54.70%, range = 0–83%) and seven of 10 participants showed this pattern from the dominant to the nondominant language (*M* change = 39.60%, range = 0–83%; see Figure 4). Of the eight participants who were available for the 3- and 6-month follow-up, six and five individuals demonstrated a significant translation effect for cognates in both the dominant and nondominant languages, respectively. At 12 months post-treatment, four of the original seven individuals demonstrated maintenance of a cognate translation effect to the dominant language, with three of the original seven maintaining this pattern of transfer to the nondominant language.

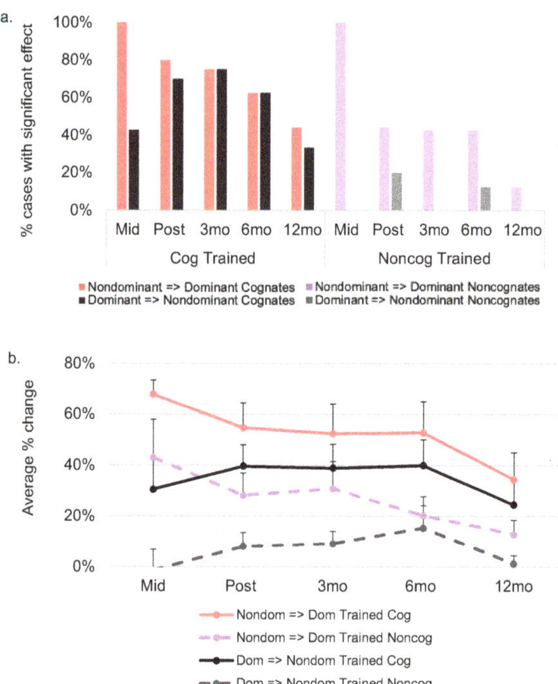

Figure 4. Cross-linguistic transfer effects by cognate status at each time point relative to pre-treatment. (**a**). Depicts the percentage of cases demonstrating a significant effect at each time point relative to pre-treatment. (**b**). Depicts the average percent change at each time point relative to pre-treatment. Performance on trained items across languages represents translation effects. At mid-treatment, seven of nine participants had received treatment in their dominant language and three had received treatment in their nondominant language; the figure shows performance for these subsets at mid-tx for trained and untrained items. See Supplementary Materials for data at the single-subject level. Mid = mid-treatment; mo = month.

With regard to cross-linguistic translation of noncognates, four of nine individuals showed a significant translation effect from their nondominant to their dominant language (M change = 28.22%, range = 0–83%) and two of 10 showed this pattern from the dominant to the nondominant language (M change = 8.10%, range = −3–47%), following both phases of treatment. A similar pattern of performance was observed at 3 and 6 months post-treatment. At 12 months post-treatment, one of the original four individuals demonstrated maintenance of a noncognate translation effect in the dominant language, with no individuals maintaining this pattern in their nondominant language.

A direct comparison of the magnitude of translation effects for cognate and noncognate items revealed a significant difference (with better translation of cognates) for four individuals from pre- to post-treatment from the nondominant to the dominant language (M difference = 25.11%, range = −17–83%; see Figure 5), and five participants showing this pattern from their dominant to nondominant language (M difference = 31.50%, range = 0–77%). At subsequent follow-ups, a gradual decline was observed in the number of participants who showed a significant difference in the magnitude of the translation effect observed between cognates and noncognates (see Figure 5).

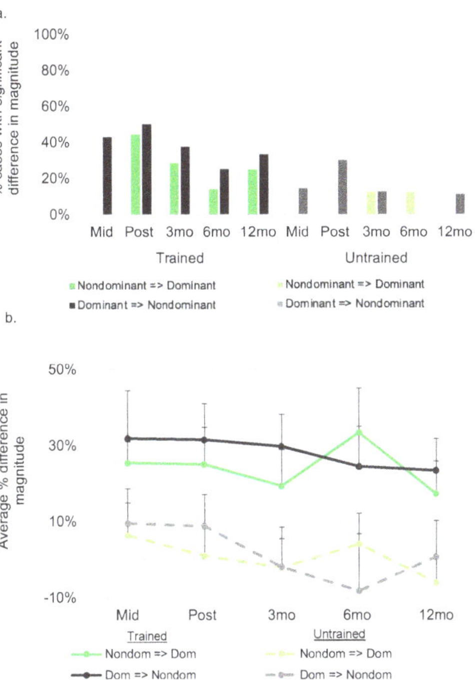

Figure 5. Cross-linguistic transfer for cognates *versus* noncognates at each time point relative to pre-treatment. (**a**). Depicts the percentage of cases demonstrating a significant difference in the magnitude of transfer between cognates and noncognates at each time point relative to pre-treatment. (**b**). Depicts the average difference in the magnitude of change between cognates and noncognates at each time point relative to pre-treatment. Performance on trained items represents translation effects. Performance on untrained items represents generalization effects. At mid-treatment, seven of nine participants had received treatment in their dominant language and two of three who had received treatment in their nondominant language had sufficient data for these contrasts; the figure shows performance for these subsets at mid-tx for trained and untrained items. See Supplemental Materials for data at the single-subject level. Mid = mid-treatment; mo = month.

3.4. Performance on Additional Outcome Measures

Paired permutation tests revealed that participants demonstrated significant improvement on the BNT at post-treatment relative to pre-treatment in the nondominant language ($t = -1.59$, $p = 0.047$; see Appendix A). Performance on this measure at other time points was not significantly different from pre-treatment, nor was performance in the dominant language at any time point relative to pre-treatment. Performance on the MMSE showed a relatively steady decline over time, with significant decline emerging at 12 months post-treatment relative to pre-treatment in the dominant language only ($t = 2.76$, $p = 0.012$). Lastly, performance on the WAB-R also showed a gradual decline over time, with significant decline noted at three months post-treatment ($t = 2.15$, $p = 0.020$) and at each subsequent follow-up (6 months post-treatment ($t = 1.75$, $p = 0.023$); one-year post-treatment ($t = 2.78$, $p = 0.006$)) in the dominant language. A similar pattern was observed in the nondominant language, but with significant decline emerging at 6 months post-treatment ($t = 2.55$, $p = 0.023$); one-year post-treatment ($t = 2.37$; $p = 0.016$).

3.5. Self and Communication Partner Assessment of Change

The mean improvement reported by all respondents (caregivers and participants combined) on the post-treatment survey was 1.17 (just above "somewhat better"). The mean rating for participants with lvPPA was 1.68 (between "somewhat better" and "better"), and for participants with svPPA, the mean rating was 0.65 (between "unchanged" and "somewhat better"). The average caregiver rating was consistent with the overall mean (1.17). The items and results from the post-treatment survey are reported in Appendix B.

4. Discussion

To our knowledge, this is the first study to systematically investigate the utility of speech-language intervention in a group of bilingual speakers with progressive aphasia. Consistent with previous studies examining treatment primarily in monolingual speakers with PPA [14,15], we hypothesized that bilingual speakers would show a robust treatment effect in both of their treated languages, with maintenance at follow-ups. We also hypothesized that generalization to untrained targets would be observed for some participants, due to the strategic nature of the intervention [14,15]. In addition, we sought to investigate whether the inclusion of cross-linguistic cognates would promote accurate translation of treated items.

4.1. Within-Language Gains and Generalization Effects

Our results indicate that bilingual speakers with mild-moderate PPA showed a significant and robust treatment effect in both of their treated languages following dual-language naming intervention. With regard to performance on matched, untrained stimuli, a greater number of participants demonstrated generalization in their dominant language at post-treatment; however, generalization was observed in the nondominant language for a smaller subset of participants. This suggests that the strategic nature of the intervention resulted in generalization to untrained items for a subset of participants, with the greatest benefit observed in the dominant language. In sum, our findings constitute further evidence that this treatment approach is beneficial for word retrieval impairments in PPA and FTD. Moreover, our results indicate that this approach is suitable for treating bilingual speakers with progressive anomia, and highlight that significant gains can be observed in both an individual's dominant and nondominant language.

4.2. Maintenance of Treatment Gains

There is pessimism in the clinical and research communities regarding not only the efficacy of treatment in individuals with progressive communication disorders but particularly the potential for maintenance of gains [101]. As such, it is crucial to document not only the immediate benefits of treatment, but also to evaluate stability of treatment effects in the face of disease progression. Many studies that report the effects of intervention in PPA have not explored performance beyond the immediate post-treatment period; however, those that have reported maintenance effects have documented stability in the follow-up period (e.g., [15,38,47,54]). Similarly, stability of treatment effects up to 12 months post-treatment was observed for the majority of our participants. As in Henry et al. (2019), participants were allowed to keep practice materials and encouraged to continue with self- guided practice following the completion of structured intervention with the clinician. In the prior study, post-treatment practice was monitored via self-report for a subset of participants and, surprisingly, a relation was not observed between amount of ongoing practice and maintenance of treatment gains. Future studies should employ methods for systematic and objective tracking of individual practice to better understand maintenance effects and the role of continued practice for individuals with PPA [15].

The maintenance effects in this study can be interpreted within the broader context of cognitive-linguistic decline observed in this cohort of bilingual speakers. Specifically, participants demonstrated gradual decline on general measures of linguistic and cognitive functioning (see Appendix A). In the context of this general progression, our findings confirm that a tailored approach to bilingual intervention results in significant improvement for trained items as well as improvement or stability in confrontation naming more broadly (as noted on the BNT, see Supplementary Materials). These findings indicate a possible protective benefit for the targeted behavior following treatment.

4.3. Cross-Linguistic Translation Effects

We observed that the majority of individuals showed a significant cognate translation effect (i.e., ability to name cognate items in the untrained language) following both phases of treatment, with fewer individuals showing an effect following the initial phase of treatment (see the Supplementary Materials for results following the initial treatment phase). For approximately half of participants, the translation effect at post-treatment was significantly greater in magnitude for cognates relative to noncognates. Cognate translation effects were generally maintained up to six months post-treatment (consistent with within-language generalization observed in our prior study [15]), with fewer individuals demonstrating a sustained benefit one year post-treatment. We note a couple of interesting patterns that emerged from our data. First, the two individuals who did not show a cognate translation effect (in at least one linguistic direction) obtained the lowest cognitive screening score (MMSE; lv3) or naming score (BNT; sv3) at pre-treatment. This observation suggests that an individual's potential to benefit from inclusion of cognates may be mediated by severity of cognitive and/or language deficits. This is also consistent with the finding that the most notable decrease in cognate translation ability (for individuals who originally showed a cognate translation effect) occurred between the 6 and 12-month follow-up visits (i.e., with increasing severity of cognitive-linguistic deficits).

Pre- and postmorbid language history variables, such as order of acquisition and frequency of use, may also influence translation effects (e.g., [3–5,68]). In PPA and other neurodegenerative disorders, nonparallel patterns of language decline [6,99–101] have been reported, which may influence frequency of language use and moderate treatment outcomes across languages. In the future, larger samples will allow us to better understand the relation between overall severity and translation effects, and to explore the possible interaction of severity indices and language history variables.

Given that the distribution of participants who received treatment in the dominant vs. nondominant language during the initial phase is unbalanced in this study ($n = 3$ received treatment in nondominant language in the initial phase), the following preliminary observations should be interpreted with caution and require replication in a larger sample utilizing a balanced design. Following the first treatment phase, a greater proportion of participants showed a cognate translation effect from the nondominant to dominant language (i.e., three of three participants who were treated in the nondominant language in the initial phase and three of seven who were treated in the dominant language in the initial phase). Findings following the initial phase of treatment are consistent with (1) patterns observed in healthy bilingual speakers (e.g., [62,66]) and with (2) transfer and translation patterns observed in stroke-induced aphasia (e.g., [65,82]), wherein ease of translation may be facilitated from the weaker to stronger language. Following both phases of treatment, the cross-linguistic translation effects for cognate items were bidirectional in our cohort, (i.e., eight of 10 from the nondominant to the dominant language and seven of 10 participants from the dominant to the nondominant language). This may indicate that the treatment approach, which targets both semantic and phonological bases of word retrieval, strengthened cross-linguistic activation between translation equivalents with phonological similarities. Together, results indicate that treatment in the nondominant language (or treatment in the dominant language followed by treatment in the nondominant language) resulted in robust translation effects for cognate items. We reiterate that this observation should be interpreted with caution due to the small number of participants, heterogeneity in clinical profile, and crucially, the fact that fewer participants received treatment in their nondominant language in the initial phase of this study ($n = 3$).

It is important to note that noncognates also have the potential to benefit cross-linguistically from this intervention, due to the targeted analysis of semantic features, which are shared across languages [58,102,103]. Nonetheless, far fewer individuals showed a significant noncognate translation effect after the initial phase of treatment (i.e., three of three from the nondominant to the dominant language and zero of seven from the dominant to the nondominant language) or following both phases of treatment (i.e., four of nine from their nondominant to their dominant language and two of 10 from their dominant to their nondominant language). The diminished translation effects for noncognates relative to cognates may be driven by a lack of phonological similarity. This is corroborated by findings from healthy bilingual speakers, which suggest that the combination of shared conceptual representations and phonology leads to the well-documented cognate facilitation effect (e.g., [56,57,60–67,104]).

4.4. Treatment and Translation Effects by PPA Variant

Following both phases of treatment, we observed that individuals with either sv or lvPPA showed a significant treatment effect irrespective of language dominance. With regard to generalization to untrained items, a slightly greater number of individuals within each variant showed generalization in the dominant language (sv = 3 vs. 1; lv = 4 vs. 3).

Although we did not have specific hypotheses regarding treatment and translation effects on the basis of the PPA variant, in this section, we note patterns that emerged in this study. With respect to cross-linguistic translation effects, individuals with svPPA and those with lvPPA demonstrated evidence of cognate translation effects. By contrast, a subset (four of nine from their nondominant to their dominant language and two of 10 from their dominant to their nondominant language) of individuals with lvPPA and no individuals with svPPA demonstrated significant translation effects for noncognates. This pattern may be explained by the different underlying deficits contributing to naming impairment in each variant. In lvPPA, semantic processing is relatively spared, and cognate facilitation is likely a result of improved access to or assembly of phonology resulting from repeated practice of target items. In the case of noncognates, the translation effects in some lvPPA cases may be attributed to the strategic nature of the intervention, which requires individuals to use residual semantic and word form knowledge in attempts to self-cue. In lvPPA, translation of noncognates was greatest from the nondominant to the dominant language. As has been reported in bilingual AD (e.g., [105,106]), it may be the case that the dominant language is more resistant to decline in bilingual speakers with lvPPA [107], and perhaps more likely to benefit from translation effects. This pattern might also reflect reliance upon the dominant language to access semantic knowledge [58].

In svPPA, learning has been characterized as rigid, with generalization reported less frequently (e.g., [37–39,43,108,109]). In addition, phonological processing is relatively spared [107] and cognate translation effects may be facilitated by strengthening of semantic representations for trained items, with similarities in phonology boosting activation for these word forms across languages. Given that learning tends to be more rigid in svPPA, it is not surprising that significant translation effects for noncognates (where spared phonological processing would not confer the same benefit) were not observed.

4.5. Additional Considerations

This study provides evidence that lexical retrieval treatment is an efficacious intervention for bilingual speakers with PPA in the mild-to-moderate range of severity. All individuals in this study were seen via a telehealth platform, which allowed for the inclusion of individuals living throughout the United States, as well as internationally. Telehealth holds promise as an assessment and treatment modality, enabling clinicians to reach individuals who may face barriers to accessing treatment, including ethnically and racially diverse groups who experience barriers to service provision more generally (e.g., [110–112]). In the future, advocacy for broad reimbursement of these services will be crucial to exploiting this treatment modality. In addition, future research should continue to broaden the evidence base for telehealth interventions intended for individuals with PPA beyond the mild-to-moderate range in order to maximize communication across the continuum of disease severity.

This study had several limitations. First, although this is, to our knowledge, the largest intervention study of bilingual speakers with PPA and FTD to date, the sample size is a limiting factor. Future studies will benefit from larger samples in order to investigate patterns of response to treatment, including cross-linguistic effects. This will also allow for the investigation of different patterns on the basis of language distance (i.e., how similar language pairs are to one another), as well as the consideration of language history variables (e.g., age of acquisition, frequency of use). It is also important to note that our results represent findings from language pairs that share cross-linguistic cognates. For individuals who speak language pairs that do not share cognates, our results suggest that the strategic component of this intervention may encourage generalization to untrained items and that cross-linguistic transfer is possible for noncognate items (particularly for individuals with lvPPA).

The treatment approach used in this study was selected due to its established benefit in monolingual speakers with PPA and due to its emphasis on training procedures that draw upon both semantic and phonological mechanisms supporting naming. From this study, it is not possible to discern whether semantic versus phonological stimulation is more crucial for within-language outcomes and transfer effects in this population. Future research may employ facilitation studies to investigate whether particular components of intervention are especially supportive of translation and generalization effects in bilingual speakers with progressive anomia. In addition, there is a need to investigate the potential for generalized improvement from naming intervention to connected speech, as such effects would further characterize the ecological validity of naming interventions administered to individuals with PPA.

5. Conclusions

There is a growing literature base addressing the treatment of progressive disorders of language. This work is crucial, as the global community anticipates a rapidly growing aging population, and consequently, an increase in the number of individuals presenting with neurodegenerative disorders (e.g., [113,114]). Simultaneously, we anticipate a growing bilingual population [115,116]. Although previous work has established a strong foundation for speech-language treatment research in monolingual speakers with neurodegenerative disorders, much work is needed to address the optimization of these approaches for bilingual speakers. At the same time, careful consideration of assessment and treatment methods is needed to ensure the use of culturally tailored approaches, as bilingual speakers often comprise culturally and ethnically diverse groups [12,117].

Our results indicate that bilingual speakers with PPA and FTD significantly improved their word retrieval for trained items assigned to each of their languages, with maintenance observed up to 6 or 12 months post-treatment. In addition, our findings indicate that monolingual clinicians may be able to select cross-linguistic cognates as a means to support gains across languages for words trained in a single language (i.e., "two for the price of one"). This has ramifications for service delivery in the U.S., where a majority of clinicians are monolingual English speakers. In the context of results from previous studies investigating treatment outcomes for PPA, our results offer complementary support and confirm that tailored behavioral intervention should be the standard of clinical care for linguistically diverse individuals with progressive aphasia.

Supplementary Materials: The following are available online at https://www.mdpi.com/article/10.3390/brainsci11111371/s1, Table S1: Results of Statistical Analyses at the Single-Subject Level for all Participants, Text S1: Results Following the Initial Phase of Treatment and Cross-Linguistic Generalization Effects to Untrained Items.

Author Contributions: Conceptualization, S.M.G., E.D.P. and M.L.H.; methodology, S.M.G., E.D.P., M.L.G.-T. and M.L.H.; formal analysis, S.M.G.; investigation, S.M.G., N.K., H.M. and R.N.; resources, B.B., M.L.G.-T. and M.L.H.; data curation, S.M.G., N.K., H.M. and R.N.; writing—original draft preparation, S.M.G.; writing—review and editing, S.M.G., E.D.P., M.L.G.-T., B.B. and M.L.H.; visualization, S.M.G.; supervision, M.L.H.; funding acquisition, S.M.G., E.D.P. and M.L.H. All authors have read and agreed to the published version of the manuscript.

Funding: This work was funded by the National Institute on Deafness and Other Communication Disorders (NIDCD) Grants, F31DC016229 (awarded to S.M.G.), R01 DC016291 and R03 DC013403 (awarded to M.L.H.), National Institute of Neurological Disorders and Stroke Grant R01 NS050915 and NIDCD Grant K24 DC015544 (awarded to M.L.G.T.).

Institutional Review Board Statement: The study was approved by the institutional review board at The University of Texas at Austin which governs human subjects research at UT Austin (protocol no. 2014-02-0133).

Informed Consent Statement: Written informed consent has been obtained from the patient(s) enrolled in this study.

Data Availability Statement: The data presented in this study are available in the supplemental materials.

Acknowledgments: We thank Jussara Vitorino for assistance with data collection; Viola Green for assistance with French translation; Lindsay Palmer, Liz Chinchilla, Sydney Hamilton, Shea De Brito, Naseem Shafaei and Liron Shlesinger for fidelity ratings, and Heather Dial and John G. Hixon for support with statistical analysis. Finally, we wish to thank all of our participants with primary progressive aphasia for the time and effort that they have devoted to this research.

Conflicts of Interest: The authors declare no conflict of interest.

Appendix A

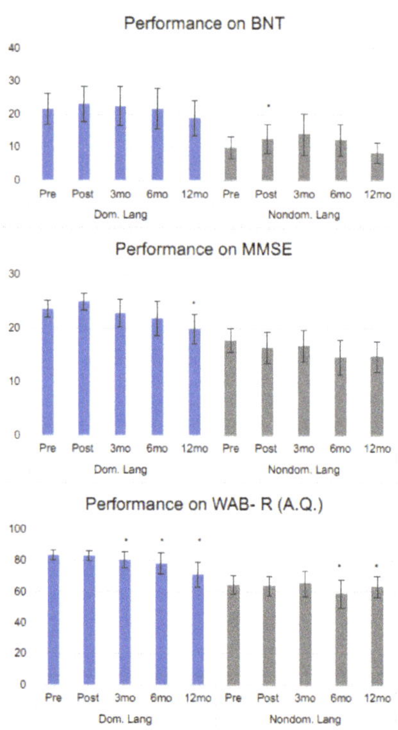

Figure A1. Performance from pre-treatment to each subsequent timepoint on a subset of cognitive and linguistic measures by language dominance. * Note. Standard error bars included for descriptive purposes. Significance determined via paired permutation tests. One-tailed tests used for BNT from pre to post-treatment; two-tailed tests used for all other timepoints/measures ($p < 0.05$). Dom. = dominant language, Nondom. = nondominant language, BNT = Boston Naming Test; MMSE = Mini-Mental State Exam, WAB-R = Western Aphasia Battery- Revised.

Appendix B

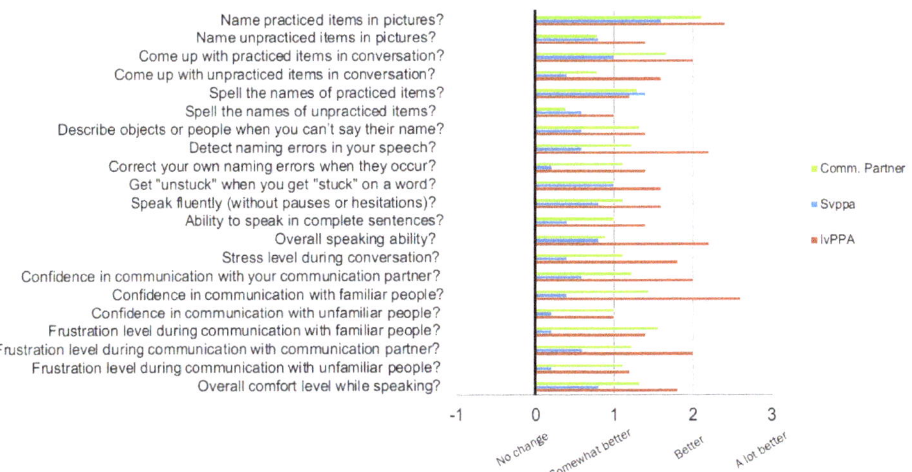

Figure A2. Survey Responses: "Compared to pre-treatment how would you rate your ability to … ".

References

1. Grosjean, F. *Bilingual: Life and Reality*; Harvard University Press: Cambridge, MA, USA, 2010. [CrossRef]
2. Marian, V.; Shook, A. The Cognitive Benefits of Being Bilingual. *Cerebrum Dana Forum Brain Sci.* **2012**, *2012*, 13.
3. Faroqi-Shah, Y.; Frymark, T.; Mullen, R.; Wang, B. Effect of treatment for bilingual individuals with aphasia: A systematic review of the evidence. *J. Neurolinguist.* **2010**, *23*, 319–341. [CrossRef]
4. Kohnert, K. Cross-Language Generalization following Treatment in Bilingual Speakers with Aphasia: A Review. *Semin. Speech Lang.* **2009**, *30*, 174–186. [CrossRef]
5. Ansaldo, A.I.; Saidi, L.G. Aphasia Therapy in the Age of Globalization: Cross-Linguistic Therapy Effects in Bilingual Aphasia. *Behav. Neurol.* **2014**, *2014*, 603085. [CrossRef] [PubMed]
6. Sandberg, C.W.; Zacharewicz, M.; Gray, T. Bilingual Abstract Semantic Associative Network Training (BAbSANT): A Polish-English case study. *J. Commun. Disord.* **2021**, *93*, 106143. [CrossRef]
7. Peñaloza, C.; Scimeca, M.; Gaona, A.; Carpenter, E.; Mukadam, N.; Gray, T.; Shamapant, S.; Kiran, S. Telerehabilitation for Word Retrieval Deficits in Bilinguals With Aphasia: Effectiveness and Reliability as Compared to In-person Language Therapy. *Front. Neurol.* **2021**, *12*, 598. [CrossRef]
8. Aziz, M.A.A.; Razak, R.A.; Garraffa, M. Targeting Complex Orthography in the Treatment of a Bilingual Aphasia with Acquired Dysgraphia: The Case of a Malay/English Speaker with Conduction Aphasia. *Behav. Sci.* **2020**, *10*, 109. [CrossRef]
9. Costa, A.S.; Jokel, R.; Villarejo, A.; Llamas-Velasco, S.; Domoto-Reilley, K.; Wojtala, J.; Reetz, K.; Machado, Á. Bilingualism in Primary Progressive Aphasia. *Alzheimer Dis. Assoc. Disord.* **2019**, *33*, 47–53. [CrossRef] [PubMed]
10. Meyer, A.M.; Snider, S.F.; Eckmann, C.B.; Friedman, R.B. Prophylactic treatments for anomia in the logopenic variant of primary progressive aphasia: Cross-language transfer. *Aphasiology* **2015**, *29*, 1062–1081. [CrossRef]
11. Rumbaut, R.G.; Massey, D.S. Immigration & Language Diversity in the United States. *Daedalus* **2013**, *142*, 141–154. [CrossRef]
12. Santhanam, S.P.; Parveen, S. Serving Culturally and Linguistically Diverse Clients: A Review of Changing Trends in Speech-Language Pathologists' Self-efficacy and Implications for Stakeholders. *Clin. Arch. Commun. Disord.* **2018**, *3*, 165–177. [CrossRef]
13. ASHA. *Demographic Profile of ASHA Members Providing Bilingual Services.* 2020. Available online: https://www.asha.org/siteassets/surveys/demographic-profile-bilingual-spanish-service-members.pdf (accessed on 15 October 2021).
14. Henry, M.; Rising, K.; DeMarco, A.; Miller, B.; Gorno-Tempini, M.; Beeson, P. Examining the value of lexical retrieval treatment in primary progressive aphasia: Two positive cases. *Brain Lang.* **2013**, *127*, 145–156. [CrossRef]
15. Henry, M.L.; Hubbard, H.I.; Grasso, S.M.; Dial, H.R.; Beeson, P.M.; Miller, B.L.; Gorno-Tempini, M.L. Treatment for Word Retrieval in Semantic and Logopenic Variants of Primary Progressive Aphasia: Immediate and Long-Term Outcomes. *J. Speech Lang. Hear. Res.* **2019**, *62*, 2723–2749. [CrossRef]
16. Beeson, P.M.; King, R.M.; Bonakdarpour, B.; Henry, M.L.; Cho, H.; Rapcsak, S.Z. Positive Effects of Language Treatment for the Logopenic Variant of Primary Progressive Aphasia. *J. Mol. Neurosci.* **2011**, *45*, 724–736. [CrossRef] [PubMed]
17. Mesulam, M.-M. Slowly progressive aphasia without generalized dementia. *Ann. Neurol.* **1982**, *11*, 592–598. [CrossRef]
18. Gorno-Tempini, M.L.; Hillis, A.E.; Weintraub, S.; Kertesz, A.; Mendez, M.; Cappa, S.F.; Ogar, J.M.; Rohrer, J.D.; Black, S.; Boeve, B.F.; et al. Classification of primary progressive aphasia and its variants. *Neurology* **2011**, *76*, 1006–1014. [CrossRef] [PubMed]

19. Montembeault, M.; Brambati, S.M.; Gorno-Tempini, M.L.; Migliaccio, R. Clinical, Anatomical, and Pathological Features in the Three Variants of Primary Progressive Aphasia: A Review. *Front. Neurol.* **2018**, *9*, 692. [CrossRef]
20. Henry, M.L.; Gorno-Tempini, M.L. The logopenic variant of primary progressive aphasia. *Curr. Opin. Neurol.* **2010**, *23*, 633–637. [CrossRef]
21. Rohrer, J.; Rossor, M.; Warren, J.D. Alzheimer's pathology in primary progressive aphasia. *Neurobiol. Aging* **2012**, *33*, 744–752. [CrossRef]
22. Spinelli, E.G.; Mandelli, M.L.; Miller, Z.A.; Santos-Santos, M.A.; Wilson, S.M.; Agosta, F.; Grinberg, L.T.; Huang, E.J.; Trojanowski, J.Q.; Meyer, M.; et al. Typical and atypical pathology in primary progressive aphasia variants. *Ann. Neurol.* **2017**, *81*, 430–443. [CrossRef]
23. Hodges, J.R.; Patterson, K. Semantic dementia: A unique clinicopathological syndrome. *Lancet Neurol.* **2007**, *6*, 1004–1014. [CrossRef]
24. Iaccarino, L.; Crespi, C.; Della Rosa, P.A.; Catricalà, E.; Guidi, L.; Marcone, A.; Tagliavini, F.; Magnani, G.; Cappa, S.; Perani, D. The Semantic Variant of Primary Progressive Aphasia: Clinical and Neuroimaging Evidence in Single Subjects. *PLoS ONE* **2015**, *10*, e0120197. [CrossRef]
25. Binney, R.J.; Henry, M.; Babiak, M.; Pressman, P.S.; Santos-Santos, M.A.; Narvid, J.; Mandelli, M.L.; Strain, P.J.; Miller, B.L.; Rankin, K.P.; et al. Reading words and other people: A comparison of exception word, familiar face and affect processing in the left and right temporal variants of primary progressive aphasia. *Cortex* **2016**, *82*, 147–163. [CrossRef]
26. Josephs, K.A.; Whitwell, J.L.; Knopman, D.S.; Boeve, B.F.; Vemuri, P.; Senjem, M.L.; Parisi, J.E.; Ivnik, R.J.; Dickson, D.W.; Petersen, R.C.; et al. Two distinct subtypes of right temporal variant frontotemporal dementia. *Neurology* **2009**, *73*, 1443–1450. [CrossRef] [PubMed]
27. Chan, D.; Anderson, V.; Pijnenburg, Y.; Whitwell, J.; Barnes, J.; Scahill, R.; Stevens, J.M.; Barkhof, F.; Scheltens, P.; Rossor, M.; et al. The clinical profile of right temporal lobe atrophy. *Brain* **2009**, *132*, 1287–1298. [CrossRef] [PubMed]
28. Henry, M.L.; Wilson, S.M.; Ogar, J.M.; Sidhu, M.S.; Rankin, K.P.; Cattaruzza, T.; Miller, B.L.; Gorno-Tempini, M.L.; Seeley, W.W. Neuropsychological, behavioral, and anatomical evolution in right temporal variant frontotemporal dementia: A longitudinal and post-mortem single case analysis. *Neurocase* **2014**, *20*, 100–109. [CrossRef]
29. Carthery-Goulart, M.T.; Silveira, A.D.C.D.; Machado, T.H.; Mansur, L.L.; Parente, M.A.D.M.P.; Senaha, M.L.H.; Brucki, S.; Nitrini, R. Nonpharmacological interventions for cognitive impairments following primary progressive aphasia: A systematic review of the literature. *Dement. Neuropsychol.* **2013**, *7*, 122–131. [CrossRef]
30. Croot, K.; Nickels, L.; Laurence, F.; Manning, M. Impairment- and activity/participation-directed interventions in progressive language impairment: Clinical and theoretical issues. *Aphasiology* **2009**, *23*, 125–160. [CrossRef]
31. Jokel, R.; Graham, N.L.; Rochon, E.; Leonard, C. Word retrieval therapies in primary progressive aphasia. *Aphasiology* **2014**, *28*, 1038–1068. [CrossRef]
32. Rising, K. Treatment for Lexical Retrieval in Primary Progressive Aphasia Lexical Retrieval Treatments in PPA. *Perspect. Neurophysiol. Neurogenic Speech Lang. Disord.* **2016**, *24*, 137–144. [CrossRef]
33. Tippett, D.C.; Hillis, A.E.; Tsapkini, K. Treatment of Primary Progressive Aphasia. *Curr. Treat. Options Neurol.* **2015**, *17*, 1–11. [CrossRef]
34. Croot, K. Treatment for Lexical Retrieval Impairments in Primary Progressive Aphasia: A Research Update with Implications for Clinical Practice. *Semin. Speech Lang.* **2018**, *39*, 242–256. [CrossRef]
35. Cadório, I.; Lousada, M.; Martins, P.; Figueiredo, D. Generalization and maintenance of treatment gains in primary progressive aphasia (PPA): A systematic review. *Int. J. Lang. Commun. Disord.* **2017**, *52*, 543–560. [CrossRef] [PubMed]
36. Cotelli, M.; Manenti, R.; Ferrari, C.; Gobbi, E.; Macis, A.; Cappa, S.F. Effectiveness of language training and non-invasive brain stimulation on oral and written naming performance in Primary Progressive Aphasia: A meta-analysis and systematic review. *Neurosci. Biobehav. Rev.* **2020**, *108*, 498–525. [CrossRef]
37. Graham, K.S.; Patterson, K.; Pratt, K.H.; Hodges, J.R. Relearning and Subsequent Forgetting of Semantic Category Exemplars in a Case of Semantic Dementia. *Neuropsychology* **1999**, *13*, 359–380. [CrossRef]
38. Heredia, C.G.; Sage, K.; Ralph, M.A.L.; Berthier, M.L. Relearning and retention of verbal labels in a case of semantic dementia. *Aphasiology* **2009**, *23*, 192–209. [CrossRef]
39. Mayberry, E.; Sage, K.; Ehsan, S.; Ralph, M.A.L. Relearning in semantic dementia reflects contributions from both medial temporal lobe episodic and degraded neocortical semantic systems: Evidence in support of the complementary learning systems theory. *Neuropsychologia* **2011**, *49*, 3591–3598. [CrossRef]
40. Jokel, R.; Rochon, E.; Anderson, N. Errorless learning of computer-generated words in a patient with semantic dementia. *Neuropsychol. Rehabil.* **2010**, *20*, 16–41. [CrossRef]
41. Savage, S.; Ballard, K.; Piguet, O.; Hodges, J.R. Bringing words back to mind—Improving word production in semantic dementia. *Cortex* **2013**, *49*, 1823–1832. [CrossRef]
42. Savage, S.A.; Piguet, O.; Hodges, J.R. Cognitive Intervention in Semantic Dementia Maintaining Words over Time. *Alzheimer Dis. Assoc. Disord.* **2015**, *29*, 55–62. [CrossRef] [PubMed]
43. Snowden, J.S.; Neary, D. Relearning of verbal labels in semantic dementia. *Neuropsychologia* **2002**, *40*, 1715–1728. [CrossRef]
44. Suarez-Gonzalez, A.; Heredia, C.G.; Savage, S.A.; Gil-Néciga, E.; García-Casares, N.; Franco-Macías, E.; Berthier, M.L.; Caine, D. Restoration of conceptual knowledge in a case of semantic dementia. *Neurocase* **2014**, *21*, 309–321. [CrossRef] [PubMed]

45. Suárez-González, A.; A Savage, S.; Caine, D. Successful short-term re-learning and generalisation of concepts in semantic dementia. *Neuropsychol. Rehabil.* **2016**, *28*, 1095–1109. [CrossRef]
46. Meyer, A.M.; Tippett, D.C.; Friedman, R.B. Prophylaxis and remediation of anomia in the semantic and logopenic variants of primary progressive aphasia. *Neuropsychol. Rehabil.* **2016**, *28*, 352–368. [CrossRef]
47. Croot, K.; Raiser, T.; Taylor-Rubin, C.; Ruggero, L.; Ackl, N.; Wlasich, E.; Danek, A.; Scharfenberg, A.; Foxe, D.; Hodges, J.R.; et al. Lexical retrieval treatment in primary progressive aphasia: An investigation of treatment duration in a heterogeneous case series. *Cortex* **2019**, *115*, 133–158. [CrossRef]
48. Lavoie, M.; Bier, N.; Laforce, R., Jr.; Macoir, J. Improvement in functional vocabulary and generalization to conversation following a self-administered treatment using a smart tablet in primary progressive aphasia. *Neuropsychol. Rehabil.* **2020**, *30*, 1224–1254. [CrossRef]
49. Krajenbrink, T.; Croot, K.; Taylor-Rubin, C.; Nickels, L. Treatment for spoken and written word retrieval in the semantic variant of primary progressive aphasia. *Neuropsychol. Rehabil.* **2018**, *30*, 915–947. [CrossRef] [PubMed]
50. Dressel, K.; Huber, W.; Frings, L.; Kümmerer, D.; Saur, D.; Mader, I.; Hüll, M.; Weiller, C.; Abel, S. Model-oriented naming therapy in semantic dementia: A single-case fMRI study. *Aphasiology* **2010**, *24*, 1537–1558. [CrossRef]
51. Jokel, R.; Anderson, N. Quest for the best: Effects of errorless and active encoding on word re-learning in semantic dementia. *Neuropsychol. Rehabil.* **2012**, *22*, 187–214. [CrossRef] [PubMed]
52. Newhart, M.; Davis, C.; Kannan, V.; Heidler-Gary, J.; Cloutman, L.; Hillis, A.E. Therapy for naming deficits in two variants of primary progressive aphasia. *Aphasiology* **2009**, *23*, 823–834. [CrossRef]
53. Grasso, S.M.; Shuster, K.M.; Henry, M.L. Comparing the effects of clinician and caregiver-administered lexical retrieval training for progressive anomia. *Neuropsychol. Rehabil.* **2017**, *29*, 866–895. [CrossRef] [PubMed]
54. Meyer, A.M.; Tippett, D.C.; Turner, R.S.; Friedman, R.B. Long-Term maintenance of anomia treatment effects in primary progressive aphasia. *Neuropsychol. Rehabil.* **2019**, *29*, 1439–1463. [CrossRef] [PubMed]
55. Beales, A.; Cartwright, J.; Whitworth, A.; Panegyres, P.K. Exploring generalisation processes following lexical retrieval intervention in primary progressive aphasia. *Int. J. Speech-Lang. Pathol.* **2016**, *18*, 299–314. [CrossRef] [PubMed]
56. Kroll, J.; Stewart, E. Category Interference in Translation and Picture Naming: Evidence for Asymmetric Connections between Bilingual Memory Representations. *J. Mem. Lang.* **1994**, *33*, 149–174. [CrossRef]
57. Kroll, J.F.; Michael, E.; Tokowicz, N.; Dufour, R. The development of lexical fluency in a second language. *Second. Lang. Res.* **2002**, *18*, 137–171. [CrossRef]
58. Kroll, J.F.; Van Hell, J.G.; Tokowicz, N.; Green, D.W. The Revised Hierarchical Model: A critical review and assessment. *Biling. Lang. Cogn.* **2010**, *13*, 373–381. [CrossRef]
59. van Hell, J.G.; de Groot, A.M. Sentence context modulates visual word recognition and translation in bilinguals. *Acta Psychol.* **2008**, *128*, 431–451. [CrossRef] [PubMed]
60. Christoffels, I.; De Groot, A.; Kroll, J. Memory and language skills in simultaneous interpreters: The role of expertise and language proficiency. *J. Mem. Lang.* **2006**, *54*, 324–345. [CrossRef]
61. De Groot, A.M.B. Determinants of word translation. *J. Exp. Psychol. Learn. Mem. Cogn.* **1992**, *18*, 1001–1018. [CrossRef]
62. DeGroot, A.; Dannenburg, L.; Vanhell, J. Forward and Backward Word Translation by Bilinguals. *J. Mem. Lang.* **1994**, *33*, 600–629. [CrossRef]
63. de Groot, A.M.B.; Nas, G.L.J. Lexical Representation of Cognates and Noncognates Compound Bilinguals. *J. Mem. Lang.* **1991**, *123*, 90–123. [CrossRef]
64. Costa, A.; Caramazza, A.; Sebastian-Galles, N. The Cognate Facilitation Effect: Implications for Models of Lexical Access. *J. Exp. Psychology Learn. Mem. Cogn.* **2000**, *26*, 1283–1296. [CrossRef]
65. Costa, A.; Santesteban, M.; Caño, A. On the facilitatory effects of cognate words in bilingual speech production. *Brain Lang.* **2005**, *94*, 94–103. [CrossRef]
66. Sáchez-Casas, R.M.; García-Albea, J.E.; Davis, C.W. Bilingual lexical processing: Exploring the cognate/non-cognate distinction. *Eur. J. Cogn. Psychol.* **1992**, *4*, 293–310. [CrossRef]
67. Rosselli, M.; Ardila, A.; Jurado, M.B.; Salvatierra, J.L. Cognate facilitation effect in balanced and non-balanced Spanish–English bilinguals using the Boston Naming Test. *Int. J. Biling.* **2014**, *18*, 649–662. [CrossRef]
68. Murray, L.L. Bilingual aphasia treatment: Clinical recommendations regarding secondary language treatment, cross-language transfer, and the use of language brokers await additional research. *Evid.-Based Commun. Assess. Interv.* **2014**, *9*, 1–6. [CrossRef]
69. Hameau, S.; Köpke, B. Cross-Language Transfer for Cognates in Aphasia Therapy with Multilingual Patients: A Case Study. *Aphasie Verwandte Geb.* **2015**, *3*, 13–19.
70. Marangolo, P.; Rizzi, C.; Peran, P.; Piras, F.; Sabatini, U. Parallel recovery in a bilingual aphasic: A neurolinguistic and fMRI study. *Neuropsychology* **2009**, *23*, 405–409. [CrossRef] [PubMed]
71. Miertsch, B.; Meisel, J.M.; Isel, F. Non-treated languages in aphasia therapy of polyglots benefit from improvement in the treated language. *J. Neurolinguist.* **2009**, *22*, 135–150. [CrossRef]
72. Kiran, S.; Sandberg, C.; Gray, T.; Ascenso, E.; Kester, E. Rehabilitation in Bilingual Aphasia: Evidence for Within- and Between-Language Generalization. *Am. J. Speech-Lang. Pathol.* **2013**, *22*, S298–S309. [CrossRef]
73. Kiran, S.; Iakupova, R. Understanding the relationship between language proficiency, language impairment and rehabilitation: Evidence from a case study. *Clin. Linguist. Phon.* **2011**, *25*, 565–583. [CrossRef]

74. Ansaldo, A.I.; Saidi, L.G.; Ruiz, A. Model-driven intervention in bilingual aphasia: Evidence from a case of pathological language mixing. *Aphasiology* **2009**, *24*, 309–324. [CrossRef]
75. Edmonds, L.A.; Kiran, S. Effect of Semantic Naming Treatment on Crosslinguistic Generalization in Bilingual Aphasia. *J. Speech Lang. Hear. Res.* **2006**, *49*, 729–748. [CrossRef]
76. Junqué, C.; Vendrell, P.; Vendrell-Brucet, J.M.; Tobena, A. Differential recovery in naming in bilingual aphasics. *Brain Lang.* **1989**, *36*, 16–22. [CrossRef]
77. Kiran, S.; Roberts, P.M. Semantic feature analysis treatment in Spanish–English and French–English bilingual aphasia. *Aphasiology* **2009**, *24*, 231–261. [CrossRef]
78. Kohnert, K. Cognitive and cognate-based treatments for bilingual aphasia: A case study. *Brain Lang.* **2004**, *91*, 294–302. [CrossRef] [PubMed]
79. Goral, M.; Rosas, J.; Conner, P.S.; Maul, K.K.; Obler, L.K. Effects of language proficiency and language of the environment on aphasia therapy in a multilingual. *J. Neurolinguist.* **2012**, *25*, 538–551. [CrossRef]
80. Meinzer, M.; Obleser, J.; Flaisch, T.; Eulitz, C.; Rockstroh, B. Recovery from aphasia as a function of language therapy in an early bilingual patient demonstrated by fMRI. *Neuropsychologia* **2007**, *45*, 1247–1256. [CrossRef] [PubMed]
81. Kurland, J.; Falcon, M. Effects of cognate status and language of therapy during intensive semantic naming treatment in a case of severe nonfluent bilingual aphasia. *Clin. Linguist. Phon.* **2011**, *25*, 584–600. [CrossRef]
82. Bedore, L.M.; Peña, E.D.; García, M.; Cortez, C. Scoring: When Does It Make a Difference? *Lang. Speech Hear. Serv. Sch.* **2005**, *36*, 188–201. [CrossRef]
83. Kohnert, K.J.; Hernandez, A.E.; Bates, E. Bilingual Performance on the Boston Naming Test: Preliminary Norms in Spanish and English. *Brain Lang.* **1998**, *65*, 422–440. [CrossRef] [PubMed]
84. Umbel, V.M.; Pearson, B.Z.; Fernandez, M.C.; Oller, D.K. Measuring Bilingual Children's Receptive Vocabularies. *Child Dev.* **1992**, *63*, 1012–1020. [CrossRef] [PubMed]
85. Folstein, M.F.; Folstein, S.E.; McHugh, P.R. "Mini-mental state": A practical method for grading the cognitive state of patients for the clinician. *J. Psychiatr. Res.* **1975**, *12*, 189–198. [CrossRef]
86. Dial, H.R.; A Hinshelwood, H.; Grasso, S.M.; Hubbard, H.I.; Gorno-Tempini, M.-L.; Henry, M. Investigating the utility of teletherapy in individuals with primary progressive aphasia. *Clin. Interv. Aging* **2019**, *14*, 453–471. [CrossRef]
87. Kiran, S.; Peña, E.; Bedore, L.; Sheng, L. Evaluating the Relationship between Category Generation and Language Use and Proficiency. Presented at the Donostia Workshop on Neurobilingualism, San Sebastian, Spain, 30 September–2 October 2010.
88. Beeson, P.M.; Egnor, H. Combining treatment for written and spoken naming. *J. Int. Neuropsychol. Soc.* **2006**, *12*, 816–827. [CrossRef] [PubMed]
89. Coltheart, M. The MRC Psycholinguistic Database. *Q. J. Exp. Psychol. Sect. A* **1981**, *33*, 497–505. [CrossRef]
90. Davies, M. The Corpus of Contemporary American English as the first reliable monitor corpus of English. *Lit. Linguist. Comput.* **2010**, *25*, 447–464. [CrossRef]
91. Marian, V.; Bartolotti, J.; Chabal, S.; Shook, A. CLEARPOND: Cross-Linguistic Easy-Access Resource for Phonological and Orthographic Neighborhood Densities. *PLoS ONE* **2012**, *7*, e43230. [CrossRef]
92. Davies, M. Corpus Del Español: Two Billion Words, 21 Countries. 2006. Available online: http://www.corpusdelespanol.org/web-dial/ (accessed on 15 October 2021).
93. Duchon, A.; Perea, M.; Sebastian-Galles, N.; Martí, M.A.; Carreiras, M. EsPal: One-stop shopping for Spanish word properties. *Behav. Res. Methods* **2013**, *45*, 1246–1258. [CrossRef]
94. New, B.; Pallier, C.; Brysbaert, M.; Ferrand, L. Lexique 2: A new French lexical database. *Behav. Res. Methods Instrum. Comput.* **2004**, *36*, 516–524. [CrossRef]
95. Davies, M. Corpus Do Português: One Billion Words, 4 Countries. Available online: http://www.corpusdoportugues.org/web-dial/ (accessed on 15 October 2021).
96. Kilgarriff, A.; Baisa, V.; Bušta, J.; Jakubíček, M.; Kovář, V.; Michelfeit, J.; Rychlý, P.; Suchomel, V. The Sketch Engine: Ten years on. *Lexicography* **2014**, *1*, 7–36. [CrossRef]
97. Rasooli, M.S.; Kouhestani, M.; Moloodi, A. Development of a Persian Syntactic Dependency Treebank. In Proceedings of the NAACL-HLT, Atlanta, GA, USA, 9–13 June 2013; pp. 306–314.
98. Dial, H.; Martin, R. Evaluating the relationship between sublexical and lexical processing in speech perception: Evidence from aphasia. *Neuropsychologia* **2017**, *96*, 192–212. [CrossRef]
99. Kertesz, A. *Western Aphasia Battery—Revised*; Pearson: San Antonio, TX, USA, 2012.
100. Kaplan, E.; Goodglass, H.; Weintraub, S. *Boston Naming Test*; Pro-Ed: Austin, TX, USA, 2001.
101. Paul, N.; Mehrhoff, J. Descriptive Analysis: Survey of Direct and Indirect Interventions for Persons with Dementia-Based Communication Disorders. *Perspect. Neurophysiol. Neurogenic Speech Lang. Disord.* **2015**, *25*, 125–141. [CrossRef]
102. Francis, W.S. Cognitive integration of language and memory in bilinguals: Semantic representation. *Psychol. Bull.* **1999**, *125*, 193–222. [CrossRef]
103. Francis, W.S. Shared core meanings and shared associations in bilingual semantic memory: Evidence from research on implicit memory. *Int. J. Biling.* **2018**, *24*, 464–477. [CrossRef]
104. Hoshino, N.; Kroll, J.F. Cognate effects in picture naming: Does cross-language activation survive a change of script? *Cognition* **2008**, *106*, 501–511. [CrossRef]

105. Mendez, M.F.; Perryman, K.M.; Pontón, M.O.; Cummings, J.L. Bilingualism and Dementia. *J. Neuropsychiatry Clin. Neurosci.* **1999**, *11*, 411–412. [CrossRef] [PubMed]

106. Ivanova, I.; Salmon, D.P.; Gollan, T.H. Which Language Declines More? Longitudinal versus Cross-sectional Decline of Picture Naming in Bilinguals with Alzheimer's Disease. *J. Int. Neuropsychol. Soc.* **2014**, *20*, 534–546. [CrossRef] [PubMed]

107. Henry, M.L.; Wilson, S.M.; Babiak, M.C.; Mandelli, M.L.; Beeson, P.M.; Miller, Z.A.; Gorno-Tempini, M.L. Phonological Processing in Primary Progressive Aphasia. *J. Cogn. Neurosci.* **2016**, *28*, 210–222. [CrossRef]

108. Graham, K.S.; Patterson, K.; Pratt, K.H.; Hodges, J.R. Can repeated exposure to "forgotten" vocabulary help alleviate word-finding difficulties in semantic dementia? An illustrative case study. *Neuropsychol. Rehabil.* **2001**, *11*, 429–454. [CrossRef]

109. Jokel, R.; Rochon, E.; Leonard, C. Treating anomia in semantic dementia: Improvement, maintenance, or both? *Neuropsychol. Rehabil.* **2006**, *16*, 241–256. [CrossRef] [PubMed]

110. Lind, M.; Simonsen, H.G.; Ribu, I.S.B.; Svendsen, B.A.; Svennevig, J.; De Bot, K. Lexical access in a bilingual speaker with dementia: Changes over time. *Clin. Linguist. Phon.* **2017**, *32*, 353–377. [CrossRef]

111. Ou, L.; Przybilla, M.; Koniar, B.; Whitley, C.B. RTB lectin-mediated delivery of lysosomal α-L-iduronidase mitigates disease manifestations systemically including the central nervous system. *Mol. Genet. Metab.* **2018**, *123*, 105–111. [CrossRef]

112. Mahendra, N.; Spicer, J. Access to Speech-Language Pathology Services for African-American Clients with Aphasia: A Qualitative Study Racial and Ethnic Disparities in Stroke Care. *Div. 14 Newsl.* **2014**, *21*, 53–62.

113. Prince, M.; Ali, G.-C.; Guerchet, M.; Prina, M.; Albanese, E.; Wu, Y.-T. Recent global trends in the prevalence and incidence of dementia, and survival with dementia. *Alzheimer's Res. Ther.* **2016**, *8*, 1–13. [CrossRef]

114. Prince, M.; Acosta, D.; Albanese, E.; Arizaga, R.; Ferri, C.P.; Guerra, M.; Huang, Y.; Jacob, K.; Jiménez-Velázquez, I.Z.; Rodriguez, J.L.; et al. Ageing and dementia in low and middle income countries–Using research to engage with public and policy makers. *Int. Rev. Psychiatry* **2008**, *20*, 332–343. [CrossRef] [PubMed]

115. Ortman, J.M.; Shin, H.B. Language Projections: 2010 to 2020. Presented at the Annual Meetings of the American Sociological Association, Las Vegas, NV, USA, 20–23 August 2011.

116. Zeigler, K.; Camarota, S.A. *67.3 Million in the United States Spoke a Foreign Language at Home in 2018*; Center for Immigration Studies: Washington, DC, USA, 2019. Available online: https://cis.org/Report/673-Million-United-States-Spoke-Foreign-Language-Home-2018 (accessed on 15 October 2021).

117. Grandpierre, V.; Milloy, V.; Sikora, L.; Fitzpatrick, E.; Thomas, R.; Potter, B. Barriers and facilitators to cultural competence in rehabilitation services: A scoping review. *BMC Health Serv. Res.* **2018**, *18*, 1–14. [CrossRef]

Article

Cognitive Intervention Strategies Directed to Speech and Language Deficits in Primary Progressive Aphasia: Practice-Based Evidence from 18 Cases

Thais Helena Machado [1,2,3,*], Maria Teresa Carthery-Goulart [4,5,6], Aline Carvalho Campanha [2] and Paulo Caramelli [1,2]

1   Programa de Pós-Graduação em Ciências Aplicadas à Saúde do Adulto, Faculdade de Medicina, Universidade Federal de Minas Gerais, Belo Horizonte 30130-100, MG, Brazil; caramelli@ufmg.br
2   Grupo de Pesquisa em Neurologia Cognitiva e do Comportamento, Departamento de Clínica Médica, Faculdade de Medicina, Universidade Federal de Minas Gerais, Belo Horizonte 30130-100, MG, Brazil; aline.campanha@gmail.com
3   Av Prudente de Morais, 290-Sala 1106, Belo Horizonte 30380-002, MG, Brazil
4   Grupo de Estudos em Neurociência da Linguagem e Cognição, Núcleo Interdisciplinar de Neurociência Aplicada, Centro de Matemática, Computação e Cognição da Universidade Federal do ABC, São Bernardo do Campo 09210-580, SP, Brazil; teresa.carthery@gmail.com
5   Grupo de Neurologia Cognitiva e do Comportamento, Divisão de Clínica Neurológica, Hospital das Clínicas da Faculdade de Medicina da Universidade de São Paulo, São Paulo 05403-000, SP, Brazil
6   INCT-ECCE (Instituto Nacional de Ciência e Tecnologia sobre Comportamento, Cognição e Ensino), Rodovia Washington Luís, Km 235, São Carlos 13565-905, SP , Brazil
*   Correspondence: thaismachado@yahoo.com

Citation: Machado, T.H.; Carthery-Goulart, M.T.; Campanha, A.C.; Caramelli, P. Cognitive Intervention Strategies Directed to Speech and Language Deficits in Primary Progressive Aphasia: Practice-Based Evidence from 18 Cases. *Brain Sci.* **2021**, *11*, 1268. https://doi.org/10.3390/brainsci11101268

Academic Editors: Jordi A. Matias-Guiu, Robert Jr Laforce and Rene L. Utianski

Received: 11 August 2021
Accepted: 18 September 2021
Published: 25 September 2021

Publisher's Note: MDPI stays neutral with regard to jurisdictional claims in published maps and institutional affiliations.

Abstract: Background: Practice-based evidence can inform and support clinical decision making. Case-report series about the implementation of programs in real-world clinical settings may contribute to verifying the effectiveness of interventions for treating PPA in specific contexts, as well as illustrating challenges that need to be overcome. Objective: To describe and provide practice-based evidence on the effectiveness of four cognitive rehabilitation programs designed for individuals with PPA and directed to speech and language impairments, which were implemented in a specialized outpatient clinic. Methods: Multiple single-case study. Eighteen individuals with different subtypes of PPA were each assigned to one out of four training programs based on comprehensive speech and language assessments. The treatments targeted naming deficits, sentence production, speech apraxia, and phonological deficits. Pre- and post-treatment assessments were undertaken to compare trained and untrained items. Gains were generalized to a different task in the first two types of intervention (naming and sentence production). A follow-up assessment was conducted 1–8 months after treatment among 7 participants. Results: All individuals presented better performance in the trained items at the post-test for each rehabilitation program accomplished, demonstrating that learning of the trained strategies was achieved during the active phase of treatment. For 13 individuals, statistical significance was reached; while for five, the results were maintained. Results about untrained items, generalization to other tasks, and follow-up assessments are presented. Conclusions: The positive results found in our sample bring some practice-based evidence for the benefits of speech and language treatment strategies for clinical management of individuals with PPA.

Keywords: primary progressive aphasia; treatment; speech and language therapy; intervention; cognitive rehabilitation

1. Introduction

Primary progressive aphasia (PPA) is characterized by gradual deterioration of language with relative preservation of other cognitive functions and functional independence, except for situations in which language is critical [1,2].

The international consensus for diagnosing PPA [2] defined three clinical variants (agrammatic/nonfluent, semantic, and logopenic). Around 20 to 35% of individuals with PPA do not fit into these three main variants and are named non-classified or mixed PPA (mxPPA) cases [3,4].

Symptom onset may occur before the age of 65, with a devastating effect on functionality. In the absence of effective pharmacological treatments [5], there is increasing interest in other approaches, particularly behavioral interventions, focusing on communication aspects or specific speech/language deficits. Relative preservation of other cognitive functions, including episodic memory [2,6], enables implementation of SLT, given that individuals with PPA are usually aware of their difficulties and can engage more independently in the activities proposed, with lower demand for support from caregivers, compared with subjects with predominantly episodic memory impairment, who have greater difficulty in learning new content.

Non-pharmacological interventions in PPA can be classified into those directed to the deficits (e.g., anomia, agrammatism, phonological working memory or speech apraxia) and functional interventions (environmental modifications and compensatory strategies). Positive results were reported in most studies but, to be recommended, treatments require further investigation regarding their effectiveness [7]. Most evidence derives from case studies or series [7,8] and randomized-controlled studies with larger samples are needed in order to increase the level of evidence, as there is no consensus regarding types and duration of interventions. However, compared with dementia syndromes (e.g., Alzheimer's disease), PPA is a rare condition with heterogeneous clinical profiles. Implementation of randomized controlled studies requires a multicenter effort to unify assessment and treatment protocols in order to gather evidence from larger samples. On the other hand, SLT practice requires individualized intervention plans that are adjusted to the context, resources and individuals' and families' preferences. Reporting the results from treatments that were implemented can also contribute to the level of practice-based evidence. Practice-based evidence can inform and support clinical decision making and is obtained from several sources, including case reports or case series in "real-life" clinical settings [9].

1.1. Research-Based Evidence on Non-Pharmacological Treatments for PPA

1.1.1. Interventions Directed to Lexical Retrieval/Semantic Deficits

Lexical retrieval and/or semantic deficits are common features of PPA syndromes and may predominate over other language or cognitive impairments for long periods. Subsequently, communication becomes markedly affected by word production and comprehension deficits, difficulties in sentence production (agrammatism or paragrammatism) and/or in syntactic comprehension. Speech may also be affected by apraxia of speech and dysarthria [1].

Lexical retrieval treatment is the most widely applied approach [7,10], independently of the clinical variant, given that individuals with PPA usually manifest anomia or word misuse (i.e., lexical retrieval and semantic deficits) with greater or lesser severity. The goal of this treatment is to restore and maintain retrieval of core vocabulary items for as long as possible.

Subjects with PPA are able to relearn target vocabulary during the active phase of treatment and to maintain gains for varying periods after the intervention [11–17]. Learning may be generalized to untrained stimuli [13,15,16,18–20]; however, these findings are still inconsistent. Rising [21] and Croot [8] reported immediate treatment gains in most individuals, and maintenance of gains (months to years) in some individuals with ongoing treatment.

Beales et al. [22] showed that relearning was the most prominent mechanism of change in PPA, followed by stimulation. Reorganization and cognitive relay were less observed. Given the progressive nature of PPA and the urgency of maintaining the preserved vocabulary, perhaps only items that are relevant to daily life should be included in treatment sets [10,18].

Meyer et al. [23] used the term prophylaxis for stimuli that were consistently named correctly (prophylaxis items) and the term remediation for those that were consistently named incorrectly (remediation items) at baseline. Studies on treatment for anomia in PPA have typically focused on remediation of words that could not be named at baseline, rather than on prophylaxis of words that could be named. However, prophylactic treatment may also have positive effects. Reilly [24] defended maintenance of known words over reacquisition of forgotten knowledge regarding semantic treatment paradigms. Studies investigating maintenance of treatment gains have suggested that retrieval accuracy can be maintained (prophylaxis items) or improved (remediation items) with long-term treatment (six months or more) [17,23–27].

Volkmer et al. [28] explained that PPA requires a "staging approach", in which "impairment-based interventions" (focusing on remediation and rehabilitation) should be implemented at early stages, while compensatory strategies (with the goal of developing strategies to facilitate completion of a particular task) should be implemented after restoration has failed and language skills are lost.

1.1.2. Interventions Targeting Speech and Sentence Production

These interventions are offered to nonfluent subjects and are aimed at syntax training with different approaches, as shown below. Schneider et al. [29] examined the effectiveness of verbal plus gestural treatment on acquisition and generalization of verb tenses in sentence production in one individual and showed that improvement in the production of sentences was achieved through using trained verb tenses.

Andrade-Calderón et al. [30] analyzed the effects of intensive speech therapy in a nfPPA case. The subject received weekly speech therapy with combined stimulation strategies relating to different components for language processing. Syntactic tasks were applied, comprising construction of sentences based on combinations of worked stimuli and on changing the gender/number/tense of structural elements. The subject showed slight improvements in prosody, fluency and spontaneous speech content, and significant improvements in repetition, reading aloud, and oral-phonatory praxis. This therapy also had a positive impact on other cognitive processes.

A constraint-induced treatment approach implemented with two nfvPPA subjects resulted in improved production of grammatical structures, with maintenance of gains observed at two months post-treatment [31].

Studies on nfvPPA subjects were also directed to speech apraxia [32] and have shown reduction in speech errors through training on text reading.

1.1.3. Interventions Directed to Phonological Deficits

Phonological deterioration starting from a phonological short-term memory deficit characterizes lvPPA. While most individuals with lvPPA mention lexical retrieval problems as their main deficit, some of them are concerned with spelling and short-term memory deficits.

With the premise that the phonological loop is a working memory component, spelling and repetition activities are positive resources used in phonological interventions. Two studies on spelling showed positive results, with learning of phoneme-to-grapheme and phoneme-to-word correspondences [33,34]. In addition, to improve fluency in nfPPA, Louis et al. [35] trained three subjects using a remediation protocol that included auditory exercises that were specifically designed to tackle phonological processing. All participants improved their performance in trained and untrained tasks (generalization to the cookie theft picture and functional communication).

The objective of the present study was to explore intervention techniques for specific language and speech deficits in PPA in a specialized outpatient clinic. Four intervention programs were implemented based on strategies that had shown positive effects in previous studies ([5,7] for reviews), and these were directed to anomia, agrammatism, speech apraxia, and phonological deficits. We investigated the effectiveness of programs in order to generate practice-based evidence and describe the challenges for implementation of these programs in a real-world clinical setting.

2. Materials and Methods

2.1. Subjects

Over a three-year period, we recruited a convenience sample of individuals with newly diagnosed PPA who were referred to this study by physicians or members of the interdisciplinary team of the Behavioral and Cognitive Neurology Outpatient Clinic of Hospital das Clínicas, Federal University of Minas Gerais, in Belo Horizonte, Brazil. This research was conducted in accordance with the Helsinki Declaration, and participants and their families signed an informed consent statement that was approved by the university's Ethics Committee.

All subjects included in the study presented diagnosis of probable PPA according to current diagnostic guidelines [2] and had undergone neurological examination and cognitive, speech, and language assessments. These participants were classified into one of three PPA variants (3 nfvPPA, 5 svPPA, and 5 lvPPA), or as mxPPA (5 subjects) presentations. The mean age was 66.3 years, 9 subjects were women, and mean educational length was 14.5 years. The duration of symptoms was 2.1 years.

These individuals with PPA were in mild-to-moderate stages of the syndrome. Their severity of impairment was determined qualitatively. Those who were able to establish functional communication with no need of cues from the therapist were considered to be mild cases. Those who needed support from the therapist, either by simplifying speech to facilitate comprehension or by providing cues to facilitate oral production were considered to be at the moderate stage. Participants with significant functional communication difficulties, such as those unable to give an oral response, or who displayed unintelligible speech were considered to be severe cases.

The inclusion criteria involved a minimum literacy level (at least two years of formal education) and agreement to complete the treatment cycle, be evaluated and undergo post-evaluation. The exclusion criteria involved severe hearing and/or visual deficits, and severe motor or language deficits that would impact the implementation of the programs.

2.2. Methods

The subjects were seen at the Behavioral and Cognitive Neurology Outpatient Clinic. The first stage consisted of a medical evaluation (the team included neurologists, geriatricians and psychiatrists), followed by evaluation by a speech therapist and a neuropsychologist. An overall cognitive assessment and a neuropsychological evaluation were used to assist in making the clinical diagnosis and to identify the degree of preservation of non-linguistic cognitive abilities. With these assessments and neuroimaging examinations, the study team assessed the clinical diagnoses and invited participants.

The cognitive and language evaluation for PPA diagnosis and characterization varied among the cases and included some of the tests listed below.

A.   Overall cognitive assessment and neuropsychological evaluation:

    (1).   Mini-mental state examination (MMSE) [36,37];

    (2).   Dementia rating scale (DRS) [38–40].

B.   Language assessment:

    (1).   Auditory comprehension tests: word and sentence comprehension tests from the Montreal-Toulouse battery (MTL) [41,42] and/or the Boston diagnostic aphasia examination (BDAE) [43,44] and/or the Cambridge semantic memory research battery (CSMRB) [45–47] and/or the token test [48,49] and/or the Trog-2 test [50–53];

    (2).   Visual confrontation naming tests: Boston naming test (BDAE) and/or CSMRB;

    (3).   Repetition—words, non-words, and phrases of MTL or BDAE;

    (4).   Reading words and non-words—HFSP reading aloud test [54];

(5).   Writing words and non-words—HFSP writing to dictation test [54–56];

(6).   Reading comprehension: subtests from MTL and BDAE;

(7).   Verbal fluency tasks:

-   Semantic category (animals) [57,58];
-   Phonemic fluency (FAS) [59];

(8).   Recognition and naming of famous faces [60];

(9).   Oral discourse—description of the cookie theft picture [43,61] and correction criteria suggested by Croisile et al. [62];

(10).   Word definition—CSMRB;

(11).   Camels and cactus test of semantic association [63,64];

(12).   Speech praxis protocol [65]; oral agility and oral discourse (BDAE).

Reading, writing, object knowledge and motor speech were assessed qualitatively. For reading and writing assessments, we used the list of words and pseudowords that was developed as part of the HFSP research project. This list was devised in order to study acquired dyslexia and dysgraphia across different written systems and contains words and pseudowords. We analyzed error types in two ways: (a) regularizations (irregular words or commonly used foreign words (e.g., "pizza") were read or written by applying grapheme-to-phoneme or phoneme-grapheme conversion rules, supported by the auditory representation of the stimuli instead of orthographic memory); (b) phonological or graphemic paralexias and paragraphias (additions, omissions and substitutions, indicative of the dysfunction of grapheme–phoneme conversion mechanisms or working memory deficits).

A different selection of tests for language assessment was applied to each subject. The results from these tests supported classification of the type of aphasia according to the semantic or syntactic losses that were identified, for example. Through this, the most evident difficulties could be identified in order to decide which type of program each individual should be referred to. Some of the tools used for language assessment were not validated for use in Portuguese but were translated, adapted, and applied to a group of cognitively healthy controls. The studies conducted on the versions of the language tests used in the current study are referenced above.

2.3. Study Design

This was a multiple single-case study consisting of four stages: (1) complete language assessment and pre-test (trained/untrained items); (2) speech and language intervention (four different types); (3) post-test (trained and untrained items) and, for the naming and sentence production interventions, subjects were also assessed in another task in order to address generalization; (4) follow up, which was conducted 1–8 months after completion of the program.

After the language assessment, each participant was allocated to a cognitive intervention program directed towards a specific language-speech impairment. This program was individualized and was chosen considering: (1) speech and language deficits (the most severe or apparent impairment); and (2) complaints and communication needs, in order to achieve functional adaptation. The functional deficits were identified during the clinical interview with the patients and their families. The best approach was then defined as a consensus with at least two speech therapists.

For the intervention phase, three speech and language therapists adapted four semi-structured programs that was designed for PPA deficits. Rehabilitation materials were personalized, and intervention programs were adjusted to the severity of deficits. The subjects were invited to participate in a 24-session program, but this length of program was not always possible, and the number of sessions was adjusted (see general procedures) to account for any particular mobility issues (for example, whether the individual was living in the city where the clinic was located, could afford transportation to the clinic, or needed a caregiver to accompany him/her, etc.).

The post-test stage consisted of reevaluation on the trained and untrained items (one week after the last rehabilitation session) to assess the effects of the treatment. In addition, for the interventions that focused on naming and on sentence production, generalization was also assessed in a different language task.

The degree of maintenance of the gains (follow up) was assessed in a subgroup of the subjects, at the time when they returned for a clinical consultation. The participants were retested on trained and untrained items. Due to time constraints, in two cases only the trained items were tested.

General Procedures

The clinical evaluations and intervention programs were performed by licensed speech and language therapists at the Behavioral and Cognitive Neurology Outpatient Clinic of Hospital das Clínicas, Federal University of Minas Gerais, in Belo Horizonte, Brazil.

The participants were offered 24 sessions of 50 min each, implemented over four months (twice a week). The programs were adjusted according to each individual's or family's time constraints in this context of a real clinical setting. In some cases, the subjects were just temporarily visiting the region, to look for a diagnosis in a specialized clinic, and treatment had to be implemented within a period of only two weeks. In other cases, the families committed to a three-month period, while in yet other cases we were able to extend the intervention and see the subjects for 12 months. Because of the heterogeneity of the duration of the treatments provided, as well as the decision to use a tailor-made approach in designing the therapy, we applied a multiple single-case study design in order to report on the effectiveness of the therapies.

Regardless of the type of intervention, the participants were encouraged to practice at home, and we offered training to the caregivers to support this practice (they were trained to assist when the subjects required help), but not all individuals and caregivers were able to follow this procedure. The stimuli sent to their homes were the same as those trained in the sessions. Practice at home was encouraged throughout the treatment; however, it was not formally monitored.

2.4. Interventions

2.4.1. Intervention Focusing on Naming

This treatment was based on Senaha et al. [66] and was aimed at naming deficits (either due to semantic memory deterioration or to lexical retrieval deficits). Its main goal was to improve or maintain individuals' performance in a set of core vocabulary items that could support their communication needs, with a remediation or prophylaxis approach, respectively. The items to be trained were selected for each subject considering: (1) specific needs and relevance to daily life; and (2) relative preservation of semantic knowledge of that item. Items were selected after interviewing the participant, spouse or frequent communication partner before the first week of the study. Before starting the rehabilitation program, the participants' families were involved in the selection of relevant words for the training. The criterion was their relevance to daily communication. The trained and untrained sets included both correctly named and incorrectly named stimuli that were presented in the pre-test. The only requirement was that the patient was seen to retain some semantic knowledge about the item in the pre-test (i.e., the ability to describe the context within which that item is usually seen, or its function, etc.). The sets included items from different semantic categories. The items consisted predominantly of picturable nouns, proper nouns, adjectives and verbs, as required, depending on the participants' communication needs. The number of items to be trained varied among the participants and was adjusted to their motivation for intervention (i.e., the amount of time that they could dedicate to daily practices). The training consisted of looking at meaningful pictures or photos of objects or people and trying to name them. The subjects were discouraged from guessing (i.e., following the principles of errorless learning) and were encouraged to check the written corresponding names at the back of each card in case they were not sure.

Then, they were asked to read the names aloud and build a meaningful sentence to use that word in context. When the subject was unable to produce this sentence on his own, the therapist elaborated it and asked for repetition. This last step was included in the training routine and differed from the procedure used by Senaha et al. [52]. As the participants' naming performance improved, the last letters/syllables were gradually erased from the back of the card until only the first letter remained as a written graphic cue that induced correct naming of each stimulus.

After selecting the training set, another set of items was prepared by two or three speech and language therapists for each subject (control set). These included items of the same grammatical category, of similar familiarity and picture complexity as in the training set. The subjects' performance regarding the trained and untrained items was assessed twice in all cases (one week before and after the intervention). Some participants had a third evaluation (follow-up). The trained items were individualized, but the untrained items were selected from a set of stimuli that the speech and language therapists used for their interventions, which were matched as much as possible to the trained set, according to psycholinguistic parameters (grammatical class, familiarity and visual complexity). The subjects' comprehension and preservation of some semantic knowledge of the stimuli in the sets was assessed indirectly through qualitative analysis on the responses to naming in the pre-test and the consensual decisions of the speech and language therapists, based on clinical judgment. Retention of basic semantic knowledge of the items in the lists was demonstrated through the ability to provide at least a basic description or show with gestures how to use the item or the context in which it is usually found.

The effectiveness of the intervention was evaluated by comparing the numbers of correct responses before and after the treatment from the trained and untrained (control) items. The generalization was evaluated by comparing the individuals' performances in another task (semantic verbal fluency), before and after the treatment.

2.4.2. Intervention Focusing on Sentence Production

This intervention was based on Bock and Levelt's model of sentence production. In our study, the two participants assigned to this treatment received an intervention targeting the positional level. We aimed at verb inflections for production of accurate simple sentences with the structure "subject-verb-object". We targeted the verb due to its central role in the sentence.

Twenty regular, familiar and high-frequency verbs were selected for the set of training and control stimuli, based on daily routines (examples: to get up, to eat, to cook, to shop, to work, and to go to sleep). We used a set of 40 written sentences with a gap to be filled by a verb in the present or in the past tense (an adverb at the beginning of the sentence cued the verb tense, i.e., "every day" or "yesterday"). Each verb was practiced in both tenses with a model provided by the therapist (repetition). The therapist provided the model aloud (adverb + subject + inflected verb) and the subject was asked to read the sentence and reproduce the verb form in the correct position and inflection (where there was a blank). Then, the therapist asked the subject to produce the full sentence again without reading support. A second drill consisted in providing a written prompt (adverb + subject + verb in the infinitive form) and ask the subject to produce the full sentence. Errors were discouraged; if necessary, the subject could use the written material (i.e., errorless learning). This procedure was repeated until the subject was able to produce the complete sentence accurately from the adverb, subject and verb prompt (e.g., from the prompt "Yesterday + to eat", the subject should produce "Yesterday I ate a sandwich"). Models and cues were gradually removed until the subjects were able to produce and speak the sentence aloud accurately.

Another 40 sentences with 20 different regular, familiar, and high frequency verbs were used as control set.

The effectiveness of the intervention was evaluated by comparing sentence production before and after the treatment, comparing gains in trained and untrained items. Discourse production from the cookie theft picture was used to look at the transference of the training to discourse.

2.4.3. Intervention Focusing on Speech Production

Based on Henry et al. [32], we implemented a treatment method using structured oral reading as a tool for improving the production of multisyllabic words (two or more syllables). This was directed towards individuals presenting apraxia of speech. During the treatment sessions, the subjects were trained in self-detection and correction of speech errors while reading one text aloud (the training involved rereading of the same text over the sessions). The treatment approach involved the following steps:

- The subject was required to read aloud a selected text. When he/she produced a word incorrectly (with one or more speech sound errors), he/she was asked to stop reading and practice that word (target).
- The subject produced the word syllable-by-syllable many times until he/she reached the correct articulation (appropriate prosody and speed of speech). If the target was a multisyllabic word, it was underlined in the text and lines were drawn dividing the word into constituent syllables. Single-syllable words were repeated until correctly produced in isolation.

After success in producing the word in isolation, the subjects were asked to read the sentence again in order to achieve correct word production in sentence context. If the word was again produced erroneously, the subjects were asked to repeat the previous steps, until the entire sentence was produced correctly.

Two different texts were applied for training (one for each participant), considering that their educational levels were different. The simplest had 120 multisyllabic words and the most complex had 319. The untrained texts had 95 and 179 multisyllabic words, respectively.

For homework, the subjects were encouraged to train on the text used in the session.

The effectiveness of the intervention was evaluated by comparing accuracy in the production of trained and untrained multisyllabic words (pre- and post-intervention). The pre-test intervention measurements considered the number of errors that the subjects made in the first reading of the text. We compared their performance in the trained text with their performance in an untrained text.

2.4.4. Intervention Focusing on Phonological Awareness and Verbal Working Memory

Spelling of words requires temporary storage of the sequence of letters in working memory (graphemic buffer) while the individual letters are being written or spelled out aloud. Moreover, spelling of familiar words in dictation involves recognition of the spoken word (access to the stored phonological representation of the word) and access to the correct spelling of the word (the stored orthographic lexical representation) [56]. Therefore, spelling is used as a strategy for phonological treatment focusing on phonological awareness and verbal working memory, in cases of aphasia [33,67,68].

Phonological deterioration, starting from a phonological short-term memory deficit, characterizes lvPPA. Whereas most individuals with this syndrome mention lexical retrieval problems as their main deficit, some are more concerned with spelling deterioration and short-term memory deficit. Given that there were few studies on lvPPA and, to our knowledge, none reported any treatment addressing phonological deficits and spelling, we developed a protocol based on the study of Louis et al. [35], while also combining some strategies used in individuals with post-stroke aphasia.

The training consisted of activities at the syllable and phonemic levels, along with oral and written spelling. Twenty regular words were selected for the training/control stimuli set. In every session, the subjects practiced the spelling of each word through dictation. If there was an error in the spelling, the therapist guided the subject to read his/her production aloud, so that the subject could try to identify the error and write and/or spell the word aloud again. If the word was misspelled again, visual support was provided (written word) and the subject was asked to copy the word. Other activities were also practiced in the sessions: forming words from a group of syllables or phonemes (synthesis), identifying the number of syllables and phonemes in words (analysis), identifying rhymes and alliterations and manipulating syllables and phonemes to form new words.

Another set of 36 regular words were used as controls.

The effectiveness of the intervention was evaluated by comparing accuracy in spelling pre- and post-intervention. We compared the performance in trained and untrained words.

3. Data Analysis and Statistics

The treatment effects were analyzed for each subject using the Wilcoxon signed rank test to determine whether there were any differences in the numbers of correct responses from the trained and untrained stimuli sets from before and to after treatment. We used JAMOVI version 1.6, [69] for the statistical analyses. Since nonparametric tests do not include confidence interval values or effect sizes, we reported estimates generated through paired t tests and Cohen's d effect sizes in order to estimate the internal validity of the study. However, those measurements should not be considered for generalization purposes. The statistical significance level was set at 0.05 and we reported 95% confidence intervals.

4. Results

Thirty-two subjects were referred to the study within the three-year recruitment period. Six subjects with severe language deficits were excluded and eight subjects did not complete the intervention program. In relation to these non-adherent cases, five were svPPA, two nfvPPA, and one mxPPA. The reasons for dropping out from the treatment program included: frustration; anxiety and discouragement due to their own difficulties; illness in the family; unwillingness to do activities at home; distance between the home and the outpatient unit; and feeling that the treatment was not solving the problem.

Eighteen individuals with different PPA variants participated in the study. Table 1 shows the demographic and clinical characterization of the participants and Table 2 shows their performance in formal language tests. For all of them, Portuguese was their first language. None of them had any visual or hearing impairments. All of them were at the mild or moderate stages of the syndrome, and all of them were allocated to one out of the four types of intervention, as mentioned previously. Seven undertook the follow-up assessment. All subjects had at least one cognitive screening and none of them manifested impairment in other major cognitive domains that could significantly interfere with language.

Out of the 18 individuals with PPA who were included in the study, 3 met the criteria for nfvPPA (participants C15, C16, and C17), 5 for lvPPA (participants C6, C7, C8, C11, and C12), 5 for svPPA (participants C1, C2, C3, C4 and C5) and 5 for mxPPA (participants C9, C10, C13, C14 and C18).

Ten participants received an intervention focused on naming (C1, C2, C3, C4, C5, C6, C7, C8, C9 and C10); two received therapy for sentence production (C15 and C16); two for speech production (C17 and C18) and four for phonological awareness and verbal working memory (C11, C12, C13 and C14) (Table 3).

Table 1. Sociodemographic and clinical characterization of the patients.

	Sex	Age (Years)	Schooling (Years)	Disease Duration at Treatment Onset (Years)	PPA Variant	Handedness	Brain Atrophy Pattern	MMSE—DRS Mattis
C1	F	60	22	2	Sv	right-handed	Bilateral T	26/30–116/144
C2	M	62	16	2	Sv	right-handed	Bilateral T (more prominent on the left)	24/30–117/144
C3	M	57	21	1.25	Sv	right-handed	Bilateral T	26/30
C4	M	65	20	2	Sv	right-handed	Left FTP	25/30
C5	F	68	16	2	Sv	right-handed	Bilateral T (more prominent on the left)	28/30
C6	F	56	15	1	Lv	left-handed	Bilateral PO	25/30–127/144
C7	M	62	16	2.5	Lv	right-handed	Left posterior P	24/30
C8	F	80	4	2.5	Lv	right-handed	Left posterior TP	10/30
C9	F	67	11	1	Mx	right-handed	Left FTP	27/30–134/144
C10	M	57	11	1	Mx	right-handed	Left FTP	29/30–131/144
C11	F	76	8	1	Lv	right-handed	Left posterior TPO	19/30
C12	F	60	15	4	Lv	right-handed	Right posterior TP	21/30–126/144
C13	M	69	16	2	Mx	right-handed	Bilateral FT (more prominent on the left)	25/30–113/144
C14	F	65	19	3	Mx	right-handed	Bilateral P (more prominent on the left)	29/30
C15	M	66	15	3	Nf	right-handed	Left FT	28/30
C16	M	70	4	1	Nf	right-handed	Left FTP	17/30–115/144
C17	M	75	7	2	Nf	right-handed	Bilateral T	25/30–131/144
C18	F	78	25	4	Mx	left-handed	Volume reduction expected for age	30/30

Note: Legend: F = female; M = male; sv = semantic variant; lv = logopenic variant; nf = nonfluent variant; mx = non-classified/mixed; T = temporal lobe; P = parietal lobe; PO = parietal occipital lobe; TP = temporal parietal lobe; FT = frontal temporal lobe; FTP = frontal temporal parietal lobe; TPO = temporal parietal occipital lobe; MMSE = mini-mental state examination; DRS Mattis = dementia rating scale.

Table 2. Performance of the subjects in formal language tests.

	Naming	Verbal Fluency			Repetition Boston Test				Sentence Comprehension		Reading	Writing	Object Knowledge	Motor Aspects of Speech
	Boston Naming ($n = 60$)	Semantic—Animals	Phonemic—F.A.S.	Words ($n = 10$)	Sentences with High-Frequency Words ($n = 8$)	Sentences with Low-Frequency Words ($n = 8$)	TROG ($n = 80$)	Token Test ($n = 57$)						
C1	30	14	27	10	8	8	75	53			Preserved	Preserved	Preserved	Preserved
C2	30	10	26	9	8	8	NA	51			Isolated phonemic paralexias	Graphic paragraphia-substitutions and regularization of foreign words	Preserved	Speech apraxia
C3	10	10	30	10	8	7	NA	NA			Surface dyslexia	Surface dysgraphia	Moderate-severe impairment	Preserved
C4	28	10	13	10	5	3	68	39			Dyslexia with regularization and semantic paralexias	Surface dysgraphia	Mild	Preserved
C5	4	3	29	10	7	8	74	49			Dyslexia with regularization	Surface dysgraphia and phonological paragraphia	Severe	Preserved
C6	41	13	33	10	7	5	68	NA			Isolated phonemic paralexias-inversion	Isolated paragraphia, spelling changes and graphic omission	Severe	Speech apraxia
C7	19	9	9	10	4	4	NA	NA			Phonological dyslexia	Phonological dysgraphia	Preserved	Preserved
C8	15	5	6	10	1	0	NA	NA			Phonological and regularization errors (low education)	Phonological and regularization errors (low education)	Mild impairment	Preserved
C9	36	18	26	10	8	6	67	47			Preserved	Dysgraphia, phonological and graphemic paragraphias, regularizations of foreign words	Mild	Preserved

Table 2. *Cont.*

	Naming	Verbal Fluency			Repetition Boston Test		Sentence Comprehension		Reading	Writing	Object Knowledge	Motor Aspects of Speech
	Boston Naming ($n = 60$)	Semantic—Animals	Phonemic—F.A.S.	Words ($n = 10$)	Sentences with High-Frequency Words ($n = 8$)	Sentences with Low-Frequency Words ($n = 8$)	TROG ($n = 80$)	Token Test ($n = 57$)				
C10	48	8	16	10	8	7	79	57	Preserved	Regularizations of foreign words	Preserved	Preserved
C11	40	10	13	10	6	5	52	39	Phonemic paralexias-inversion, omission and substitution	Dysgraphia with phonological paragraphia, spelling changes, graphemic and syllabic omission	Mild	Speech apraxia
C12	33	14	21	9	6	6	62	51	Phonological dyslexia	Phonological dysgraphia (phonological paragraphias and graphemic omission)	Preserved	Speech apraxia
C13	24	10	5	8	6	4	NA	30	Morphological and phonological paralexias (mainly) and lexicalization	Graphemic paragraphia-omission and phonological paragraphia	Mild	Preserved
C14	4	13	15	9	4	4	54	41	Phonemic paralexias-omission	Regularization of foreign words graphemic paragraphia-omission and addition and phonological paragraphia	Preserved	Speech apraxia
C15	35	5	2	9	6	4	51	22	Phonemic paralexias-inversion, omission and substitutions and regularization of foreign words	Regularization of foreign words spelling changes	Mild	Speech apraxia

Table 2. *Cont.*

	Naming	Verbal Fluency			Repetition Boston Test				Sentence Comprehension		Reading	Writing	Object Knowledge	Motor Aspects of Speech
	Boston Naming ($n = 60$)	Semantic—Animals	Phonemic—F.A.S.	Words ($n = 10$)	Sentences with High-Frequency Words ($n = 8$)	Sentences with Low-Frequency Words ($n = 8$)	TROG ($n = 80$)	Token Test ($n = 57$)						
C16	21	3	3	7	0	1	38	12			Phonemic paralexias-omission and substitutions and regularization of foreign words	Graphemic and phonological paragraphias and lexicalization	Severe	Speech apraxia
C17	35	11	16	9	6	1	76	49			Preserved	Regularization of foreign words and spelling changes	Mild	Speech apraxia
C18	40	11	35	10	8	5	76	48			Preserved	Preserved	Preserved	Speech apraxia

Note: Legend: reading and writing (Boston test and HFSP protocol); motor aspects of speech (speech praxis protocol and Boston test); object knowledge (Cambridge semantic memory research battery). NA = not available.

Table 3. Results from the intervention programs.

Type of Treatment	Subjects	Number of Sessions	Trained/Treated Items						Untrained Items					
			Number of Items	Baseline Accuracy	Post-Intervention Accuracy	p	Confidence Interval—Compared	Effect Size (Estimate) (Cohen's d)	Number of Items	Baseline Accuracy	Post-Intervention Accuracy	p	Confidence Interval—Compared	Effect Size (Estimate) (Cohen's d)
Naming	C1	20	177	88	166	<0.01	$-0.51/-0.366'$	0.885	60	30	35	<0.05	$-0.155/-0.011'$	0.299
	C2	8	60	18	36	<0.01	$-0.419/-0.180'$	0.649	60	34	32	0.346	$-0.0134/0.080'$	0.184
	C3	5	43	0	38	<0.01	$-0.983/-0.784'$	2.725	60	10	6	0.072	0.0016/0.132	0.265
	C4	16	86	63	77	<0.01	$-0.2424/-0.083'$	0.438	60	28	30	0.346	$-0.080/0.013'$	0.184
	C5	8	139	89	139	<0.01	$-0.44/-0.27'$	0.747	60	4	9	0.037	$-0.15/-0.011'$	0.299
	C6	14	92	37	46	<0.01	$-0.160/-0.360'$	0.327	60	45	41	<0.01	$-0.542/-0.224'$	0.625
	C7	12	80	48	64	<0.01	$-0.290/-0.110$	0.497	20	11	11	NS	NS	NS
	C8	16	30	0	27	<0.01	$-1.014/-0.786'$	2.95	30	0	21	<0.01	$-0.874/-0.526'$	1.50
	C9	7	140	108	140	<0.01	$-0.29/-0.158'$	0.542	60	36	43	<0.01	$-0.20/-0.033'$	0.360
	C10	11	147	137	147	<0.01	$-0.10/-0.02'$	0.269	60	48	50	0.34	$-0.08/0.0013'$	0.184

Table 3. *Cont.*

Type of Treatment	Subjects	Trained/Treated Items							Untrained Items					
		Number of Sessions	Number of Items	Baseline Accuracy	Post-Intervention Accuracy	p	Confidence Interval—Compared	Effect Size (Estimate) (Cohen's d)	Number of Items	Baseline Accuracy	Post-Intervention Accuracy	p	Confidence Interval—Compared	Effect Size (Estimate) (Cohen's d)
Phonological awareness and verbal working memory	C11	12	20	6	7	1	−0.154/0.054'	0.224	36	17	14	0.149	−0.011/0.178'	0.297
	C12	19	20	15	16	1	−0.154/0.054'	0.224	36	31	30	1	−0.028/0.084'	0.167
	C13	19	20	6	7	1	−0.155/0.054'	0.224	36	27	18	<0,01	0.101/0.398'	0.569
	C14	23	20	14	16	0.34	−0.244/0.044'	0.325	36	21	23	0.346	−0.134/0.023'	0.239
Sentence production	C15	6	20	17	20	0.149	−0.321/0.0215'	0.409	20	14	20	0.020	−0.520/−0.800'	0.638
	C16	10	20	2	14	<0.01	0.835/−0.364'	1.194	20	1	7	0.020	−0.520/−0.800'	0.638
Speech production	C17	10	120	108	117	<0.01	−0.123/−0.027'	0.284	95	12	47	<0.01	−0.467/−0.269'	0.760
	C18	24	319	277	319	<0.01	−0.169/−0.094'	0.389	179	120	140	<0.01	−0.158/−0.065'	0.354

Note: Legend: NS = not significant.

Table 4. Results from the generalization across subjects.

Type of Treatment	Subjects	Generalization to Others Tasks						
		Verbal Fluency—Animals		Discourse Production from the Cookie Theft Picture		p	Confidence Interval—Compared	Effect Size (Estimate) (Cohen's d)
		Pre	Post	Pre	Post			
Naming	C1	14	18	NU	NU	0.072	−0.435/−0.009'	0.519
	C2	14	10	NU	NU	0.073	0.0150/0.0556'	0.609
	C3	12	12	NU	NU	NS	NS	NS
	C4	10	5	NU	NU	0.037	0.0786/0.691'	0.760
	C5	3	9	NU	NU	0.02	−0.775/−0.148'	0.889
	C6	13	14	NU	NU	1	−0.226/0.829'	0.267
	C7	8	9	NU	NU	1	−0.245/0.0907'	0.277
	C8	5	4	NU	NU	1	−0.0907/0.245'	0.277
	C9	18	14	NU	NU	0.072	0.009/0.435'	0.519
	C10	8	8	NU	NU	NS	NS	NS

Table 4. *Cont.*

Type of Treatment	Subjects	Generalization to Others Tasks						
		Verbal Fluency—Animals		Discourse Production from the Cookie Theft Picture		*p*	Confidence Interval—Compared	Effect Size (Estimate) (Cohen's d)
		Pre	Post	Pre	Post			
Sentence production	C15	NU	NU	8	10	0.346	−0.194/0.0343′	0.289
	C16	NU	NU	4	1	0.149	−0.0169/0.257′	0.362

Note: Legend: NU = not undertaken; NS = not significant; Pre = Pre-intervention; Post = Post-intervention.

Table 5. Results from the follow-up assessments.

	Subjects	Follow Up—Trained Items							Follow Up—Untrained Items					
		Time Interval (Months)	Number of Items	Post-Intervention Accuracy	Follow-Up Accuracy	*p*	Confidence Interval—Compared	Effect Size (Cohen's d)	Number of Items	Post-Intervention Accuracy	Follow-Up Accuracy	*p*	Confidence Interval—Compared	Effect Size (Cohen's d)
Naming	C1	6	177	166	163	0.149	−0.017/0.279	0.131	60	35	35	NS	NS	NS
	C7	8	80	64	48	<0.001	0.263–0.728	0.497	20	11	9	0.163	−0.044/0.244	0.325
	C9	6	140	140	125	<0.001	0.174/0.515	0.345	60	43	42	1	−0.170/0.050	0.129
	C10	6	147	147	144	0.149	−0.018/0.306	0.144	60	50	50	NS	NS	NS
Phonological awareness and verbal working memory	C11	2	20	7	6	0.330	−0.223/0.665	0.224	36	14	NU	NU	NU	NU
	C13	4	20	7	5	0.346	−0.129/0.771	0.325	36	18	NU	NU	NU	NU
Sentence production		1	20	20	20	NS	NS	NS	20	20	20	NS	NS	NS

Note: Legend: NU = not undertaken; NS = not significant.

4.1. Intervention Focusing on Naming

As shown in Table 3, all the subjects improved significantly with regard to trained items. However, the estimated effect sizes varied from large (C1, C3, and C8) to medium (C2, C5, C9) and small (C4, C6, C7, and C10). The set of trained stimuli varied among the participants: for C3 and C8, an intervention of remediation was implemented in which only items that participants failed to name at the baseline were trained. For the other subjects, prophylaxis items were also included. The number of pictures selected for the training varied among the subjects, depending on the severity of the deficit and the acceptance and motivation to engage in the treatment.

Four participants presented significantly improved performance regarding untrained stimuli: C1 and C5 (svPPA); C8 (lvPPA); and C9 (mxPPA) (Table 4). C1 and C8 received a remediation program in which their pre-test performance was very different between trained and untrained stimuli. The implications of this design for the interpretation of therapy gains are addressed in the discussion.

Generalization to a different task was observed only in C1 and C5 (both svPPA), with marginal significance in C1. Five subjects kept the same level of performance and three declined, but not significantly.

Follow up was conducted in four cases (C1, C7, C9, and C10). Two participants maintained the treatment results and two worsened significantly with regard to trained items. For the untrained items, all subjects maintained the results (Table 5).

4.2. Intervention Focusing on Sentence Production

As shown in Table 3, two subjects received this treatment. C15 received prophylaxis treatment and presented no significant change in the trained items (but there was an increase in the correctness of the trained items). There was a significant improvement in the untrained items. This strategy was also implemented in another task, with improvement in the cookie theft picture, but without statistical significance (Table 4).

In contrast, subject C16 received remediation treatment and improved significantly in trained and untrained items with large and medium effects, respectively. However, the strategy was not transferred to discourse, such that there was a significant decline in relation to the cookie theft picture.

A follow-up assessment was undertaken in relation to one participant (C15), one month after the end of the intervention, with maintenance of treatment results, both for trained and for untrained items (Table 5).

4.3. Intervention Focusing on Speech Production

As shown in Table 3, both subjects who participated in this intervention improved significantly in relation to the trained and untrained texts. Thus, they presented significant reductions in articulatory errors in multisyllabic words. These participants did not perform tests to assess generalization for other tasks and neither of them returned for the follow-up evaluation.

4.4. Intervention Focusing on Phonological Awareness and Verbal Working Memory

The four subjects who took part in this training did not present any significant improvement in spelling after the intervention, either for trained or untrained items. However, correct responses to trained items numerically increased among all the subjects, whereas the number of correct responses to untrained items decreased for three of them and increased for C14.

Follow-up of trained items was possible for C11 and C13. Both participants demonstrated maintenance of the treatment results (Table 5).

5. Discussion

This study investigated the implementation and effectiveness of four different interventions for PPA. We used a client-centered approach in which treatments were offered considering the subjects' main difficulties and concerns, and with individualized relevant stimuli for training. To our knowledge, this is one of the largest case series reporting language intervention results in PPA, and it has strong ecological validity in that it reports on work conducted in a public specialized outpatient clinic. We adjusted the programs to several individual variables involving patients and their caregivers, which is expected to happen in real clinical contexts. Motivation and engagement with treatment were also considered. Thus, some subjects received a more prophylactic form of treatment, whereas others received treatments with more items that involved "relearning" or "reacquiring".

We acknowledge that the high variability of treatments compromises the generalization and replicability of our results. In addition, as the list of trained and untrained stimuli were not strictly matched according to psycholinguistic parameters or to pre-treatment performance, there are important limitations on interpreting the results from generalization. Therefore, our conclusions and discussions should be considered at the level of "practice-based evidence" [9], in which we observed benefits from SLT in a large sample of PPA subjects. We proposed different interventions addressing not only naming and lexical retrieval, but also other language and speech impairments.

Practice-based evidence can also be demonstrated through case studies of individuals with PPA with gains after intervention [9]. Moreover, the ASHA report of the Joint Coordinating Committee on Evidence-Based Practice [70] argues for the importance of the initial investigation evidence, even when it does not meet rigorous quality standards. That report also mentioned principles of evidence-based practice followed by speech and language therapists that were considered in the present study: client-focused care approach, clear communication to aid the client's weight clinical alternatives, pursuit of consensus decisions, and top-notch clinical care.

5.1. Intervention Focusing on Naming

In our study, all ten subjects (5 svPPA, 3 lvPPA and 2 mxPPA) who underwent this type of intervention improved significantly in relation to the treated items and four also significantly improved in relation to untreated items. Other studies have had similar results and have demonstrated that individuals with PPA are able to relearn target vocabulary during the active phase of treatment [11–17] and that learning can be generalized to untrained stimuli [13,15,16,18–20]. However, the latter result is not consistent across studies. Among our subjects, two svPPA subjects presented generalizations for other language activities (semantic verbal fluency). Our results corroborate the results in the literature [10,71], in that they show that generalization is particularly difficult to achieve in the semantic variant, given that in situations of degraded semantic knowledge, learning is rigid and context dependent. Patients with more evident therapy gains received a remediation program in which pre-test performance was very different for trained and untrained stimuli. In a repeated-measurement design, extreme results tend to regress to the mean. In our study, this statistical phenomenon may have inflated the improvement in treated items, compared with untreated items. Despite this limitation, the gains were clinically significant and confirm the results from previous studies, thus supporting practice-based evidence of a benefit from behavioral interventions addressing naming deficits in PPA.

Four participants underwent a follow-up evaluation, on average six months after the end of the intervention. Two maintained the treatment results and two worsened significantly in relation to trained items. In the untrained items, all subjects maintained their results. Our findings differ partly from those of the systematic review of Cadório et al. [71], which included 25 papers on semantic therapy in different PPA subtypes, encompassing 51 subjects in total. Those authors stated that generalization was more difficult to achieve in the semantic variant (as seen in most of these subjects), compared with the nonfluent and logopenic variants. However, the lack of strict control of psycholinguistic variables, as well as the differences in programs (remediation vs. prophylaxis), limits the interpretation of generalization and maintenance findings from the present study. On the other hand, the personal relevance of the stimuli selected and involvement of the individual with PPA in this selection are factors that may have contributed to the success of the language therapy in the present study, in the same way as in other reports [10,12,14,72].

Similarly to the present study, Croot [8] also studied lexical retrieval treatment among individuals with heterogeneous clinical presentations of PPA. The heterogeneous nature of the sample allowed to observe a range of treatment outcomes and adherence patterns under the same treatment protocol and to describe disease and participant factors associated with these outcomes.

5.2. Intervention Focusing on Sentence Production

NfPPA usually presents with mixed symptoms of motor and cognitive-linguistic deficits. Studies on treatments for this variable are less common than on treatments for svPPA. The results show that approaches that focus on the deficit (agrammatism, phonological skills, and speech apraxia, for example) are beneficial to individuals with PPA.

Our two participants who underwent this type of intervention (both nfPPA) improved in relation to both treated and untreated items, thus corroborating previous studies [29–31,35,73]. Regarding untrained items, both of them improved significantly.

Only one subject presented generalization for other language activities (cookie theft picture description), with better sentence construction in relation to the pre-test. Schneider et al. [29] and Louis et al. [35] also showed generalization of results for items and untrained material. Cadório et al. [71] show that generalization is easier to achieve in this group of subjects than in relation to the semantic subtype.

One participant underwent a follow-up evaluation one month after the end of the intervention, with maintenance of the results. Among the follow-up studies, only Hameister et al. [31] reported that learning was maintained after the end of therapy.

5.3. Intervention Focusing on Speech Production

Few studies have implemented interventions to improve fluency in nfPPA.

Structured oral reading proved to be an efficient and effective means of addressing multisyllabic word production in speech apraxia associated with nfPPA. In the study by Henry et al. [32], one participant showed a reduction in speech errors during the reading of novel text. Similarly, the two subjects in our sample who underwent this intervention (one nfPPA and one mxPPA) improved significantly in relation to both treated and untreated items.

5.4. Intervention Focusing on Phonological Awareness and Verbal Working Memory

Among the four subjects (two lvPPA and two mxPPA) treated with this type of intervention, none presented any significant improvement in spelling after the intervention, in relation either to trained or to untrained items. However, all four of them showed numerical increases in the correct responses relating to trained items, whereas three showed decreases relating to untrained items and only subject C14 showed an increase in this regard.

It is noteworthy that maintenance signs of the same level of function in progressive disorders should be seen as a success. Moreover, in these cases it is important to slow down the progression and maintain the communication abilities of subjects [74].

Regarding follow up, two participants were reassessed, with maintenance of treatment results, but without statistical significance. This was comparable with the results of Beeson et al. [75] and Henry et al. [16], but different from Rapp and Glucroft [34], who demonstrated worsened results in the follow-up reassessment.

5.5. General Remarks about Treatments and Concluding Comments

We have reported on treatment results for a case series of individuals with PPA. We now discuss some challenges and limitations of our study and other factors of relevance to interventions directed towards speech and language deficits in PPA.

There are few studies on PPA treatment in low and middle-income countries. Like in other studies on this topic, our sample was not large (although larger than in other studies that recruited individuals in the same clinic) and the participants' characteristics varied considerably even within the same variant of PPA. However, given the relatively low prevalence of PPA, treatment studies on this population usually involve a small number of participants [10].

Another related matter is adherence to treatment. In our sample, eight subjects dropped out of the study before the post-test: five svPPA, two nfvPPA, and one mxPPA. The average number of sessions that they attended was 17.5. Their reasons for dropping out from the treatment comprised frustration, anxiety, and discouragement with their own difficulties, illness in the family, unwillingness to do activities at home, distance between the consultation office and their home and a perception that their speech and language difficulties were not being "solved".

Jokel et al. [74] stated that many individuals who participate in a group intervention program find it rewarding and positive. Nonetheless, our results show that this finding is not consistent across different samples. Furthermore, there may be a publication bias such that patients who do not adhere to treatments are not included in publications. In our experience, individual treatment was not always motivating and generated frustration and anxiety among some subjects who were aware of their progressing condition, deficits and prognosis from treatment.

Information about participant adherence to treatment requirements is rarely reported in research studies. Taylor-Rubin et al. [76] studied adherence to treatment in the clinical setting in PPA and mentioned that treatment generally requires the person with PPA and their caregiver to play an active role in initiating and continuing the daily home practice. We believe that personalization of therapeutic material and identification with it favors adherence to the rehabilitation program. Thus, the individuals' involvement in the selection of stimuli may have been a factor contributing to the success of language therapy in the present study and in other reports [12,72]. In our case series, all the stimuli were personalized, with the aim of improving adherence and achieving better functional results. The use of meaningful materials would favor the stronger use of these materials to support functional communication and indirectly increase participation levels. The goal of rehabilitation is to empower people with cognitive impairment and dementia such that they can participate in everyday life in their families and communities in meaningful ways [77].

Taylor-Rubin et al. [76] discussed personal intrinsic factors (such as depression and mood) and treatment-related extrinsic factors (such as time required and duration), along with social factors, which are a combination of intrinsic and extrinsic factors, and how these relate to adherence to treatment. Their results suggested that commencement of treatment while the person with PPA is in the early stages of disease progression may improve adherence and increase the possibility of positive treatment outcomes. However, according to our experience in the public healthcare system, patients take too long to have their first consultation (for reasons discussed below), which limits the chances of always beginning the treatment in the initial stage of the disease.

The initial severity of deficits and the length of time since the onset of symptoms affects the response to treatment, although it is difficult to establish how this occurs. It is coherent to think that the longer the disease duration is, the greater the linguistic impairment will be and hence the greater the treatment limitations. There is considerable inconsistency in reporting the time that has elapsed post-onset and severity levels in the literature on treatments, since the onset of symptoms is not easily defined. Another effort towards treating individuals with PPA consists of interpreting the response to treatment in the context of disease progression, given that a situation of little or no change in the language skills treated may represent a positive outcome, in comparison with the expected decline [21,74].

A significant number of potential participants could not be included in our study because they were severely impaired. One good alternative for these individuals and their families would be orientation and interaction groups, for the exchanging of experiences and counseling, as proposed by Jokel et al. [74] and by Mooney, Beale and Fried-Oken [78]. It is noteworthy that only seven participants returned for the follow-up assessment. In addition to the difficulty in carrying out follow-up treatment studies on individuals with neurodegenerative diseases, due to their cognitive decline [33], the structure of the public care system in Brazil with specialized clinics usually located in large cities, operating within universities, gives rise to further difficulty. Given that several subjects were not residents of the city where the study was conducted and, instead, were there only for diagnosis and intervention therapy cycles, it became more difficult to have them return for follow-up on a regular basis. The initial schedule envisaged carrying out all reassessments three months after the end of treatment. However, this interval varied between one and six months, due to personal, social, and family issues. This factor was referenced in other contexts. Volkmer et al. [28] mentioned barriers to the provision of speech and language therapy services. They argued that many people with PPA are never referred to speech and language therapy services in the first place, due to the lack of evidence that these interventions give clinically meaningful benefit in PPA, and due to the limited specific speech and language therapy services available.

That is also the reason for some of the very short-term cycles of interventions reported in this study. We believe that interventions need to be patient-centered and tailored. Ideally, cognitive rehabilitation programs should be long-term, in line with the progression of the disease and the changing needs of the subjects. However, many individuals do not have access to cognitive intervention clinics or cannot afford treatments. In contrast, some public services need to deal with high demand from patients and cannot provide long-term follow up. For these contexts, brief cycles of intervention and follow up can be an alternative.

We believe that one important contribution of this paper was that it allowed us to share our clinical experience in implementing interventions among PPA subjects. We reported the results from programs and strategies that could be implemented by speech and language therapists as part of a more comprehensive rehabilitation program. Short cycles can be implemented in contexts where patients lack access to full care and to interventions that can be implemented by caregivers. Conversely, in more complete care settings, therapists may combine different strategies according to the needs of the individual.

Other options for interventions with promising preliminary results are being studied. These include neuromodulation, computer-based approaches, the use of social media and electronic devices, and home-based interventions [23,79–81]. They may offer more treatment options, even for the most serious cases. For this study, we considered only behavioral approaches that were already reported in the literature, with the aims of increasing the number of published cases and making the level of evidence stronger.

Behavioral interventions in PPA showed improvement of the targeted language function. However, not all of them showed generalizable and long-lasting effects. Tippett et al. [82] pointed out some reasons that would account for these findings: heterogeneity of symptoms and pathological processes, reflected by the different PPA variants, different stages of disease progression at baseline, and variable rates of decline among participants and studies. Moreover, the trained items were individualized in this study, but untrained items were selected from the speech and language therapist's materials. Thus, the trained and untrained items were not well matched according to the psycholinguistic criteria. Hence, generalization must be considered with caution. It is important to consider the use of more balanced sets (trained and untrained) in future studies. Similarly, the direct treatment gains in the pre- and post-design (for treated items) need to be interpreted with caution for each individual, since we do not know how stable the pre- and post-scores were. Multiple-baseline assessments would provide a better design for the study and must be implemented in future research.

Generalization of treatment gains for untrained tasks may be related to the nature of the intervention and to the use of episodic/autobiographical information [83]. The maintenance of the results achieved in the training does not seem to be influenced by the PPA subtype, but by other factors, such as continuous practice, duration of treatment, and frequency of sessions [71]. All the item sets exhibited a decline in accuracy from the end of treatment to the follow-up evaluation, which was consistent with the degenerative nature of PPA.

Some participants reported having a subjective perception of improvement in functionality regarding communication at the end of treatment. However, as we did not have any means of objective assessment for analyzing this information, we did not include this observation as part of our results. In future studies, we intend to objectively quantify this information.

Another limitation related to the lack of control over practice at home. Differences between participants may have contributed to different treatment results. The absence of supervision of the control stimuli in the patients' daily life should also be considered as a limitation of the work, since this could potentially interfere with the results.

We recognize that the absence of a control group is a limitation, but we point out that it is a small sample and heterogeneous as to the types of deficits, which makes it difficult to compare patients with and without rehabilitation.

Lastly, we can highlight that this study addressed some important matters: 1. Our study reported on a range of interventions targeted to the individuals' communication needs; 2. Different treatments were selected for different individuals, determined by the participants' language symptoms, and not by their PPA variant; 3. Our study had stronger ecological validity because it was implemented in a clinical context and because the number of subjects who did not adhere to therapy and the reasons for this were also reported.

Although PPA is a progressive disorder, both the immediate effects of treatment and, in some cases, the maintenance results, were positive. The results from our study show the effectiveness of specific behavioral interventions even at "low dose" (short-term intervention cycles).

Author Contributions: Conceptualization, T.H.M., M.T.C.-G. and P.C.; methodology, T.H.M., M.T.C.-G. and P.C.; validation, T.H.M., M.T.C.-G., A.C.C. and P.C.; formal analysis, M.T.C.-G.; investigation, T.H.M., M.T.C.-G. and A.C.C.; resources, T.H.M., M.T.C.-G. and A.C.C.; data curation, T.H.M., M.T.C.-G. and A.C.C.; writing—original draft preparation, T.H.M. and M.T.C.-G.; writing—review and editing, T.H.M., M.T.C.-G. and P.C.; visualization, T.H.M., M.T.C.-G., A.C.C. and P.C.; supervision, M.T.C.-G. and P.C.; project administration, P.C. All authors have read and agreed to the published version of the manuscript.

Funding: This research received no external funding.

Institutional Review Board Statement: The study was conducted according to the guidelines of the Declaration of Helsinki and approved by Research Ethics Committee (COEP) of the Federal University of Minas Gerais (UFMG) under certificate n° 2.018.855 on 17 April 2017.

Informed Consent Statement: Informed consent was obtained from all subjects involved in the study.

Data Availability Statement: The data presented in this study are available on request from the corresponding author.

Acknowledgments: P.C. is funded by CNPq, Brazil (grant for research productivity). M.T.C.-G. is funded by the Federal University of ABC (UFABC), CAPES (001), FAPESP and INCT-ECCE (Instituto Nacional de Ciência e Tecnologia sobre Comportamento, Cognição e Ensino), with support from the Brazilian National Research Council (CNPq, Grant # 465686/2014-1), the Coordination of Superior Level Staff Improvement (CAPES, Grant # 88887.136407/2017-00), and the São Paulo Research Foundation (Grant # 2014/50909-8), Brazil.

Conflicts of Interest: The authors declare no conflict of interest related to the present work.

References

1. Mesulam, M.M.; Grossman, M.; Hillis, A.; Kertesz, A.; Weintraub, S. The core and halo of primary progressive aphasia and semantic dementia. *Ann. Neurol. Off. J. Am. Neurol. Assoc. Child Neurol. Soc.* **2003**, *54*, S11–S14. [CrossRef]
2. Gorno-Tempini, M.L.; Hillis, A.E.; Weintraub, S.; Kertesz, A.; Mendez, M.; Cappa, S.F.; Ogar, J.M.; Rohrer, J.; Black, S.; Boeve, B.F. Classification of primary progressive aphasia and its variants. *Neurology* **2011**, *76*, 1006–1014. [CrossRef] [PubMed]
3. Senaha, M.L.H.; Caramelli, P.; Brucki, S.; Smid, J.; Takada, L.T.; Porto, C.S.; César, K.G.; Matioli, M.N.P.; Soares, R.T.; Mansur, L.L. Primary progressive aphasia: Classification of variants in 100 consecutive Brazilian cases. *Dement. Neuropsychol.* **2013**, *7*, 110–121. [CrossRef]
4. Wicklund, M.R.; Duffy, J.R.; Strand, E.A.; Machulda, M.M.; Whitwell, J.L.; Josephs, K.A. Quantitative application of the primary progressive aphasia consensus criteria. *Neurology* **2014**, *82*, 1119–1126. [CrossRef] [PubMed]
5. Croot, K.; Nickels, L.; Laurence, F.; Manning, M. Impairment- and activity/participation-directed interventions in progressive language impairment: Clinical and theoretical issues. *Aphasiology* **2009**, *23*, 125–160. [CrossRef]
6. Marshall, C.R.; Hardy, C.J.; Volkmer, A.; Russell, L.L.; Bond, R.L.; Fletcher, P.D.; Clark, C.N.; Mummery, C.J.; Schott, J.M.; Rossor, M.N. Primary progressive aphasia: A clinical approach. *J. Neurol.* **2018**, *265*, 1474–1490. [CrossRef] [PubMed]
7. Carthery-Goulart, M.T.; Silveira, A.D.C.D.; Machado, T.H.; Mansur, L.L.; Parente, M.A.D.M.P.; Senaha, M.L.H.; Brucki, S.M.D.; Nitrini, R. Nonpharmacological interventions for cognitive impairments following primary progressive aphasia: A systematic review of the literature. *Dement. Neuropsychol.* **2013**, *7*, 122–131. [CrossRef]
8. Croot, K. Treatment for lexical retrieval impairments in primary progressive aphasia: A research update with implications for clinical practice. In Proceedings of the Seminars in Speech and Language, Hsinchu, Taiwan, 4–5 October 2018; pp. 242–256.
9. Ruggero, L.; Croot, K.; Nickels, L. How Evidence-Based Practice (E3BP) Informs Speech-Language Pathology for Primary Progressive Aphasia. *Am. J. Alzheimer Dis. Other Dement.* **2020**, *35*, 1533317520915365. [CrossRef] [PubMed]
10. Jokel, R.; Graham, N.L.; Rochon, E.; Leonard, C. Word retrieval therapies in primary progressive aphasia. *Aphasiology* **2014**, *28*, 1038–1068. [CrossRef]
11. Bozeat, S.; Patterson, K.; Hodges, J. Relearning object use in semantic dementia. *Neuropsychol. Rehabil.* **2004**, *14*, 351–363. [CrossRef]
12. Bier, N.; Macoir, J.; Gagnon, L.; Van der Linden, M.; Louveaux, S.; Desrosiers, J. Known, lost, and recovered: Efficacy of formal-semantic therapy and spaced retrieval method in a case of semantic dementia. *Aphasiology* **2009**, *23*, 210–235. [CrossRef]
13. Dewar, B.-K.; Patterson, K.; Wilson, B.A.; Graham, K.S. Re-acquisition of person knowledge in semantic memory disorders. *Neuropsychol. Rehabil.* **2009**, *19*, 383–421. [CrossRef] [PubMed]
14. Heredia, C.G.; Sage, K.; Ralph, M.A.L.; Berthier, M.L. Relearning and retention of verbal labels in a case of semantic dementia. *Aphasiology* **2009**, *23*, 192–209. [CrossRef]
15. Mayberry, E.J.; Sage, K.; Ehsan, S.; Ralph, M.A.L. Relearning in semantic dementia reflects contributions from both medial temporal lobe episodic and degraded neocortical semantic systems: Evidence in support of the complementary learning systems theory. *Neuropsychologia* **2011**, *49*, 3591–3598. [CrossRef]
16. Henry, M.L.; Rising, K.; DeMarco, A.T.; Miller, B.L.; Gorno-Tempini, M.L.; Beeson, P.M. Examining the value of lexical retrieval treatment in primary progressive aphasia: Two positive cases. *Brain Lang.* **2013**, *127*, 145–156. [CrossRef] [PubMed]
17. Savage, S.A.; Ballard, K.J.; Piguet, O.; Hodges, J.R. Bringing words back to mind–Improving word production in semantic dementia. *Cortex* **2013**, *49*, 1823–1832. [CrossRef]
18. Jokel, R.; Rochon, E.; Anderson, N.D. Errorless learning of computer-generated words in a patient with semantic dementia. *Neuropsychol. Rehabil.* **2010**, *20*, 16–41. [CrossRef]
19. Montagut, N.; Sánchez-Valle, R.; Castellví, M.; Rami, L.; Molinuevo, J.L. Reaprendizaje de vocabulario. Análisis comparativo entre un caso de demencia semántica y enfermedad de Alzheimer con afectación predominante del lenguaje. *Rev. Neurol.* **2010**, *50*, 152–156. [CrossRef]
20. Jokel, R.; Anderson, N.D. Quest for the best: Effects of errorless and active encoding on word re-learning in semantic dementia. *Neuropsychol. Rehabil.* **2012**, *22*, 187–214. [CrossRef]
21. Rising, K. Treatment for lexical retrieval in primary progressive aphasia. *Perspect. Neurophysiol. Neurogen. Speech Lang. Disord.* **2014**, *24*, 137–144. [CrossRef]
22. Beales, A.; Whitworth, A.; Cartwright, J. A review of lexical retrieval intervention in primary progressive aphasia and Alzheimer's disease: Mechanisms of change, generalisation, and cognition. *Aphasiology* **2018**, *32*, 1360–1387. [CrossRef]
23. Meyer, A.M.; Getz, H.R.; Brennan, D.M.; Hu, T.M.; Friedman, R.B. Telerehabilitation of anomia in primary progressive aphasia. *Aphasiology* **2016**, *30*, 483–507. [CrossRef]
24. Reilly, J. How to constrain and maintain a lexicon for the treatment of progressive semantic naming deficits: Principles of item selection for formal semantic therapy. *Neuropsychol. Rehabil.* **2016**, *26*, 126–156. [CrossRef]
25. Meyer, A.M.; Tippett, D.C.; Friedman, R.B. Prophylaxis and remediation of anomia in the semantic and logopenic variants of primary progressive aphasia. *Neuropsychol. Rehabil.* **2018**, *28*, 352–368. [CrossRef]
26. Meyer, A.M.; Tippett, D.C.; Turner, R.S.; Friedman, R.B. Long-term maintenance of anomia treatment effects in primary progressive aphasia. *Neuropsychol. Rehabil.* **2018**, *29*, 1439–1463. [CrossRef] [PubMed]

27. Rogalski, E.J.; Saxon, M.; McKenna, H.; Wieneke, C.; Rademaker, A.; Corden, M.E.; Borio, K.; Mesulam, M.-M.; Khayum, B. Communication Bridge: A pilot feasibility study of Internet-based speech–language therapy for individuals with progressive aphasia. *Alzheimer Dement. Transl. Res. Clin. Interv.* **2016**, *2*, 213–221. [CrossRef] [PubMed]

28. Volkmer, A.; Rogalski, E.; Henry, M.; Taylor-Rubin, C.; Ruggero, L.; Khayum, R.; Kindell, J.; Gorno-Tempini, M.L.; Warren, J.D.; Rohrer, J.D. Speech and language therapy approaches to managing primary progressive aphasia. *Pract. Neurol.* **2020**, *20*, 154–161. [CrossRef] [PubMed]

29. Schneider, S.L.; Thompson, C.K.; Luring, B. Effects of verbal plus gestural matrix training on sentence production in a patient with primary progressive aphasia. *Aphasiology* **1996**, *10*, 297–317. [CrossRef]

30. Andrade-Calderón, P.; Salvador-Cruz, J.; Sosa-Ortiz, A.L. Positive impact of speech therapy in progressive non-fluent aphasia. *Acta Colomb. Psicol.* **2015**, *18*, 101–114. [CrossRef]

31. Hameister, I.; Nickels, L.; Abel, S.; Croot, K. "Do you have mowing the lawn?"—Improvements in word retrieval and grammar following constraint-induced language therapy in primary progressive aphasia. *Aphasiology* **2017**, *31*, 308–331. [CrossRef]

32. Henry, M.; Meese, M.; Truong, S.; Babiak, M.; Miller, B.; Gorno-Tempini, M. Treatment for apraxia of speech in nonfluent variant primary progressive aphasia. *Behav. Neurol.* **2013**, *26*, 77–88. [CrossRef]

33. Tsapkini, K.; Hillis, A.E. Spelling intervention in post-stroke aphasia and primary progressive aphasia. *Behav. Neurol.* **2013**, *26*, 55–66. [CrossRef]

34. Rapp, B.; Glucroft, B. The benefits and protective effects of behavioural treatment for dysgraphia in a case of primary progressive aphasia. *Aphasiology* **2009**, *23*, 236–265. [CrossRef]

35. Louis, M.; Espesser, R.; Rey, V.; Daffaure, V.; Di Cristo, A.; Habib, M. Intensive training of phonological skills in progressive aphasia: A model of brain plasticity in neurodegenerative disease. *Brain Cogn.* **2001**, *46*, 197–201. [CrossRef]

36. Folstein, M.F.; Folstein, S.E.; McHugh, P.R. "Mini-mental state": A practical method for grading the cognitive state of patients for the clinician. *J. Psychiatr. Res.* **1975**, *12*, 189–198. [CrossRef]

37. Brucki, S.; Nitrini, R.; Caramelli, P.; Bertolucci, P.H.; Okamoto, I.H. Sugestões para o uso do mini-exame do estado mental no Brasil. *Arq. Neuro-Psiquiatr.* **2003**, *61*, 777–781. [CrossRef] [PubMed]

38. Mattis, S. *Dementia Rating Scale: Professional Manual*; Psychological Assessment Resources: Lutz, FL, USA, 1988.

39. Porto, S.; Charchat-Fichman, H.; Caramelli, P.; Bahia, V.; Nitrini, R. Dementia Rating Scale-DRS-in the diagnosis of patients with Alzheimer's dementia. *Arq. Neuropsiquiatr.* **2003**, *61*, 339–345. [CrossRef]

40. Foss, M.P.; Carvalho, V.A.D.; Machado, T.H.; Reis, G.C.D.; Tumas, V.; Caramelli, P.; Nitrini, R.; Porto, C.S. Mattis Dementia Rating Scale (DRS): Normative data for the Brazilian middle-age and elderly populations. *Dement. Neuropsychol.* **2013**, *7*, 374–379. [CrossRef] [PubMed]

41. Nespoulous, J.; Lecours, A.; Lafond, D.; Parente, M. *Protocole Montréal-Tolouse MT-86 D'examen Linguistique de L'aphasie-Version Beta*; Laboratoire Théophile-Alajouanine: Montréal, QC, Canada, 1986.

42. Parente, M.; Ortiz, K.; Soares, E.; Scherer, L.; Fonseca, R.; Joanette, Y.; Lecours, A.; Nespoulous, J. *Bateria Montreal-Toulouse de Avaliação da Linguagem-Bateria MTL-Brasil*; Vetor Editora: São Paulo, Brazil, 2016.

43. Goodglass, H.; Kaplan, E. *The Assessment of Aphasia and Related Disorders*; Lippincott Williams & Wilkins: Philadelphia, PA, USA, 1983.

44. Ratnavalli, E. Progress in the last decade in our understanding of primary progressive aphasia. *Ann. Indian Acad. Neurol.* **2010**, *13*, S109. [CrossRef] [PubMed]

45. Hodges, J.R.; Patterson, K.; Oxbury, S.; Funnell, E. Semantic dementia: Progressive fluent aphasia with temporal lobe atrophy. *Brain* **1992**, *115*, 1783–1806. [CrossRef] [PubMed]

46. Adlam, A.-L.R.; Patterson, K.; Bozeat, S.; Hodges, J.R. The Cambridge Semantic Memory Test Battery: Detection of semantic deficits in semantic dementia and Alzheimer's disease. *Neurocase* **2010**, *16*, 193–207. [CrossRef] [PubMed]

47. Carthery-Goulart, M.T.; Estequi, J.G.; Areza-Fegyveres, R.; Silveira, A.C.; César, K.; Senaha, M.L.H.; Brucki, S.; Nitrini, R. Dissociação Entre Seres Vivos e Artefatos: Investigação de Efeito Categoria Específica no Processamento de Substantivos na Bateria de Memória Semântica de Cambridge. *Rev. Psicol. Pesqui.* **2013**, *7*, 108–120. [CrossRef]

48. De Renzi, A.; Vignolo, L.A. Token test: A sensitive test to detect receptive disturbances in aphasics. *Brain A J. Neurol.* **1962**, *85*, 665–678. [CrossRef] [PubMed]

49. Moreira, L.; Schlottfeldt, C.G.; Paula, J.J.D.; Daniel, M.T.; Paiva, A.; Cazita, V.; Coutinho, G.; Salgado, J.V.; Malloy-Diniz, L.F. Estudo normativo do Token Test versão reduzida: Dados preliminares para uma população de idosos brasileiros. *Arch. Clin. Psychiatry* **2011**, *38*, 97–101. [CrossRef]

50. Bishop, D.V. *Test for the Reception of Grammar, Age and Cognitive Performance*; Research Centre, University of Manchester: Manchester, UK, 1983.

51. Bishop, D.V. *Test for Reception of Grammar: TROG-2*, 2nd ed.; Harcourt Assessment: San Antonio, TX, USA, 2003.

52. Oliveira, R. *Compreensão Oral de Sentenças em Idosos Cognitivamente Saudáveis: Caracterização e Investigação de Sua Relação com Outros Aspectos do Funcionamento Cognitivo e com Fatores Sócio-Demográficos*; UFABC: Santo André, Brazil, 2013.

53. Pereira, M.; Goulart, M.; Mansur, L.; Lopes, D.; Negrão, E.; Agonilha, D. Tradução e adaptação do teste de recepção gramatical TROG-2 para o português brasileiro. In Proceedings of the Anais-Congresso Brasileiro de Fonoaudiologia, Salvador, Bahia, Brazil, 21–24 October 2009.

54. Parente, M.; Hosogi, M.; Delgado, A.; Lecours, A. Protocolo de Leitura Para o Projeto HFSP (Human Frontier Science Program). 1992. Available online: https://www.hfsp.org/ (accessed on 10 September 2021).
55. Carthery, M.T. Caracterização dos Distúrbios de Escrita na Doença de Alzheimer. Ph.D. Thesis, Universidade de São Paulo (USP), São Paulo, Brazil, 2000.
56. Carthery, M.T.; de Mattos Pimenta Parente, M.; Nitrini, R.; Bahia, V.S.; Caramelli, P. Spelling tasks and Alzheimer's disease staging. *Eur. J. Neurol.* **2005**, *12*, 907–911. [CrossRef] [PubMed]
57. Caramelli, P.; Carthery-Goulart, M.T.; Porto, C.S.; Charchat-Fichman, H.; Nitrini, R. Category fluency as a screening test for Alzheimer disease in illiterate and literate patients. *Alzheimer Dis. Assoc. Disord.* **2007**, *21*, 65–67. [CrossRef]
58. Fichman, H.C.; Fernandes, C.S.; Nitrini, R.; Lourenço, R.A.; Paradela, E.M.D.P.; Carthery-Goulart, M.T.; Caramelli, P. Age and educational level effects on the performance of normal elderly on category verbal fluency tasks. *Dement. Neuropsychol.* **2009**, *3*, 49–54. [CrossRef]
59. Machado, T.H.; Fichman, H.C.; Santos, E.L.; Carvalho, V.A.; Fialho, P.P.; Koenig, A.M.; Fernandes, C.S.; Lourenço, R.A.; Paradela, E.M.D.P.; Caramelli, P. Normative data for healthy elderly on the phonemic verbal fluency task-FAS. *Dement. Neuropsychol.* **2009**, *3*, 55–60. [CrossRef]
60. Carmo, C.F.; Machado, T.H.; Amaral-Carvalho, V.; Santos, E.L.; Beato, R.G.; Caramelli, P. A Brazilian protocol to assess recognition of famous faces. In Proceedings of the VIII RPDA—Reunião de Pesquisadores em Doença de Alzheimer e Desordens Relacionadas, São Paulo, Brazil, 19 December 2011; p. 39.
61. Alves, D.C.; de Paula Souza, L.A. Performance de moradores da grande São Paulo na descrição da prancha do Roubo de Biscoitos. *Rev. CEFAC* **2005**, *7*, 13–20.
62. Croisile, B.; Ska, B.; Brabant, M.-J.; Duchene, A.; Lepage, Y.; Aimard, G.; Trillet, M. Comparative study of oral and written picture description in patients with Alzheimer's disease. *Brain Lang.* **1996**, *53*, 1–19. [CrossRef]
63. Bozeat, S.; Ralph, M.A.L.; Patterson, K.; Garrard, P.; Hodges, J.R. Non-verbal semantic impairment in semantic dementia. *Neuropsychologia* **2000**, *38*, 1207–1215. [CrossRef]
64. Salmazo-Silva, H.; Parente, M.A.D.M.P.; Rocha, M.S.; Baradel, R.R.; Cravo, A.M.; Sato, J.R.; Godinho, F.; Carthery-Goulart, M.T. Lexical-retrieval and semantic memory in Parkinson's disease: The question of noun and verb dissociation. *Brain Lang.* **2017**, *165*, 10–20. [CrossRef]
65. Darley, F.L.; Aronson, A.E.; Brown, J.R. *Alteraciones Motrices del Habla*; Editorial Médica Panamericana: Madrid, Spain, 1978.
66. Senaha, M.L.H.; Brucki, S.M.D.; Nitrini, R. Rehabilitation in semantic dementia: Study of the effectiveness of lexical reacquisition in three patients. *Dement. Neuropsychol.* **2010**, *4*, 306–312. [CrossRef] [PubMed]
67. Folk, J.R.; Rapp, B.; Goldrick, M. The interaction of lexical and sublexical information in spelling: What's the point? *Cogn. Neuropsychol.* **2002**, *19*, 653–671. [CrossRef] [PubMed]
68. Luzzatti, C.; Colombo, C.; Frustaci, M.; Vitolo, F. Rehabilitation of spelling along the sub-word-level routine. *Neuropsychol. Rehabil.* **2000**, *10*, 249–278. [CrossRef]
69. The Jamovi Project. Jamovi (Version 1.6) [Computer Software]. 2021. Available online: https://www.jamovi.org (accessed on 10 September 2021).
70. Robey, R.; Apel, K.; Dollaghan, C.; Ellmo, W.; Hall, N.; Helfer, T.; Lonsbury-Martin, B. *Report of the Joint Coordinating Committee on Evidence-Based Practice*; ASHA: Rockville, MD, USA, 2004.
71. Cadório, I.; Lousada, M.; Martins, P.; Figueiredo, D. Generalization and maintenance of treatment gains in primary progressive aphasia (PPA): A systematic review. *Int. J. Lang. Commun. Disord.* **2017**, *52*, 543–560. [CrossRef] [PubMed]
72. Snowden, J.S.; Neary, D. Progressive anomia with preserved oral spelling and automatic speech. *Neurocase* **2003**, *9*, 27–43. [CrossRef]
73. Henry, M.L.; Hubbard, H.I.; Grasso, S.M.; Mandelli, M.L.; Wilson, S.M.; Sathishkumar, M.T.; Fridriksson, J.; Daigle, W.; Boxer, A.L.; Miller, B.L. Retraining speech production and fluency in non-fluent/agrammatic primary progressive aphasia. *Brain* **2018**, *141*, 1799–1814. [CrossRef]
74. Jokel, R.; Meltzer, J. Group intervention for individuals with primary progressive aphasia and their spouses: Who comes first? *J. Commun. Disord.* **2017**, *66*, 51–64. [CrossRef]
75. Beeson, P.M.; King, R.M.; Bonakdarpour, B.; Henry, M.L.; Cho, H.; Rapcsak, S.Z. Positive effects of language treatment for the logopenic variant of primary progressive aphasia. *J. Mol. Neurosci.* **2011**, *45*, 724–736. [CrossRef]
76. Taylor-Rubin, C.; Croot, K.; Nickels, L. Adherence to lexical retrieval treatment in Primary Progressive Aphasia and implications for candidacy. *Aphasiology* **2019**, *33*, 1182–1201. [CrossRef]
77. Clare, L. Rehabilitation for people living with dementia: A practical framework of positive support. *PLoS Med.* **2017**, *14*, e1002245. [CrossRef] [PubMed]
78. Mooney, A.; Beale, N.; Fried-Oken, M. Group communication treatment for individuals with PPA and their partners. In Proceedings of the Seminars in Speech and Language, Hsinchu, Taiwan, 4–5 October 2018; pp. 257–269.
79. Bier, N.; Brambati, S.; Macoir, J.; Paquette, G.; Schmitz, X.; Belleville, S.; Faucher, C.; Joubert, S. Relying on procedural memory to enhance independence in daily living activities: Smartphone use in a case of semantic dementia. *Neuropsychol. Rehabil.* **2015**, *25*, 913–935. [CrossRef]
80. Evans, W.S.; Quimby, M.; Dickey, M.W.; Dickerson, B.C. Relearning and retaining personally-relevant words using computer-based flashcard software in primary progressive aphasia. *Front. Hum. Neurosci.* **2016**, *10*, 561. [CrossRef]

81. Tsapkini, K.; Webster, K.T.; Ficek, B.N.; Desmond, J.E.; Onyike, C.U.; Rapp, B.; Frangakis, C.E.; Hillis, A.E. Electrical brain stimulation in different variants of primary progressive aphasia: A randomized clinical trial. *Alzheimer Dement. Transl. Res. Clin. Interv.* **2018**, *4*, 461–472. [CrossRef]
82. Tippett, D.C.; Hillis, A.E.; Tsapkini, K. Treatment of primary progressive aphasia. *Curr. Treat. Options Neurol.* **2015**, *17*, 362. [CrossRef]
83. Henry, M.L.; Hubbard, H.I.; Grasso, S.M.; Dial, H.R.; Beeson, P.M.; Miller, B.L.; Gorno-Tempini, M.L. Treatment for word retrieval in semantic and logopenic variants of primary progressive aphasia: Immediate and long-term outcomes. *J. Speech Lang. Hear. Res.* **2019**, *62*, 2723–2749. [CrossRef] [PubMed]

brain sciences

Article

Application of Machine Learning to Electroencephalography for the Diagnosis of Primary Progressive Aphasia: A Pilot Study

Carlos Moral-Rubio [1], Paloma Balugo [2], Adela Fraile-Pereda [2], Vanesa Pytel [3], Lucía Fernández-Romero [3], Cristina Delgado-Alonso [3], Alfonso Delgado-Álvarez [3], Jorge Matias-Guiu [3], Jordi A. Matias-Guiu [3,*,†] and José Luis Ayala [1,†]

1   Department of Computer Arquitecture and Automation, Faculty of Informatics, Universidad Complutense de Madrid, 28040 Madrid, Spain; carlosmoralrubio@gmail.com (C.M.-R.); jayala@ucm.es (J.L.A.)
2   Department of Neurophysiology, Institute of Neuroscience, Hospital Clínico San Carlos (IdISCC), Universidad Complutense de Madrid, 28040 Madrid, Spain; pbalugo@gmail.com (P.B.); adefraile@gmail.com (A.F.-P.)
3   Department of Neurology, Institute of Neuroscience, Hospital Clínico San Carlos (IdISCC), Universidad Complutense de Madrid, 28040 Madrid, Spain; vanesa.pytel@gmail.com (V.P.); lucia_28028@hotmail.com (L.F.-R.); cristinadelgado1409@gmail.com (C.D.-A.); alfonso.delgado.alvarez@hotmail.com (A.D.-Á.); matiasguiu@gmail.com (J.M.-G.)
*   Correspondence: jordimatiasguiu@hotmail.com or jordi.matias-guiu@salud.madrid.org; Tel.: +34-676933312
†   Jordi A. Matias-Guiu and José Luis Ayala contributed equally to this work as senior authors.

Citation: Moral-Rubio, C.; Balugo, P.; Fraile-Pereda, A.; Pytel, V.; Fernández-Romero, L.; Delgado-Alonso, C.; Delgado-Álvarez, A.; Matias-Guiu, J.; Matias-Guiu, J.A.; Ayala, J.L. Application of Machine Learning to Electroencephalography for the Diagnosis of Primary Progressive Aphasia: A Pilot Study. *Brain Sci.* **2021**, *11*, 1262. https://doi.org/10.3390/brainsci11101262

Academic Editors: Stephen D. Meriney and Heather Bortfeld

Received: 12 August 2021
Accepted: 23 September 2021
Published: 24 September 2021

Publisher's Note: MDPI stays neutral with regard to jurisdictional claims in published maps and institutional affiliations.

Abstract: Background. Primary progressive aphasia (PPA) is a neurodegenerative syndrome in which diagnosis is usually challenging. Biomarkers are needed for diagnosis and monitoring. In this study, we aimed to evaluate Electroencephalography (EEG) as a biomarker for the diagnosis of PPA. **Methods.** We conducted a cross-sectional study with 40 PPA patients categorized as non-fluent, semantic, and logopenic variants, and 20 controls. Resting-state EEG with 32 channels was acquired and preprocessed using several procedures (quantitative EEG, wavelet transformation, autoencoders, and graph theory analysis). Seven machine learning algorithms were evaluated (Decision Tree, Elastic Net, Support Vector Machines, Random Forest, K-Nearest Neighbors, Gaussian Naive Bayes, and Multinomial Naive Bayes). **Results.** Diagnostic capacity to distinguish between PPA and controls was high (accuracy 75%, F1-score 83% for kNN algorithm). The most important features in the classification were derived from network analysis based on graph theory. Conversely, discrimination between PPA variants was lower (Accuracy 58% and F1-score 60% for kNN). **Conclusions.** The application of ML to resting-state EEG may have a role in the diagnosis of PPA, especially in the differentiation from controls. Future studies with high-density EEG should explore the capacity to distinguish between PPA variants.

Keywords: electroencephalography; resting-state; primary progressive aphasia; biomarkers machine learning; K-Nearest Neighbors; frontotemporal dementia; Alzheimer's disease; graph theory

1. Introduction

Primary progressive aphasia (PPA) is a clinical syndrome secondary to the neurodegeneration of language brain regions and networks [1]. There are currently three main variants of PPA recognized in the literature: non-fluent (nfvPPA), semantic (svPPA), and logopenic variants (lvPPA) [2]. Diagnosis of PPA in the early stages is usually challenging. On the one hand, very mild word-finding difficulties may be present in aging, and the insidious onset of PPA symptoms limit an early identification [3]. On the other hand, there is a certain overlap between the PPA variants, especially between nfvPPA and lvPPA [4,5]. Neuropsychological batteries and language assessments are usually time consuming, and a high level of expertise is necessary for an adequate interpretation of the clinical findings. Although some novel and brief cognitive tests are being developed [6,7], neuroimaging and cerebrospinal fluid biomarkers are usually performed to confirm PPA diagnosis and the specific variant in each case. Structural magnetic resonance imaging (MRI) have shown adequate values of sensitivity and specificity for the diagnosis of svPPA, but diagnostic properties for the other variants are poorer [8,9]. Other more advanced MRI sequences show different patterns between PPA variants, but are not generally applicable routinely. Regarding positron emission tomography imaging, the 18F-FDG tracer has shown adequate values for diagnosis of the three variants of PPA, especially svPPA and lvPPA [10]. Amyloid tracers may distinguish between patients with amyloid deposition (generally associated with lvPPA) or not, but it does not have enough sensitivity to discriminate between subtypes of non-Alzheimer's disease variants. Novel tracers, such as tau tracers, are still under investigation and are not usually available beyond research settings [11]. Consequently, the combination of several tools is often necessary to conduct an adequate diagnosis. However, some of these techniques are not available in all clinical settings, which jeopardizes the equality of opportunities. In recent years, there is an increasing interest in an accurate diagnosis of neurodegenerative disorders. Furthermore, early diagnosis may imply early access to language therapies, which have shown positive effects in PPA [12,13]. In addition, the classification of PPA into three clinical variants improves the prediction of the underlying pathology [14]. Thus, novel and cost-effective biomarkers are necessary for early detection and differential diagnosis between PPA variants.

One of the key processes in neurodegenerative disorders comprises the alterations in brain activity and network disruptions [15]. There are several methods for measuring brain activity, with differences in the spatiotemporal resolution and applicability. Some methods, such as single-unit recordings, have high spatial and temporal precision, but are invasive and are not applicable to large networks and clinical practice. Among the non-invasive methods, functional MRI, magnetoencephalography and electroencephalography (EEG) permit the assessment of brain activity across the entire brain [16]. These methods are generally well tolerated and applicable in clinical practice, and evaluate brain activity with a resolution on the scale of millimeters and centimeters. This means that each voxel of a conventional MRI or a channel of an EEG reflect the activity of thousands of neurons and billions of synapses [17]. In comparison to functional MRI, EEG shows lower spatial but higher temporal resolution. However, both techniques are regarded as useful for the assessment of brain activity and connectivity. As advanced computational algorithms promise to improve signal processing and filtering, noninvasive recording devices are increasingly being investigated and applied. Some approaches record neural potentials from the scalp, and depending on the intensity of recording, they can capture the activity of thousands of neurons. Multiple layers obstruct information transmission from the cerebral cortex to the scalp, resulting in signal amplitudes and spatial resolution that are reduced. The electrodes are also sensitive to external interferences such as eye movements, face movements, chewing, or swallowing, among others [18].

EEG is a widely available technique very useful for the diagnosis of epileptic disease. In the last years, quantitative EEG has also been confirmed as a helpful biomarker in the assessment of several neurodegenerative disorders [19]. These studies suggest a potential clinical application of EEG in the assessment of neurodegenerative disorders, either in the differential diagnosis between them or with other non-neurodegenerative causes, including psychiatric conditions [20]. Data regarding the application of EEG signal in PPA are scarce. In a recent study [21], three patients with nfvPPA and five with primary progressive apraxia of speech (two of them also showing aphasia) underwent EEG. A theta slowing was detected in almost all patients with nfvPPA, suggesting a potential clinical application. Another recent study has detected some particular findings in the analysis of EEG microstates in 8 patients with svPPA in comparison with controls and Alzheimer's dementia [22].

Machine learning (ML) techniques may be helpful in improving the diagnostic performance of EEG, as has been shown in predicting epileptic seizures [23–26], Alzheimer's disease [27], or depression [28,29]. The rationale for the application of ML to EEG is based on the following factors. First, the visual analysis of EEG is time-consuming and requires high levels of expertise. Second, changes in neurodegenerative disorders may be less visually evident than epileptiform activity, which also limits the inter-rater reliability. Third, filter settings, frequency bands and criteria for thresholds are not clearly defined in the setting of neurodegenerative disorders [30].

ML for EEG analysis may be divided into two approaches: feature-based and end-to-end. On the one hand, feature-based decoding algorithms have a long track record of effectiveness in various EEG decoding challenges [20,31]. The data are often represented by handcrafted and previously selected features in this approach. End-to-end decoding algorithms, on the other hand, allow raw or minimally pre-processed data as inputs [32,33]. To date, end-to-end deep learning has gotten much interest due to its success in other disciplines of research. At least for the extraction of the features, this technique might lead to better solutions or the discovery of unexpectedly informative characteristics, and it does not involve handcrafting. In terms of learning features, end-to-end models have a reputation for being "black boxes".

In this study, we aimed to evaluate the potential of EEG as a biomarker for the diagnosis of PPA. For that purpose, the EEG raw signal was pre-processed in terms of feature transformations to enlarge the representation domain. We evaluated the diagnostic performance of EEG for the diagnosis of PPA, and the differential diagnosis between the three PPA Variants, applying ML models.

2. Materials and Methods

2.1. Participants

Forty patients with PPA were enrolled in this study. All patients met the current diagnostic criteria for PPA [1]. Patients were evaluated with a comprehensive language and neuropsychological protocol, which has been described elsewhere [34]. Structural MRI and FDG-PET were performed in all cases supporting the clinical diagnosis. Accordingly, patients were categorized as nfvPPA ($n = 18$), svPPA ($n = 10$), and lvPPA ($n = 12$). Twenty controls (control group, CG) were also included for comparison. Table 1 shows the details of the groups participating in the study.

The CG was obtained from patients that underwent EEG because of a previous history of syncope, but visual analysis of EEG, neuroimaging, and clinical follow-up were normal, excluding potential neurological disorders.

Table 1. Main clinical and demographic characteristics.

	PPA	nfvPPA	svPPA	lvPPA
Number of participants	40	18 (45%)	10 (25%)	12 (30%)
Age	68.7 ± 6.94	68.55 ± 7.29	66.80 ± 6.35	70.50 ± 6.97
Women	26 (65%)			
Years of education	13.90 ± 4.26	13.33 ± 4.41	14.20 ± 4.15	14.50 ± 4.35
Years since symptom onset	4.00 ± 2.25	4.83 ± 1.94	4.00 ± 2.98	2.75 ± 1.42
ACE-III	55.78 ± 26.59	71.76 ± 22.07	53.89 ± 15.72	48.00 ± 23.37
CDR-FTLD (Sum of boxes)	2.6 ± 1.81	2.22 ± 1.54	2.60 ± 1.67	3.16 ± 2.26

2.2. EEG Acquisition

EEGs were recorded in a resting state condition with the eyes closed and under the supervision of trained personnel. EEGs were acquired on a NicoletOne device of 32 channels, using the standard 10/20 system and referenced to A1. Time of acquisition was 20 min.

2.3. Preprocessing

Different signal transformations were applied, aiming to expand the amount of information. Signal preprocessing was performed following the pipeline implemented by [35] in EEGLAB Software (Matlab). These procedures try to minimize the external noise and artefacts that are usually present in a raw EEG signal. The following steps were conducted, which are also summarized in Figure 1:

1. Time ranges selection. Original signals are too long to be analyzed and can contain some additional noise, so we manually selected those time ranges with higher quality in the signal representation to get the most accurate and clean signal. This process also considered the labels recorded during the EEG acquisition and clinical assessment, which notify about the state of the patient, unexpected events, or activities that could impact on the signal.
2. High-pass filtering at 1 Hz. This filter was applied to remove baseline noise, remove noise introduced by sweating, and prepare the signal for ICA analysis.
3. Apply CleanLine process with the following configuration: 10 Hz of bandwidth at 50 Hz line frequency. This preprocessing step removes line noise and related harmonics from each one of the scalp channels using a novel approach, as described in [36]. For that purpose, and for each sliding window over the original data, a multi-taper FFT is applied to transform the signal to the frequency domain; after that, the complex amplitude of the desired frequency is extracted. With that information, a noise signal in the frequency domain is generated and, finally, the time-domain associated noise signal that needs to be extracted from the original one is also created.
4. Re-reference data to average. This is the most effective and easiest way to re-reference EEG data because it establishes that the summed up power across the scalp topography should sum zero. In other words, we removed the mean over all scalp channels to every single channel to make sure that all channels contribute with the same weight.
5. Low pass filter at 40 Hz. This step was applied in order to remove any possible undesired high-frequency signal that was not removed by CleanLine. Although other investigations are looking for biomarkers in higher frequency ranges of EEG signal, most recent research works are focusing in lower frequency ranges [37]. For simplicity of our analysis and control of error sources, we have limited our work to the lower frequency bands.
6. To apply ICA (Independent Component Analysis) to the signal. This method is a linear decomposition technique which aims to find the source signals from a set of mixed signals, as it occurs with EEG. Unlike PCA (Principal Component Analysis), ICA tries to retrieve those original signals that are maximally statistical independent in just one domain [38].
7. To epoch data into windows of duration equal to one second without overlapping.
8. Visual artifact rejection of epochs. As a final step, we reviewed manually all signals and all their epochs looking for artefacts or undesired signal events.

2.4. Quantitative EEG

QEEG, quantitative EEG, is the frequency domain transformation of the original EEG signal [39]. To obtain this transformed signal, the Discrete Fourier Transform method was applied over our sampled (at 500 Hz) EEG signal. Given a $x(n)$ discrete EEG signal, the definition of Fourier Transform (FT) is as follows:

$$X_k = \sum_{n=0}^{N} x_n e^{-2\pi i k n / N},\qquad(1)$$

for $0 \leq k \leq N - 1$.

This transformation gave a transformed domain that increases the representation domains of the original signal and, hence, the information provided. We divided the total frequency range into non-overlapping frequency bands:

- Delta from 1–4 Hz.
- Ipsilon from 4–8 Hz.
- Alpha from 8–14 Hz.
- Beta from 14–30 Hz.

- Gamma from 30–45 Hz.
- OoB (out of bag) for frequencies higher than 45 Hz.

2.5. Wavelet Transformation

Wavelet Transform (WT) is the decomposition of the original signal into a set of basis functions consisting of contractions, expansions, and translations [40]. This is a similar approach to FT but using wavelet functions to achieve the transformed domain.

WT of the EEG signal was obtained by filtering repeatedly until reaching the desired level of decomposition. In each repetition, we applied a low-pass filter to obtain the approximation coefficient (CA) and a high-pass filter to obtain the detailed coefficient (CD). After every filtering stage, the signal was down-sampled by half the sampling frequency of the previous level.

A total of seven sequential subdivisions were applied until a sufficient number of transformations, all correctly subdivided in frequency, was achieved. This process provided eight signals, each one assigned to a different frequency range:

- Subband 1 from 125 to 250 Hz.
- Subband 2 from 62.5 to 125 Hz.
- Subband 3 from 31.2 62.5 Hz.
- Subband 4 from 15.6 to 31.2 Hz.
- Subband 5 from 7.8 to 15.6 Hz.
- Subband 6 from 3.9 to 7.8 Hz.
- Subband 7 from 1.9 to 3.9 Hz.
- Subband 8 from 0 to 1.9 Hz.

From all the extracted signals, sub-bands 1 and 2 were removed from the pipeline because neither of them offered any information after the application of step 5 in the preprocessing pipeline section (low-pass filter at 40 Hz).

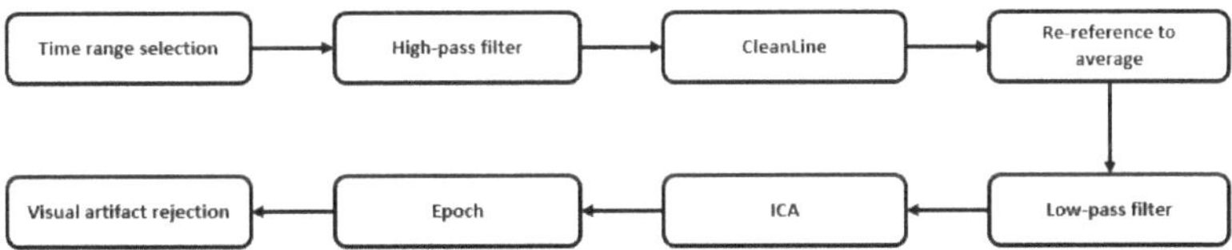

Figure 1. Preprocessing pipeline.

2.6. AutoEncoders

Autoencoders are a specific type of neural networks architecture where the input is the same as the output. They compress the input into a lower-dimensional code and then reconstruct the output from this representation. The code is a compact "summary" or "compression" of the input, also called the latent-space representation. This novel modeling technique is also exploited in dimensionality reduction problems.

This architecture was created by using an Encoder-Decoder system (Figure 2). The first part, the Encoder, used fully-connected layers in which the number of input neurons is higher than the number of output neurons, to achieve that reduction of dimensionality. The second part, the Decoder, used fully-connected layers in which the number of input neurons is lower than the number of output neurons to achieve the reconstruction of the original signal.

In addition to this vanilla configuration, there are other complex configurations with multiple hidden layers for the encoding and decoding part, some of them even add noise to the input or to the intermediate part to force the neural network as a regularization technique to a better generalization.

As stated, this type of neural networks are usually applied as a technique for dimensionality reduction, but we decided to use them as a mechanism to transform the time domain (similarly to the FT and WT). We applied the Encoder to the EEG values along time, creating a new domain of features and information representation.

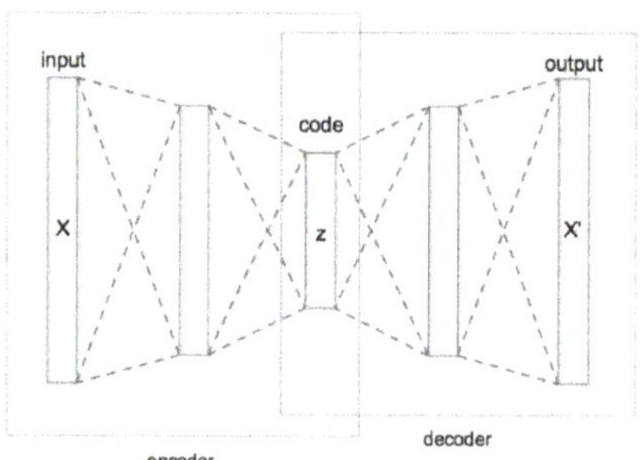

Figure 2. Encoder-Decoder architecture of AutoEncoder.

2.7. Graph Theory Analysis

Graph Theory Analysis (GTA) is a mathematical formalism used to model pairwise relations between objects. Here, we used the EEG sources from the different electrodes to generate the network that merges such data [41].

For that purpose, we generated a graph matrix, also called adjacency matrix, by calculating all pairs of partial correlations between all available channels or electrodes [42]. The absolute value operator was also applied to this matrix to achieve our final result, which in this case was an undirected weighted adjacency matrix.

Once the network was created, it was analyzed using different metrics:

- Node degree. This metric represents the number of links detected for every node.
- Path length. Mean of the shortest links present in the network.
- Clustering coefficient. Number of triangular connections in the network, divided by the theoretical maximum number of triangular connections. This variable represents the clustering capacity of the generated network.

In addition to these metrics, we also created a brain representation that shows each electrode as a node in a graph structure, and the connections that we obtained between those electrodes. An example of this representation ca be found in the Figures 3 and 4 where the connections of the CG and PPA groups, filtered by alpha frequency range (from 8 to 14 Hz), are shown, respectively.

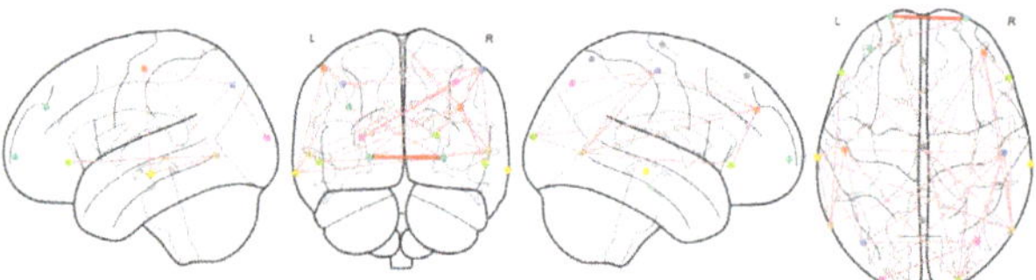

Figure 3. Network representation for CG group, including nodes, connections and their strength in alpha frequency band.

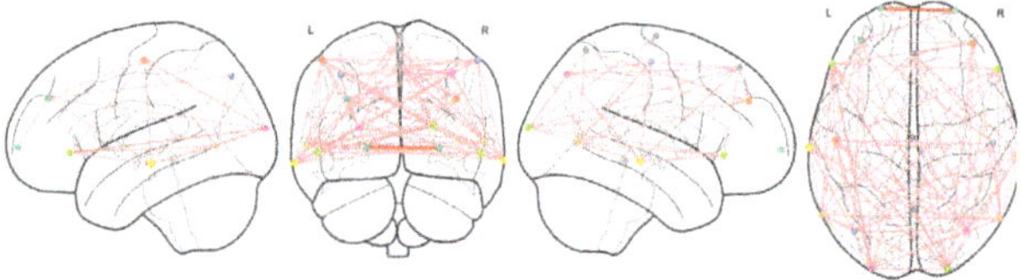

Figure 4. Network representation for PPA group, including nodes, connections and their strength in alpha frequency band.

2.8. Data Analysis

2.8.1. Binary Classification Model between PPA and CG

A classification model was generated to differentiate between CG and PPA patients based only on the extracted EEG features. Seven classification algorithms were evaluated: Decision Tree, Elastic Net (EN), Support Vector Machine (SVM), Random Forest (RF), k-Nearest Neighbors (kNN), Gaussian Naive Bayes (Gaussian NB) and Multinomial Naive Bayes (Multinomial NB).

The following pre-processing pipeline was applied to prepare the data for modelling:

- Train-test split. In this step we randomly generated train and test samples from the original dataset by applying 80% for training sample and 20% for test. This split was stratified, namely, the proportion of examples in each class is preserved into train and test samples.
- Scaling. We applied a MinMaxScaler method to each column in order to transform their range of values into the range [0, 1].
- Univariate Feature Selection. A feature selection step was applied to reduce the number of features to only 50 features from the original set (309). ANOVA F-value was computed for each column-target model and only the best 50 scores were selected.

In the training process, the selection of the best hyperparameters for each model was accomplished by a Bayes-search optimization algorithm (this optimization algorithm created a full space with all possible hyperparameter values and applied Bayes Theorem in order to find those exact values that minimized the error function). All models performed a binary classification using a 10-fold cross-validation using F1-Score metric (2) as the main metric. This metric was selected to optimize the classification problem, which was imbalanced. Precision (3), Sensitivity (4), Specificity (5) and Youden Index (6) are also displayed.

$$F1 - Score = \frac{2 * Precision * Sensitivity}{Precision + Sensitivity} \tag{2}$$

$$Precision = \frac{TruePositives}{TruePositives + FalsePositives} \tag{3}$$

$$Sensitivity = \frac{TruePositives}{TruePositives + FalseNegatives}, \tag{4}$$

$$Specificity = \frac{TrueNegatives}{TrueNegatives + FalsePositives}, \tag{5}$$

$$YoudenIndex = Sensitivity + Specificity - 1, \tag{6}$$

2.8.2. Classification Model for All Groups

Following the same pipeline described in the binary classification model, a multiclass classification model was applied in order to distinguish between nfvPPA, svPPA, lvPPA and CG. The same pre-processing steps, hyperparameter tuning techniques and cross-validation options were applied here. All models performed a multiclass classification with 4 different classes (one per each group of patients) and using F1-Score metric as their main aim.

2.8.3. Network Analysis

We transformed each EEG signal into a Network; this allowed to extract additional network metrics and enlarge the set of features per patient, but it also allowed to evaluate the differences between two brains in terms of activity according to these network metrics.

We also visualized the generated connections between EEG channels in each group of patients. This provided meaningful information about interactions of brain regions across the different groups of patients.

2.8.4. Principal Components Analysis

In order to explain the complexity of our working dataset, we applied Principal Component Analysis (PCA) to reduce the dimensionality. PCA aims to find the directions of maximum variance in high-dimensional data and projects them onto a new subspace with equal or fewer dimensions than the original one. Hence, it reduces the number of features by combining them linearly. We performed a dimensionality reduction to only two principal components. Accordingly, the visualization of all subjects as data points is allowed by looking for the linear combination of all the extracted features into these two principal components. In this way, it is possible to visualize the multi-dimensional data distribution and evaluate how mixed are the data instances in the representation space.

3. Results

3.1. Classification Model between CG and PPA

Main metrics are summarized in Table 2.

Table 2. Metrics from classification models for PPA vs. CG.

Model	F1-Score	Precision	Sensitivity	Accuracy
Decision Tree	0.38	0.39	0.38	0.42
kNN	0.83	0.78	0.88	0.75
SVM	0.58	0.72	0.86	0.58
Random Forest	0.37	0.32	0.43	0.58
Elastic Net	0.4	0.33	0.5	0.66
Gaussian NB	0.78	0.9	0.75	0.83
Multinomial NB	0.73	0.73	0.75	0.75

Seven different models were evaluated for the binary classification: Decision Tree, EN, SVM, RF, kNN, Gaussian NB and Multinomial NB (Table 2). kNN model, a non-parametric supervised classification method, achieved the best performance, showing a Sensitivity of 0.88, an F1-Score value of 0.83 and a Specificity of 0.5. The confusion matrix from the best model (kNN) is shown in Figure 5, as well as its ROC curve in Figure 6.

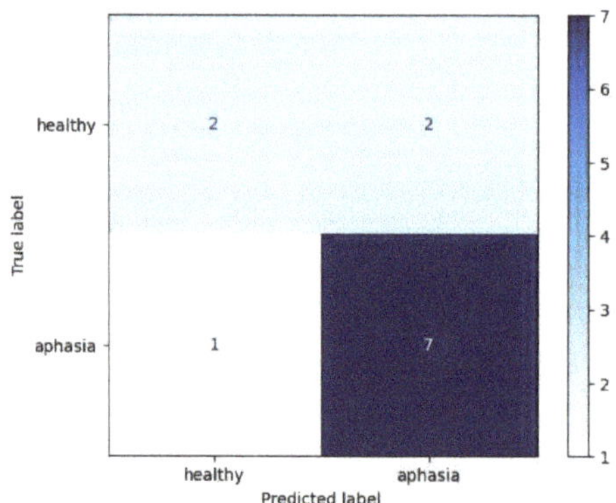

Figure 5. Confusion matrix from kNN binary classification (PPA vs. CG) model for the test set.

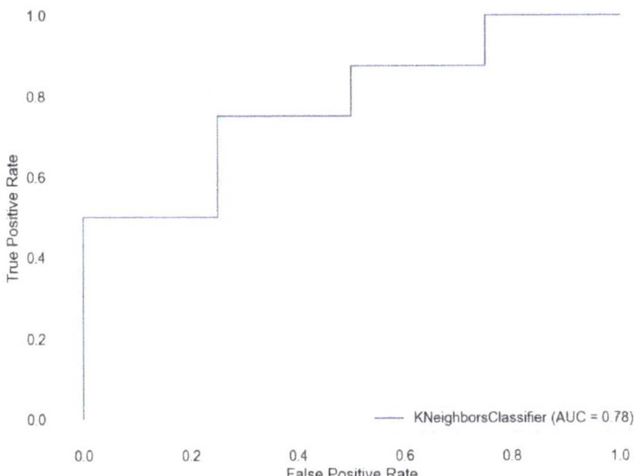

Figure 6. ROC curve from kNN binary classification (PPA vs. CG) model.

The 20 most important variables used for training are depicted in Figure 7. A Decision Tree model is included in Figure 8. Regarding the most relevant variables, all variables except one were generated by the network transformations. Specifically, 40% from Node Degree, 50% from Clustering Coefficient, 5% Path Length, and 5% qEEG. Similarly, most features used in the Decision Tree algorithm are associated with network analysis.

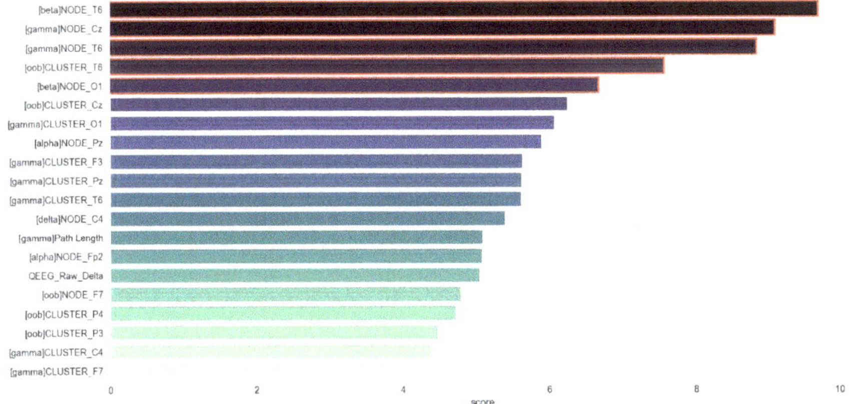

Figure 7. Representation of the 20 most important features.

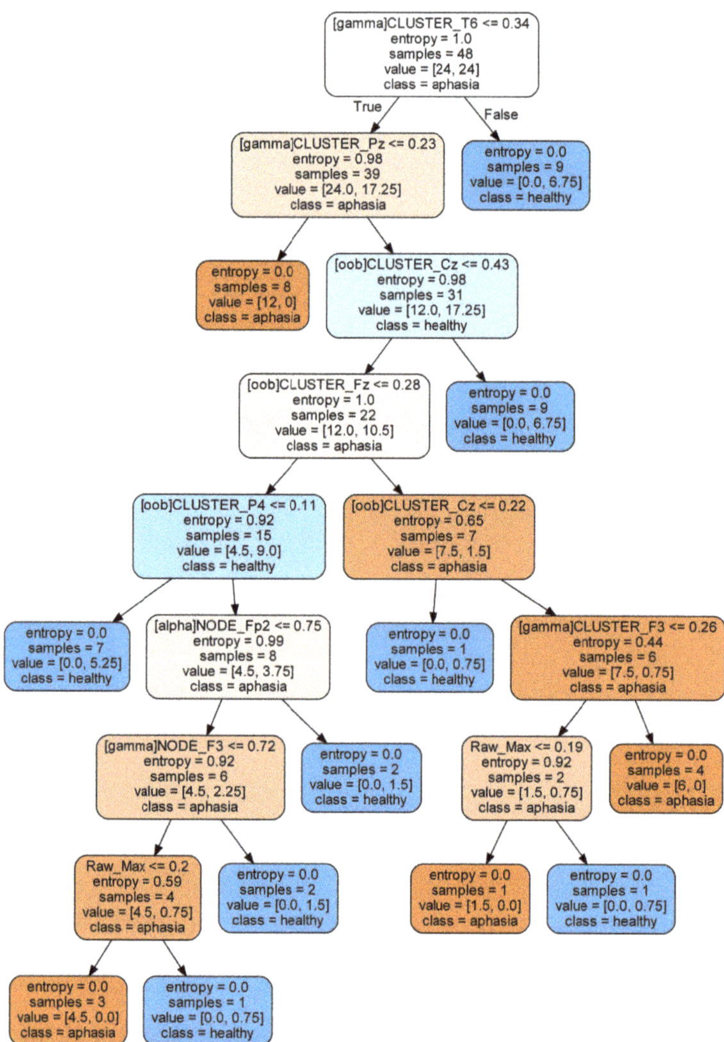

Figure 8. Representation of decisions in Decision Tree binary model. Scores represent the ANOVA F-value. Five most relevant variables are shown in red.

3.2. Classification Model between All Groups

A multiclass model (4 classes) was developed to evaluate the possibility of automatic detection among all the PPA variants and CG. The same aforementioned models were evaluated (Table 3). Again, kNN model achieved the best performance. However, Sensitivity was 0.58 and F1-Score was 0.6. Confusion matrix for this model is shown in Figure 9.

Table 3. Metrics from classification models for 4 groups (nfvPPA, svPPA, lvPPA, and CG).

Model	F1-Score	Precision	Sensitivity	Accuracy
Decision Tree	0.32	0.32	0.4	0.42
kNN	0.6	0.68	0.58	0.58
SVM	0.39	0.40	0.46	0.5
Random Forest	0.39	0.38	0.48	0.5
Elastic Net	0.34	0.31	0.33	0.38
Gaussian NB	0.27	0.25	0.31	0.33
Multinomial NB	0.2	0.2	0.25	0.25

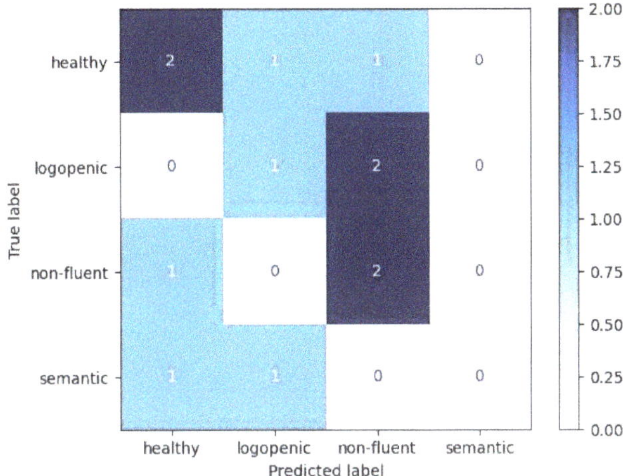

Figure 9. Confusion matrix from kNN multiclass classification model.

As in the previous analysis, Figure 10 shows the graphical representation of the Decision Tree Model. In this case, decisions are mainly based on features obtained from network analysis and Autoencoder transformations.

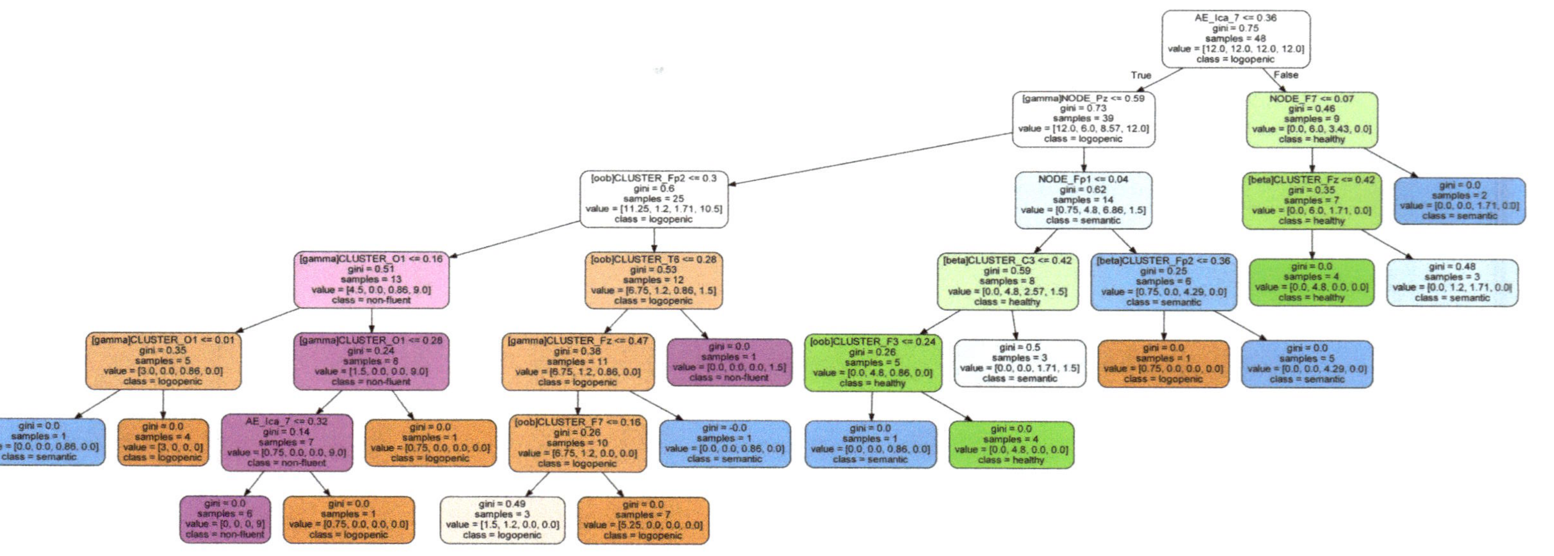

Figure 10. Representation of decisions in Decision Tree multiclass model.

4. Discussion

In this pilot study, we evaluated the diagnostic performance of a resting-state EEG obtained in clinical practice conditions for the diagnosis of PPA. We applied ML models, as they may be helpful to maximize the diagnostic capacity from many variables with no a priori hypotheses. In this regard, diagnostic performance was relatively high for the detection of patients with PPA in comparison with the control group. This suggests that there are certain EEG abnormalities that may be detected in patients with PPA. The most important features ranked by the algorithms for the classification and included in the decision trees algorithms involve mainly temporal and frontal channels in both hemispheres. Interestingly, features derived from network analysis obtained the best classification, emphasizing the role of graph theory in the analysis of EEG data [32]. These findings are consistent with recent investigations that are exploiting this new area of analysis [43,44].

Conversely, the application of EEG to the diagnosis of the specific variant of PPA did not achieve a satisfactory classification. Previous studies using quantitative data from EEG for the differential diagnosis of neurodegenerative disorders have obtained generally better results [31]. For instance, applying support vector machines, a 91% of accuracy was found to distinguish Alzheimer's disease and dementia with Lewy bodies, and 88% for Alzheimer's disease and Frontotemporal dementia [45]. Another study achieved a 93.3% of accuracy to classify between Alzheimer's disease and Frontotemporal dementia [46]. However, these studies were performed with small samples, and were not replicated in larger studies [47]. The application of EEG to PPA is probably more challenging, due to the regional overlap between PPA variants in contrast to other disorders such as Alzheimer's disease and Frontotemporal Dementia. In this regard, high-density EEG with a larger number of channels might obtain better results in the classification between PPA variants.

Our key insight is that machine learning itself can deal well with errors, qualitative and corrupted data and, more importantly for our purposes, integrate heterogeneous data from multiple domains. With this aim, our research work has enlarged the dataset to increase the information representation. The applied machine learning algorithms can jointly manage transformations from time to frequency domain, wavelet or network representation provided the setup parameters are carefully selected. In [48], a total of 49 experimental studies published from 2009 until 2020, which apply machine learning algorithms on resting-state EEG recordings from AD patients, were reviewed. These works did not evaluate the benefits of increased information representation in classification accuracy. Most of the studies focused on AD detection incorporating Support Vector Machines (SVM). Conversely, we found that classification algorithms based on distance (similar to kNN, where the function is only approximated locally and all computation is deferred until function evaluation) can improve performance.

The visualization of the multi-dimensional data used in our study, and the complexity of such dataset, has been performed with a PCA. Using this method, we found that patients are not clearly separated (Figure 11). This aspect is important for optimal performance in SVM models in the classification [49]). In contrast, kNN model, which is based on local distances, could lead to better predictions. This explains why we observed better performance of kNN model with respect to SVM model. Additionally, kNN model works better with a small number of features [50], which is our case after the application of the feature selection method. To follow this line, we replicated the same pipeline in the CG vs. PPA classification, skipping the feature selection phase. Thus, we compared the results of all models using only the best 50 features against using all generated features. As shown in Table 4, SVM model obtained better performance than kNN. However, the absolute values of the quality metrics (F1-score, precision, sensitivity and accuracy) explain the need for the feature selection process followed in our study.

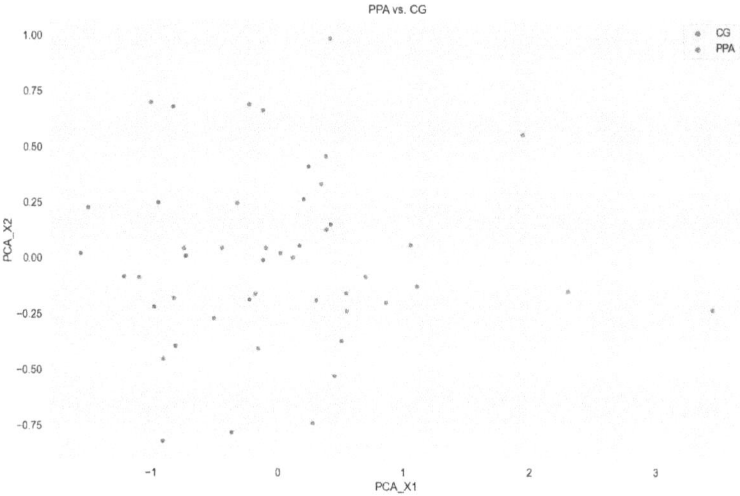

Figure 11. PPA vs. CG in PCA two dimensions.

Table 4. Metrics from classification models for PPA vs. CG using all columns (with no feature selection).

Model	F1-Score	Precision	Sensitivity	Accuracy
Decision Tree	0.49	0.50	0.50	0.50
kNN	0.46	0.6	0.38	0.42
SVM	0.50	0.56	0.56	0.50
Random Forest	0.40	0.33	0.50	0.67
Elastic Net	0.40	0.33	0.50	0.67
Gaussian NB	0.37	0.32	0.44	0.58
Multinomial NB	0.56	0.56	0.56	0.58

The trend of increased information representation may be seen in recent works like [51] (still SVM in AD), or [52] (where, apart from EEG, a wide range of diagnostic tests were included). Our approach has focused exclusively in the classification possibilities of the EEG signal for the PPA, but we expect that our results could be improved by the addition of other tests such as neuroimaging, cognitive assessment, or genetics.

Patients included in this study fulfilled the current diagnostic criteria, with both MRI and FDG-PET supporting the diagnosis. As they were generally in early stages and EEG in comparison with controls was discriminative, these findings raise the possibility to explore in future studies the role of EEG in the clinical follow-up of patients with PPA, especially in the setting of clinical trials, in which reliable, reproducible and non-invasive endpoints are necessary [53].

Our study has some limitations. First, we included 40 patients, generally in early stages but not in the first consultation. Future studies should enroll a larger sample size and specifically focusing on patients in the early stages to confirm a potential role of EEG in the detection of PPA. Second, in this study, we only applied ML to EEG data. One of the main strengths of ML is the combination of multiple sources of information. Thus, the application of ML to studies including multimodality assessments (cognitive testing, MRI, PET) may be of interest to disentangle the best tools (isolated or in combination) for diagnosis of PPA [54].

5. Conclusions and Future Work

Our study shows that the application of ML to resting-state EEG may have a role in the diagnosis of PPA, especially in the differentiation from controls. EEG may have some advantages compared with other biomarkers.

ML techniques were applied to evaluate the possibility to automatically classify EEG data from PPA patients with respect to a control group. Our work showed that a feature expansion process can increase the information representation and achieve good classification accuracy, using mainly features from the graph-network representation of the EEG signal. The capability to classify PPA variants was also evaluated. Although lower, the classification capacity is still promising and advises further development of these automatic techniques for phenotype classification from EEG signals.

We are currently increasing the sample size to improve the classification accuracy of the models. In addition, we aim to enlarge the frequency range in the input dataset (over 45 Hz) to evaluate whether higher-frequency components may help the biomarker discovery with machine learning and deep learning methods.

Author Contributions: Conceptualization, J.A.M.-G. and J.L.A.; methodology, P.B., A.D.-Á., A.F.-P., C.M.-R.; formal analysis, C.M.-R. and J.L.A.; investigation, P.B., A.F.-P., V.P., L.F.-R., C.D.-A. and J.A.M.-G.; data curation, A.D.-Á., P.B.A, A.F.-P., V.P., C.D.-A. and J.A.M.-G.; writing—original draft preparation, C.M.-R., P.B., J.L.A. and J.A.M.-G.; writing—review and editing, All; supervision, J.M.-G.; project administration, J.A.M.-G.; funding acquisition, J.M.-G., J.A.M.-G. and J.L.A. All authors have read and agreed to the published version of the manuscript.

Funding: Jordi A. Matias-Guiu is supported by Instituto de Salud Carlos III through the project INT20/00079 (co-funded by European Regional Development Fund "A way to make Europe").

Institutional Review Board Statement: The study was conducted according to the guidelines of the Declaration of Helsinki, and approved by the Institutional Review Board (or Ethics Committee) of Hospital Clinico San Carlos (protocol code 17/247, date of approval 27th June 2017).

Informed Consent Statement: Written informed consent was obtained from all subjects involved in the study.

Data Availability Statement: The data presented in this study are available on request from the corresponding author.

Acknowledgments: Jordi A Matias-Guiu is supported by Instituto de Salud Carlos III through the project INT20/00079 (co-funded by European Regional Development Fund "A way to make Europe"). We also acknowledge support from Spanish Ministry of Science and Innovation under project PID2019-110866RB-I00.

Conflicts of Interest: The authors declare that they have no conflict of interests.

Abbreviations

The following abbreviations are used in this manuscript:

ML	Machine Learning
EEG	Electroencephalogram
MRI	Magnetic Resonance Imaging
PET	Positron Emission Tomography
qEEG	quantitative EEG
MCI	Mild Cognitive Impairment
AD	Alzheimer Disease
ICA	Independent Component Analysis
PCA	Principal Component Analysis
WT	Wavelet Transform
GTA	Graph Theory Analysis
OoB	Out of Bag
CG	Control Group

PPA   Primary Progressive Aphasia
nfvPPA   Non-Fluent Primary Progressive Aphasia
svPPA   Semantic Primary Progressive Aphasia
lvPPA   Logopenic Primary Progressive Aphasia
SVM   Support Vector Machine
SD   Standard Deviation
ROC   Receiver Operating Characteristic
kNN   k-Nearest Neighbors
NB   Naive Bayes

References

1. Gorno-Tempini, M.L.; Hillis, A.E.; Weintraub, S.; Kertesz, A.; Mendez, M.; Cappa, S.F.; Ogar, J.M.; Rohrer, J.D.; Black, S.; Boeve, B.F.; et al. Classification of primary progressive aphasia and its variants. *J. Neurol.* **2011**, *76*, 1006–1014. [CrossRef]
2. Marshall, C.R.; Hardy, C.J.D.; Volkmer, A.; Russell, L.L.; Bond, R.L.; Fletcher, P.D.; Clark, C.N.; Mummery, C.J.; Schott, J.M.; Rossor, M.N.; et al. Primary progressive aphasia: A clinical approach. *J. Neurol.* **2018**, *265*, 1474–1490. [CrossRef] [PubMed]
3. Stiver, J.; Staffaroni, A.M.; Walters, S.M.; You, M.Y.; Casaletto, K.B.; Erlhoff, S.J.; Possin, K.L.; Lukic, S.; La Joie, R.; Rabinovici, G.D.; et al. The Rapid Naming Test: Development and initial validation in typically aging adults. *Clin. Neuropsychol.* **2021**, 1–22. [CrossRef] [PubMed]
4. Matías-Guiu, J.A.; Cabrera-Martín, M.N.; Moreno-Ramos, T.; Valles-Salgado, M.; Fernandez-Matarrubia, M.; Carreras, J.L.; Matías-Guiu, J. Amyloid and FDG-PET study of logopenic primary progressive aphasia: Evidence for the existence of two subtypes. *J. Neurol.* **2015**, *262*, 1463–1472. [CrossRef] [PubMed]
5. Tetzloff, K.A.; Whitwell, J.L.; Utianski, R.L.; Duffy, J.R.; Clark, H.M.; Machulda, M.M.; Strand, E.A.; Josephs, K.A. Quantitative assessment of grammar in amyloid-negative logopenic aphasia. *Brain Lang* **2018**, *186*, 26–31. [CrossRef]
6. Matias-Guiu, J.A.; Pytel, V.; Hernández-Lorenzo, L.; Patel, N.; Peterson, K.A.; Matías-Guiu, J.; Garrard, P.; Cuetos, F. Spanish Version of the Mini-Linguistic State Examination for the Diagnosis of Primary Progressive Aphasia. *J. Alzheimers Dis.* **2021**. [CrossRef] [PubMed]
7. Epelbaum, S.; Saade, Y.M.; Flamand Roze, C.; Roze, E.; Ferrieux, S.; Arbizu, C.; Nogues, M.; Azuar, C.; Dubois, B.; Tezenas du Montcel, S.; et al. A Reliable and Rapid Language Tool for the Diagnosis, Classification, and Follow-Up of Primary Progressive Aphasia Variants. *Front. Neurol.* **2020**, *11*, 571657. [CrossRef]
8. Sajjadi, S.A.; Sheikh-Bahaei, N.; Cross, J.; Gillard, J.H.; Scoffings, D.; Nestor, P.J. Can MRI Visual Assessment Differentiate the Variants of Primary-Progressive Aphasia? *AJNR Am. J. Neuroradiol.* **2017**, *38*, 954–960. [CrossRef]
9. Matias-Guiu, J.A.; Cabrera-Martín, M.N.; Matías-Guiu, J.; Carreras, J.L. FDG-PET/CT or MRI for the Diagnosis of Primary Progressive Aphasia? *AJNR Am. J. Neuroradiol.* **2017**, *38*, E63. [CrossRef]
10. Matías-Guiu, J.A.; Cabrera-Martín, M.N.; Pérez-Castejón, M.J.; Moreno-Ramos, T.; Rodríguez-Rey, C.; García-Ramos, R.; Ortega-Candil, A.; Fernandez-Matarrubia, M.; Oreja-Guevara, C.; Matías-Guiu, J.; et al. Visual and statistical analysis of ^{18}F-FDG PET in primary progressive aphasia. *Eur. J. Nucl. Med. Mol. Imaging* **2015**, *42*, 916–927. [CrossRef] [PubMed]
11. Josephs, K.A.; Martin, P.R.; Botha, H.; Schwarz, C.G.; Duffy, J.R.; Clark, H.M.; Machulda, M.M.; Graff-Radford, J.; Weigand, S.D.; Senjem, M.L.; et al. F]AV-1451 tau-PET and primary progressive aphasia. *Ann. Neurol.* **2018**, *83*, 599–611. [CrossRef] [PubMed]
12. Henry, M.L.; Hubbard, H.I.; Grasso, S.M.; Mandelli, M.L.; Wilson, S.M.; Sathishkumar, M.T.; Fridriksson, J.; Daigle, W.; Boxer, A.L.; Miller, B.L.; et al. Retraining speech production and fluency in non-fluent/agrammatic primary progressive aphasia. *Brain* **2018**, *141*, 1799–1814. [CrossRef]
13. Henry, M.L.; Hubbard, H.I.; Grasso, S.M.; Dial, H.R.; Beeson, P.M.; Miller, B.L.; Gorno-Tempini, M.L. Treatment for Word Retrieval in Semantic and Logopenic Variants of Primary Progressive Aphasia: Immediate and Long-Term Outcomes. *J. Speech Lang Hear Res.* **2019**, *62*, 2723–2749. [CrossRef] [PubMed]
14. Bergeron, D.; Gorno-Tempini, M.L.; Rabinovici, G.D.; Santos-Santos, M.A.; Seeley, W.; Miller, B.L.; Pijnenburg, Y.; Keulen, M.A.; Groot, C.; van Berckel, B.N.M.; et al. Prevalence of amyloid-β pathology in distinct variants of primary progressive aphasia. *Ann. Neurol.* **2018**, *84*, 729–740. [CrossRef] [PubMed]
15. McMackin, R.; Muthuraman, M.; Groppa, S.; Babiloni, C.; Taylor, J.P.; Kiernan, M.C.; Nasseroleslami, B.; Hardiman, O. Measuring network disruption in neurodegenerative diseases: New approaches using signal analysis. *J. Neurol. Neurosurg. Psychiatry* **2019**, *90*, 1011–1020. [CrossRef] [PubMed]
16. Vinjamuri, R. (Ed.) *Advances in Neural Signal Processing*; IntechOpen: London, UK, 2020.
17. Logothetis, N.K. What we can do and what we cannot do with fMRI. *Nature* **2008**, *453*, 869–878. [CrossRef]
18. Paszkiel, S. *Analysis and Classification of EEG Signals for Brain–Computer Interfaces*; Springer: Berlin, Germany, 2020. [CrossRef]
19. Popa, L.L.; Dragoș, Ha.; Strilciuc, Ș.; Pantelemon, C.; Mureșanu, I.; Dina, C.; Văcăraș, V.; Muresanu, D. Added Value of QEEG for the Differential Diagnosis of Common Forms of Dementia. *Clin. EEG Neurosci.* **2021**, *52*, 201–210. [CrossRef] [PubMed]
20. Metin, S.Z.; Erguzel, T.T.; Ertan, G.; Salcini, C.; Kocarslan, B.; Cebi, M.; Metin, B.; Tanridag, O.; Tarhan, N. The Use of Quantitative EEG for Differentiating Frontotemporal Dementia From Late-Onset Bipolar Disorder. *Clin. EEG Neurosci.* **2018**, *49*, 171–176. [CrossRef]

21. Utianski, R.L.; Caviness, J.N.; Worrell, G.A.; Duffy, J.R.; Clark, H.M.; Machulda, M.M.; Withwell, J.L.; Josephs, K.A. Electroencephalography in primary progressive aphasia and apraxia of speech. *Aphasiology* **2018**, *33*, 1410–1417. [CrossRef]
22. Grieder, M.; Koenig, T.; Kinoshita, T.; Utsunomiya, K.; Wahlund, L.O.; Dierks, T.; Nishida, K. Discovering EEG resting state alterations of semantic dementia. *Clin. Neurophysiol.* **2016**, *127*, 2175–2181. [CrossRef]
23. Subasi, A.; Kevric, J.; Abdullah Canbaz, M. Epileptic seizure detection using hybrid machine learning methods. *Neural Comput. Appl.* **2019**, *31*, 317–325. [CrossRef]
24. Hügle, M.; Heller, S.; Watter, M.; Blum, M.; Manzouri, F.; Dumpelmann, M.; Schulze-Bonhage, A.; Woias, P.; Boedecker, J. Early Seizure Detection with an Energy-Efficient Convolutional Neural Network on an Implantable Microcontroller. In Proceedings of the 2018 International Joint Conference on Neural Networks (IJCNN), Rio de Janeiro, Brazil, 8–13 July 2018; pp. 1–7. [CrossRef]
25. Kiral-Kornek, I.; Roy, S.; Nurse, E.; Mashford, B.; Karoly, P.; Carroll, T.; Payne, D.; Saha, S.; Baldassano, S.; O'Brien, T.; et al. Epileptic Seizure Prediction Using Big Data and Deep Learning: Toward a Mobile System. *EBioMedicine* **2018**, *27*, 103–111. [CrossRef] [PubMed]
26. Mirowski, P.; Madhavan, D.; LeCun, Y.; Kuzniecky, R. Classification of patterns of EEG synchronization for seizure prediction. *Clin. Neurophysiol.* **2009**, *120*, 1927–1940. [CrossRef]
27. Lehmann, C.; Koenig, T.; Jelic, V.; Prichep, L.; John, R.E.; Wahlund, L.; Dodge, Y.; Dierks, T. Application and comparison of classification algorithms for recognition of Alzheimer's disease in electrical brain activity (EEG). *J. Neurosci. Methods* **2007**, *161*, 342–350. [CrossRef]
28. Cai, H.; Sha, X.; Han, X.; Wei, S.; Hu, B. Pervasive EEG diagnosis of depression using Deep Belief Network with three-electrodes EEG collector. In Proceedings of the 2016 IEEE International Conference on Bioinformatics and Biomedicine, BIBM 2016, Shenzhen, China, 15–18 December 2016; pp. 1239–1246.
29. Hosseinifard, B.; Moradi, M.H.; Rostami, R. Classifying depression patients and normal subjects using machine learning techniques and nonlinear features from EEG signal. *Comput. Methods Programs Biomed.* **2013**, *109*, 339–345. [CrossRef]
30. Gemein, L.; Schirrmeister, R.; Chrabaszcz, P.; Wilson, D.; Boedecker, J.; Schulze-Bonhage, A.; Hutter, F.; Ball, T. Machine-learning-based diagnostics of EEG pathology. *NeuroImage* **2020**, *220*, 117021. [CrossRef]
31. Garn, H.; Coronel, C.; Waser, M.; Caravias, G.; Ransmayr, G. Differential diagnosis between patients with probable Alzheimer's disease, Parkinson's disease dementia, or dementia with Lewy bodies and frontotemporal dementia, behavioral variant, using quantitative electroencephalographic features. *J. Neural Transm.* **2017**, *124*, 569–581. [CrossRef] [PubMed]
32. Vecchio, F.; Miraglia, F.; Alú, F.; Orticoni, A.; Judica, E.; Cotelli, M.; Rossini, P.M. Contribution of Graph Theory Applied to EEG Data Analysis for Alzheimer's Disease Versus Vascular Dementia Diagnosis. *J. Alzheimers Dis.* **2021**, *82*, 871–879. [CrossRef]
33. Ieracitano, C.; Mammone, N.; Hussain, A.; Morabito, F.C. A Convolutional Neural Network based self-learning approach for classifying neurodegenerative states from EEG signals in dementia. In Proceedings of the 2020 International Joint Conference on Neural Networks (IJCNN), Glasgow, UK, 19–24 July 2020; pp. 1–8. [CrossRef]
34. Matias-Guiu, J.A.; Suárez-Coalla, P.; Pytel, V.; Cabrera-Martín, M.N.; Moreno-Ramos, T.; Delgado-Alonso, C.; Delgado-Álvarez, A.; Matías-Guiu, J.; Cuetos, F. Reading prosody in the non-fluent and logopenic variants of primary progressive aphasia. *Cortex* **2020**, *132*, 63–78. [CrossRef]
35. SCNN. Makoto's Preprocessing Pipeline. 2021. Available online: https://sccn.ucsd.edu/wiki/Makoto's_preprocessing_pipeline (accessed on 21 June 2021).
36. SCNN. CleanLine. 2021. Available online: https://github.com/sccn/cleanline (accessed on 21 June 2021).
37. Maturana-Candelas, A.; Gómez, C.; Poza, J.; Ruiz-Gómez, S.J.; Hornero, R. Inter-band Bispectral Analysis of EEG Background Activity to Characterize Alzheimer's Disease Continuum. *Front. Comput. Neurosci.* **2020**, *14*, 70. [CrossRef]
38. Beharelle, A.R.; Small, S.L. Chapter 64-Imaging Brain Networks for Language: Methodology and Examples from the Neurobiology of Reading. In *Neurobiology of Language*; Hickok, G., Small, S.L., Eds.; Academic Press: San Diego, CA, USA, 2016; pp. 805–814. [CrossRef]
39. Smailovic, U.; Jelic, V. Neurophysiological Markers of Alzheimer's Disease: Quantitative EEG Approach. *Neurol. Ther.* **2019**, *8*, 37–55. [CrossRef] [PubMed]
40. Cheong, L.C.; Sudirman, R.; Hussin, S. Feature extraction of EEG signal using wavelet transform for autism classification. *ARPN J. Eng. Appl. Sci.* **2015**. *10*, 8533–8540.
41. Mulders, P.; Eijndhoven, P.; Beckmann, C. *Identifying Large-Scale Neural Networks Using fMRI*; Academic Press: Cambridge, MA, USA, 2016; pp. 209–237. [CrossRef]
42. Jalili, M.; Knyazeva, M.G. Constructing brain functional networks from EEG: Partial and unpartial correlations. *J. Integr. Neurosci.* **2011**, *10*, 213–232. [CrossRef]
43. Prakash, B.; Baboo, G.K.; Baths, V. A Novel Approach to Learning Models on EEG Data Using Graph Theory Features-A Comparative Study. *Big Data Cogn. Comput.* **2021**, *5*, 39. [CrossRef]
44. Wadhera, T. Brain network topology unraveling epilepsy and ASD Association: Automated EEG-based diagnostic model. *Expert Syst. Appl.* **2021**, *186*, 115762. [CrossRef]
45. Snaedal, J.; Johannesson, G.H.; Gudmundsson, T.E.; Blin, N.P.; Emilsdottir, A.L.; Einarsson, B.; Johnsen, K. Diagnostic accuracy of statistical pattern recognition of electroencephalogram registration in evaluation of cognitive impairment and dementia. *Dement. Geriatr. Cogn. Disord.* **2012**, *34*, 51–60. [CrossRef] [PubMed]

46. Lindau, M.; Jelic, V.; Johansson, S.E.; Andersen, C.; Wahlund, L.O.; Almkvist, O. Quantitative EEG abnormalities and cognitive dysfunctions in frontotemporal dementia and Alzheimer's disease. *Dement. Geriatr. Cogn. Disord.* **2003**, *15*, 106–114. [CrossRef]
47. Caso, F.; Cursi, M.; Magnani, G.; Fanelli, G.; Falautano, M.; Comi, G.; Leocani, L.; Minicucci, F. Quantitative EEG and LORETA: Valuable tools in discerning FTD from AD? *Neurobiol. Aging* **2012**, *33*, 2343–2356. [CrossRef]
48. Tzimourta, K.D.; Christou, V.; Tzallas, A.T.; Giannakeas, N.; Astrakas, L.G.; Angelidis, P.; Tsalikakis, D.; Tsipouras, M.G. Machine Learning Algorithms and Statistical Approaches for Alzheimer's Disease Analysis Based on Resting-State EEG Recordings: A Systematic Review. *Int. J. Neural Syst.* **2021**, *31*, 2130002. [CrossRef]
49. Winters-Hilt, S.; Merat, S. SVM clustering. *BMC Bioinform.* **2007**, *8* (Suppl. 7), S18. [CrossRef]
50. Hastie, T.; Tibshirani, R.; Friedman, J. *The Elements of Statistical Learning*; Springer Series in Statistics; Springer New York Inc.: New York, NY, USA, 2001.
51. Vecchio, F.; Miraglia, F.; Alù, F.; Menna, M.; Judica, E.; Cotelli, M.; Rossini, P.M. Classification of Alzheimer's Disease with Respect to Physiological Aging with Innovative EEG Biomarkers in a Machine Learning Implementation. *J. Alzheimers Dis.* **2020**, *75*, 1253–1261. [CrossRef]
52. Gaubert, S.; Houot, M.; Raimondo, F.; Ansart, M.; Corsi, M.C.; Naccache, L.; Sitt, J.D.; Habert, M.O.; Dubois, B.; De Vico Fallani, F.; et al. A machine learning approach to screen for preclinical Alzheimer's disease. *Neurobiol. Aging* **2021**, *105*, 205–216. [CrossRef] [PubMed]
53. Pytel, V.; Cabrera-Martin, M.; Delgado-Alvarez, A.; Ayala, J.; Balugo, P.; Delgado-Alonso, C.; Yus, M.; Carreras, M.; Carreras, J.; Matias-Guiu, J.; et al. Personalized repetitive transcranial magnetic stimulation for primary progressive aphasia. *J. Alzheimers Dis.* **2021**. [CrossRef] [PubMed]
54. Matias-Guiu, J.A.; Díaz-Álvarez, J.; Cuetos, F.; Cabrera-Martín, M.N.; Segovia-Ríos, I.; Pytel, V.; Moreno-Ramos, T.; Carreras, J.L.; Matías-Guiu, J.; Ayala, J.L. Machine learning in the clinical and language characterisation of primary progressive aphasia variants. *Cortex* **2019**, *119*, 312–323. [CrossRef] [PubMed]

Review

Longitudinal Changes in Cognition, Behaviours, and Functional Abilities in the Three Main Variants of Primary Progressive Aphasia: A Literature Review

Justine de la Sablonnière, Maud Tastevin, Monica Lavoie and Robert Laforce, Jr. *

Clinique Interdisciplinaire de Mémoire, Département des Sciences Neurologiques du CHU de Québec, Faculté de Médecine, Université Laval, Quebec City, QC G1J 1Z4, Canada; justine.de-la-sablonniere.1@ulaval.ca (J.d.l.S.); maud.tastevin.1@ulaval.ca (M.T.); monica.lavoie.1@ulaval.ca (M.L.)
* Correspondence: robert.laforce@fmed.ulaval.ca; Tel.: +1-418-649-5980

Abstract: Primary progressive aphasias (PPAs) are a group of neurodegenerative diseases presenting with insidious and relentless language impairment. Three main PPA variants have been described: the non-fluent/agrammatic variant (nfvPPA), the semantic variant (svPPA), and the logopenic variant (lvPPA). At the time of diagnosis, patients and their families' main question pertains to prognosis and evolution, but very few data exist to support clinicians' claims. The objective of this study was to review the current literature on the longitudinal changes in cognition, behaviours, and functional abilities in the three main PPA variants. A comprehensive review was undertaken via a search on PUBMED and EMBASE. Two authors independently reviewed a total of 65 full-text records for eligibility. A total of 14 group studies and one meta-analysis were included. Among these, eight studies included all three PPA variants. Eight studies were prospective, and the follow-up duration was between one and five years. Overall, svPPA patients showed more behavioural disturbances both at baseline and over the course of the disease. Patients with lvPPA showed a worse cognitive decline, especially in episodic memory, and faster progression to dementia. Finally, patients with nfvPPA showed the most significant losses in language production and functional abilities. Data regarding the prodromal and last stages of PPA are still missing and studies with a longer follow-up observation period are needed.

Keywords: primary progressive aphasia; natural history; longitudinal assessment; cognitive changes; behavioural and psychological symptoms of dementia; level of functioning

Citation: de la Sablonnière, J.; Tastevin, M.; Lavoie, M.; Laforce, R., Jr. Longitudinal Changes in Cognition, Behaviours, and Functional Abilities in the Three Main Variants of Primary Progressive Aphasia: A Literature Review. *Brain Sci.* **2021**, *11*, 1209. https://doi.org/10.3390/brainsci11091209

Academic Editors: Nadine Martin and Paola Marangolo

Received: 12 July 2021
Accepted: 9 September 2021
Published: 14 September 2021

Publisher's Note: MDPI stays neutral with regard to jurisdictional claims in published maps and institutional affiliations.

1. Introduction

Primary progressive aphasias (PPA) are a group of neurodegenerative diseases that present with an insidious, progressive, and isolated impairment in language. Other cognitive functions are typically preserved for at least two years after the onset of the disease [1]. Mesulam (1982) was the first to describe six cases of progressive aphasia without accompanying signs of dementia and associated with focal perisylvian left atrophy [2]. A few years later, Snowden et al. (1989) introduced the term "semantic dementia" referring to dementia with profound loss of conceptual knowledge [3]. Afterward, Neary et al. (1998) published diagnostic criteria for progressive non-fluent aphasia and semantic dementia [4], and in 2004, a third type of PPA was described—the logopenic variant primary progressive aphasia [5]. More recently, diagnostic criteria for three main variants of PPA have been identified by Gorno-Tempini et al. (2011) [6]. The classification is based on language features and can be supported by the pattern of atrophy found on neuroimaging and pathological examination. We used the classification from Gorno-Tempini et al. (2011) as a framework for this study, but it is noteworthy that other clinical diagnoses and mixed cases exist even if not in the scope of this review (e.g., primary progressive apraxia of speech [7]).

According to the criteria of Gorno-Tempini et al. (2011), the non-fluent/agrammatic variant (nfvPPA) is characterized by the presence of agrammatism and/or apraxia of speech.

Table 1. Observational studies exploring some elements of the longitudinal changes in cognition, behaviours, and functional abilities in the main PPA variants.

Authors (Year)	Participants	Study Design	Follow-Up	Clinical Assessment	Main Results
Rogalski et al. (2011) [24]	lvPPA: $n = 6$ nfvPPA: $n = 3$ svPPA: $n = 4$ HC: $n = 27$	Prospective group study	1 FU at 2 years	(a) Cognition: Clinical judgment, behavioural scales, neuropsychological tests (b) Language: NAT, PPVT, WAB-AQ, PASS, BNT (c) Imaging: MRI	(a) Initial clinical distinctive neuropsychological patterns become blurred at follow-up. (b) Persistence of differential impairment of word comprehension in svPPA and grammatical processing in nfvPPA. For lvPPA, marked decline in naming ability. (c) No correlation between loss of cortical volume and clinical progression of aphasia. Preservation of lateralization to left hemisphere.
Hsieh et al. (2012) [25]	lvPPA: $n = 9$ nfvPPA: $n = 12$ svPPA: $n = 17$ AD: $n = 17$	Retrospective group study	Two assessments at least 12 months apart	(a) Cognition: ACE-R, FRS	(a) Faster decline in PPA than in AD, but no difference between variants. Longer time between symptoms onset and clinical diagnosis for svPPA compared to nfvPPA and AD.
Leyton et al. (2013) [26]	lvPPA: $n = 13$ svPPA: $n = 11$ HC: $n = 17$	Prospective group study	Yearly Mean duration of 3 years	(a) Cognition: MMSE, ACE-R (b) Language: Confrontation naming, single-word comprehension and repetition	(a) 3x greater decline in lvPPA for ACE-R and MMSE, the most rapid decline being in attention and visuospatial domains. lvPPA: Global impairment (meeting criteria for dementia) by 12 months. svPPA: Impairments confined to verbally mediated tasks (sparing visuospatial domain) for up to 3 years. (b) Duration of symptoms had an effect on memory and naming performances, with no differences between PPA groups.
Linds et al. (2015) [27]	nfvPPA + svPPA: $n = 13$ bvFTD: $n = 30$ AD: $n = 118$	Retrospective group study	Every year	(a) BPSD: FBI, NPI, FTLD-CDR (b) LOF: FRS	(a) Education was a predictor for ROC on the FBI-disinhibition subscale. FBI total and its sub-scale scores for apathy and disinhibition correlated with duration of illness. (b) LOF only studied at baseline.

Table 1. *Cont.*

Authors (Year)	Participants	Study Design	Follow-Up	Clinical Assessment	Main Results
Matias-Guiu et al. (2015) [28]	lvPPA: $n = 17$ nfvPPA: $n = 12$ svPPA: $n = 4$ Unclassified: $n = 2$ HC: $n = 16$	Prospective group study	Every 4 to 6 months Mean length unknown	(a) Cognition: MMSE, ACE-R (b) Language: BNT, letter/word verbal fluency, BDAE («Cookie Theft» picture), Barcelona Test (language subtests), PASS (c) LOF: IDDD, FAQ	(a) 74.3% developed a non-language symptom or deficit (PPA-plus). (b) Median time between onset and PPA-plus = 36 months (c) nfvPPA: Parkinsonism, behavioural disorder and motor neuron disease. (d) lvPPA: Memory or global impairment. (e) svPPA: Behavioural disorder. (f) Right laterality and years of education associated with lower risk of progression to PPA-plus while lvPPA is associated with higher risk.
Gómez-Tortosa et al. (2016) [29]	nfvPPA: $n = 39$ svPPA: $n = 41$	Retrospective group study	Biannual Mean length = 5 years	(a) BPSD: NPI-Q, pharmacotherapy	(a) No differences in first behavioural assessments. (b) At last assessment: svPPA: higher frequency and intensity of agitation and higher frequency of delirium/hallucinations. Greater need for antipsychotics ($p = 0.001$), 49% of patients. nfvPPA: higher frequency of depression. Greater need for antidepressants.
O'Connor et al. (2016) [30]	nfvPPA: $n = 11$ svPPA: $n = 18$	Prospective group study	Baseline and one FU at mean 1.4 years	(a) Cognition: ACE-R (b) BPSD: CBI-R (c) LOF: DAD	(a) Greater memory impairment at baseline in svPPA. (b) More stereotypical behaviour at baseline in svPPA. (c) Similar decline in functional score in both groups. svPPA: Functional and cognitive scores at baseline are predictors of functional decline. nfvPPA: Functional score at baseline is a predictor of functional decline. Functional abilities remained virtually intact up to 5 years from disease onset while behavioural changes were present from an early stage.

Table 1. *Cont.*

Authors (Year)	Participants	Study Design	Follow-Up	Clinical Assessment	Main Results
Van Langenhove et al. (2016) [31]	lvPPA: $n = 21$ nfvPPA: $n = 22$ svPPA: $n = 30$ bvFTD: $n = 33$ AD: $n = 31$	Prospective group study	1 FU at a mean of 12 months	(a) Cognition: CDR, ACE-III, CBI-R (b) BPSD: CBI-R (c) LOF: CBI-R	(a) Baseline: Memory impairment lvPPA > nfvPPA. Follow-up: Memory remains less impaired for nfvPPA. (b) Baseline: Prevalence = svPPA > nfvPPA > lvPPA. In svPPA, mostly stereotypical behaviour, empathy loss and apathy. In nfvPPA and lvPPA, mostly apathy. Follow-up: >70% developed a clinically relevant change in at least one behavioural symptom. Apparition of behaviour changes in 38 to 50% patients. Hallucinations and delusions remained rare in all groups. (c) Baseline: Similar level of impairment for daily activities across PPA except for greater impairment in everyday skills in lvPPA. Follow-up: Decline in everyday skills less pronounced in svPPA.
Ash et al. (2019) [32]	lvPPA: $n = 14$ nfvPPA: $n = 9$ svPPA: $n = 11$ bvFTD: $n = 14$ HC: $n = 36$	Prospective group study	1 FU at a mean of 26 months	(a) Cognition: MMSE, FDS, RDS, (b) Language: BDAE («Cookie Theft» picture), BNT, phonemic and semantic fluency	(a) Decline in global cognition in all variants. For nfvPPA and bvFTD, significant decline on MMSE only. (b) Decline in language production over time in all variants but more so in nfvPPA. No difference in rate of decline in language between variants. No correlation between decline in cognition and language.
Ferrari et al. (2019) [33]	lvPPA: $n = 23$ nfvPPA: $n = 26$ svPPA: $n = 19$	Retrospective group study	$M = 2.06$ years Frequency unknown	(a) Cognition: MMSE (b) BPSD: NPI (c) LOF: BADL, IADL (d) ApoE4 status	(a) Mean loss of 4 points at 1 year and 9 at 2 years. (b) No influence of BDSP on disease progression. (c) Severe functional dependency in 20% at 2.5 years. Cognitive decline in 1st year is a risk factor for functional impairment while high education is protective. (d) Cognitive decline associated with ApoE4 status. Higher prevalence of mutism in ApoE4 patients.

Table 1. *Cont.*

Authors (Year)	Participants	Study Design	Follow-Up	Clinical Assessment	Main Results
Funayama et al. (2019) [34]	lvPPA: $n = 10$	Prospective group study	Every year Duration 6 to 10 years post onset	(a) Cognition: CDR (b) Language: Standard Language Test of Aphasia (c) BPSD: NM scale	(a) Decline in CDR of 3.4 points/year, change of dementia severity every 1.7 year. 4.1 year to reach CDR 1 (mild dementia), 5.7 years to DCR 2 (moderate), and 7.3 years to CDR 3 (severe). Dementia progression parallels linguistic decline. Difficulties with using electronic appliances began 3.3 years post onset, episodic memory deficits 4 years post-onset, and topographical disorientation 5.2 years. 60% could not recognize family members, 50% with pica, 30% with mirror sign (visuospatial deficits and body schema disorder).
Cosseddu et al. (2020) [35]	nfvPPA: $n = 77$ svPPA: $n = 40$ bvFTD: $n = 286$	Retrospective and prospective group study	Every year Mean length = 3.1 years	(a) Cognition: FTLD-CDR (b) BPSD: FBI, NPI	(a) Increase in negative symptoms with disease severity in bvFTD and PPA. (b) Increase in positive symptoms until intermediate phases, followed by reduction in later phases. Positive symptoms less common in nfvPPA.
Foxe et al. (2021) [36]	lvPPA: $n = 41$ nfvPPA: $n = 44$ svPPA: $n = 62$ HC: $n = 60$	Prospective group study	FU every year	(a) Cognition: ACE-III or ACE-R, WAIS-III (b) LOF: DAD	(a) Decline in overall cognition in all three variants but twice as rapid rate in lvPPA than nfvPPA and svPPA, Faster decline across the majority of cognitive domains in lvPPA. lvPPA: Worst performance on verbal fluency and memory domains at all time points. Attention and language higher at baseline but declined faster than all other subdomains. Greater decline than svPPA in memory and language subdomains but no difference with nfvPPA. nfvPPA: Disproportionate impairment in verbal fluency at all time points compared to other domains. Faster decline for language and memory. svPPA: Greater impairments in verbal fluency, language, and memory than other subdomains. (b) Faster rate of decline for lvPPA and nfvPPA compared to svPPA. Correlation between functional and cognitive decline for all groups across all time periods. Impact of cognition on functional capacity greater for nfvPPA at most time points.

111

Clinical Assessment

All included observational studies used validated international scales to evaluate the different aspects of cognition, language, autonomy, and behavioural and psychiatric symptoms of dementia (BPSD). Cognition was the most frequently assessed clinical aspect and was included in 12 studies. The Addenbrooke's Cognitive Examination (ACE) and its subsequent versions (III and revised) [38,39], as well as the Mini-Mental State Examination (MMSE) [40], were the most frequently used tools, in six and five studies, respectively. Other cognitive evaluation tools used were the Clinical Dementia Rating Scale (CDR) and its modified version for Frontotemporal Lobar Degeneration (FTLD-CDR) in four studies [41,42]. Some of the less frequently used tools were the Cambridge Behavioural Inventory-Revised (CBI-R) [43] and the Wechsler Adult Intelligence Scale (WAIS) [44]. Some subtests of these different neuropsychological batteries were also used individually. Although language was invariably evaluated at baseline, only five studies reported longitudinal assessment of language. The Boston Naming Test (BNT) was the most frequently used test, in four studies [45]. The Progressive Aphasia Severity Scale (PASS) [46] was used in two studies. The other tests were used in only one study each and are presented in Table 1. Language domains assessed varied across studies and included phonemic and semantic fluency, confrontation naming, comprehension, reading, writing, and repetition.

Eight studies evaluated the onset and evolution of BPSD. The most frequently used test was the Neuropsychiatric Inventory (NPI), a semi-structured clinician interview of caretakers [47] in five articles. The Cambridge Behavioural Inventory-Revised (CBI-R) [43] and the Frontal Behavioural Inventory (FBI) [48] were both used in two studies each. Level of functioning in basic activities of daily living (BADL) and instrumental activities of daily living (IADL) were assessed in seven studies, directly or through clinical questionnaires. The most frequently used tool was the disability assessment for dementia (DAD) [49], in three studies. One study used the Functional Activities Questionnaire (FAQ) [50], which was recently proven to be a useful functional measure for longitudinal changes in FTD [51].

Some studies also looked at other variables not included in the scope of this review such as neuroimaging patterns and progression of atrophy, development of parkinsonian syndromes, genetics, and pharmacotherapy.

4. Discussion

We proposed herein the first comprehensive review of longitudinal changes in cognition, language, BPSD, and functional abilities in PPA. A total of 14 observational studies were included in this review, as well as a meta-analysis studying survival. General findings of PPAs will first be discussed, followed by specific findings for each variant and then survival data. Finally, limitations of the current work and the impact of the findings on clinical care and future perspectives will be addressed.

4.1. Similarities between All Three Main Variants of PPA

In all three PPA variants, studies that assessed cognition reported a decline over time. Previous studies have indeed demonstrated that even if language is primarily affected in PPA, other cognitive functions are impaired as well [5,52–54]. Regardless of the variant, two studies described a faster decline in cognition for PPAs, when compared to AD [25,34]. Indeed, Funayama et al. (2019) reported an annual rate change in the CDR sum of boxes of 3.4 ± 1.1 in their group of 10 lvPPA patients. This is a greater rate of decline than what Doody et al. (2010) previously reported in their group of 597 AD patients [55]. In Hsieh et al. (2012), the annualized rate of change was greater in all three PPA variants (9 lvPPA, 12 nfvPPA, and 17 svPPA patients) when compared to the AD group (17 patients) on the ACE-R. Over a year, the PPA patients lost on average 10 points, as compared to less than 5 by AD patients. However, these findings must be interpreted with caution since several neurocognitive tests are influenced by language abilities. Clinically, it is common to see PPA patients with lower scores on cognitive testing that do not correspond to the level of functioning on collateral history. Therefore, specific assessment and neuropsychological tools that take into account language impairments should be used with PPA patients [56]. Only one study included in this review explored the correlation between dementia progression and decline in language [32], but no such association was found. On the other hand, Funayama et al. (2019) did describe a relationship between dementia progression and language decline, although no statistical analysis was performed.

Among the few studies which evaluated changes in language, a decline was described in all variants. Ferrari et al. (2019) reported mutism in 31% of patients at 2.7 years. Although the MMSE score and fluent language at baseline were previously described as protective factors for mutism [57], these relationships were not established in the study by Ferrari et al. (2019).

There were discordant findings among the studies regarding BPSD. One study described no influence of behavioural and psychiatric symptoms at baseline on disease progression [33]. Conversely, another study found that apathy and stereotypical behaviour at baseline were predictors of functional decline for nfvPPA [30]. Linds et al. (2015) showed that education was a protective factor for disinhibition, but this finding was not indicated in other publications [27]. The main limitation regarding BPSD is the heterogeneity of the symptoms, and NPI is often limited in its description. Moreover, BPSD fluctuation over time could complicate the interpretation of the results. BPSD assessment would require a longer follow-up, with several validated scales and consideration of qualitative data for a more exhaustive list of symptoms [58].

Regarding the level of functioning in IADL and BADL, all three variants showed a decline at follow-up. In Foxe et al. (2021), over a period of four years, the mean DAD total scores with 95% confidence intervals, decreased from 82.7 (76.1–89.3) to 48 (39.5–56.4) for lvPPA patients, from 86 (79.4–92.5) to 51.3 (43.4–59) for nfvPPA patients, and from 85 (79.8–90.1) to 55.6 (50.3–60.9) in svPPA patients. One study demonstrated a direct link between the decline in functioning and cognition, with the relationship increasing over time [36]. This link was already described in a previous study of 2009 studying changes in functioning and cognition in 9 nfvPPA and 11 svPPA patients [59]. Moreover, cognition at baseline and its deterioration in the first year were predictive factors of greater functional incapacities throughout the course of the disease [30]. This is coherent with two previous studies; one showed that a higher MMSE score at baseline was a predictor of preservation of autonomy in the following years [57], while the other, conversely, correlated lower MMSE and FTLD-CDR scores at baseline with more rapid change overtime in functional measures for nfvPPA and svPPA, respectively [51]. This finding is not surprising and could be explained by the fact that patients with lower functional and cognitive scores at baseline have either a more aggressive disease or a lower cognitive reserve [60,61]. Education also seemed to be a protective factor for functional impairment [28,33]. Indeed, patients with higher education are most likely to have a higher cognitive reserve and therefore to be able to compensate longer in IADL and BADL. In their study, Ferrari et al. (2019) reported a severe functional dependency in 20% of the patients at 2.5 years. Other data from the literature reported a need for assistance in BADL in 50% of the patients at five years [57]. This is in contrast with the findings of O'Connor et al. (2016), who described a sparing of functional abilities for five years from onset. However, in this study, which included only nfvPPA and svPPA patients, only one tool was used for assessment of functioning, and one could argue that more extensive deficits would have been detected with a more comprehensive evaluation.

4.2. Non-Fluent Variant of Primary Progressive Aphasia

Four studies revealed that nfvPPA patients showed a greater decline in language production over time. In Ash et al. (2019), the decline in language production was more important for fluency and grammar, whereas in Rogalski et al. (2011), participants showed a decline in all language domains, with each of the three patients being too impaired to complete at least one of the different measures [24,32]. Similarly, Ulugut et al. (2021) found that out of eight patients who displayed mutism at follow-up, seven of them were classified as nfvPPA variant [37]. Finally, in Foxe et al. (2021), these patients showed a disproportionate impairment in verbal fluency at all time points and a faster decline in language during follow-up [36]. This decline in speech production is explained by the progression of the atrophy in the left frontal and subcortical areas, regions that are important networks for language production [62–64]. These findings are coherent with a previous study in which the nfvPPA patients were not able to complete ACE-R at one year of follow-up, due to language deterioration [59].

Regarding behavioural and psychiatric symptoms, nfvPPA patients have a tendency towards negative symptoms [36,59]. Indeed, they showed a higher frequency of depression and a greater need for antidepressants, as opposed to antipsychotics [29]. In their study, Van Langenhove et al. (2016) reported that apathy was the most prominent symptom, present in 46% of nfvPPA patients at baseline and increased to 68% at follow-up.

There also seems to be a tendency in this variant for a faster decline in the level of functioning. Hsieh et al. (2012) reported a faster decline at the Frontotemporal Dementia Rating Scale (FRS), an assessment tool measuring, among others, changes in everyday abilities such as using the phone and taking medication, in the nfvPPA group than in AD patients. Indeed, over 12 months, among the group of 12 nfvPPA patients, over 80% of patients showed a decline in the FRS. Similarly, in Foxe et al. (2021), nfvPPA patients had a faster decline in the level of functioning, compared to svPPA, with an annual rate of decline on the DAD total score of 8.7 points, compared to 7.4 points. Another study revealed that nfvPPA and lvPPA had a worse decline in daily life activities at a one-year follow-up, with the first group having the greatest impairments in self-care [31]. In line with these findings, Mioshi et al. (2009) had previously reported that nfvPPA patients showed significant changes both in BADL and IADL at follow-up [59].

4.3. Semantic Variant of Primary Progressive Aphasia

Specific findings regarding BPSD were highlighted in svPPA patients. First, they tended to show behavioural symptoms earlier in the disease course and more frequently, compared to the other variants, as well as AD and behavioural variant of frontotemporal dementia (bvFTD) [28,33,37]. Indeed, Van Langenhove et al. (2016) found that 74% of svPPA patients had behavioural changes at baseline, compared to 54% of nfvPPA patients and 47% lvPPA patients. At follow-up, the tendency remained with 80% of svPPA patients showing at least one behavioural symptom. In the study by Matias-Guiu et al. (2015), half of the svPPA patients (two out of four) developed behavioural disorders. The most frequent disturbances were stereotypical behaviour, empathy loss, and apathy. These findings are consistent with the study by O'Connor et al. (2016), which included 18 svPPA patients and found that patients with this variant displayed more stereotypical behaviour at baseline (60% vs. 9% in nfvPPA). Moreover, in Ulugut et al. (2021), 58% of svPPA patients (group of 24) eventually met diagnostic criteria for bvFTD. Compared to nfvPPA, svPPA patients also showed a higher frequency of agitation and delirium/hallucinations [29]. There was a significant difference in the severity of irritability, agitation, delirium, and apathy and a greater need for antipsychotic drugs. Increased behavioural dysfunctions in svPPA, especially disinhibition, were already underlined in the literature [65,66]. Heterogeneity in the results could, in part, be explained by a misdiagnosis of the right temporal variant of FTLD, also called the right semantic variant. It is possible that the course of the disease in the left svPPA and right semantic variant could be significantly different. A recent study showed that prosopagnosia, episodic memory impairment, and behavioural changes such as disinhibition, apathy, compulsiveness, and loss of empathy were the most common initial symptoms for the right temporal variant, whereas, during the disease course, patients developed language problems such as word-finding difficulties and anomia [67]. Distinctive symptoms of the right semantic variant, compared to the other groups, included depression, somatic complaints, motor, and mental slowness.

Interestingly, a few studies suggested that svPPA patients had a longer duration of symptoms before the diagnosis. In Van Langenhove et al. (2016), symptoms duration at baseline was 4.4 years for svPPA, compared with 2.3 and 3.5 years in nfvPPA and lvPPA, respectively. In Hseish et al. (2012), svPPA patients had a mean disease duration of 4.2 years at the time of diagnosis, compared with 2.3 and 3.9 years for nfvPPA and lvPPA. This tendency for a longer duration of symptoms in the svPPA variant, although not always statistically significant, was also reported in other studies [66,68]. This could be explained by the fact that the loss of semantic knowledge can be masked by word-finding difficulties (e.g., vague words, circumlocutions) and therefore may be overlooked by family members, which, in turn, delays recognition of the syndrome. In contrast, nfvPPA has a more striking presentation with agrammatism and halting speech. Furthermore, lvPPA variants present with impaired single-word retrieval in spontaneous speech, which is more frequently recognized by family members as an early sign of dementia.

4.4. Logopenic Variant of Primary Progressive Aphasia

Five studies revealed that the lvPPA patients showed a worse decline in global cognition, compared to other variants [26,28,31,36,37]. Foxe et al. (2021) found that lvPPA patients had a twice as rapid decline rate in overall cognition, despite performing intermediate to the other variants at baseline. Funayama et al. (2019) also showed this tendency to faster progression. In their study, lvPPA patients evaluated with the clinical dementia rating sum of boxes, had a change in dementia severity every 1.7 years and reached severe dementia (CDR 3) in 7.3 ± 1.6 years, a faster progression, compared to Alzheimer's disease [55]. In Ulugut et al. (2021), 83% of lvPPA patients (group of 18) acquired global cognitive impairment consistent with Alzheimer's disease dementia. Moreover, among cognitive skills, memory seemed to be the most frequently and severely affected ability, as demonstrated in four studies [28,31,36,37]. The 10 lvPPA patients in the Funayama et al. study (2019) showed episodic memory deficits beginning at 4.0 ± 2.0 years after onset. These findings are not surprising considering that the logopenic variant is most frequently associated with Alzheimer's pathology [8,69]. Studies that included thorough imaging analysis also showed a greater cognitive decline in lvPPA, associated with progression of brain atrophy in the regions typically damaged in AD [62]. Level of functioning and BPSD were less studied in lvPPA than in the two other variants. Foxe et al. (2021) found a faster rate of decline in the level of functioning in lvPPA in comparison to svPPA with an annual rate of decline on the DAD total score of 8.7 and 7.4 points, respectively, within their groups of 41 lvPPA and 62 svPPA. As for BPSD, they were found to be less prevalent in lvPPA than in the two other variants, with apathy being the most frequent [31].

4.5. Survival Data

The meta-analysis by Kansal et al. (2016) was the only study, to our knowledge, which addressed survival in PPA. In total, 27 studies focusing on survival and years of life lost (YLL) were included with patients presenting AD, corticobasal degeneration (CBD), progressive supranuclear palsy (PSP), and all FTLD variants (svPPA, nfvPPA, bvFTLD, and FTLD-ALS). In contrast to survival, which emphasizes life expectancy, YLL highlights premature mortality. YLL is, therefore, useful for quantifying premature deaths in policy contexts. The median survival in the svPPA variant was significantly longer than in nfvPPA (12 years versus 7.66 years). However, the mean survival for svPPA was estimated at 7.45 years, and 8.11 years for nfvPPA with no statistically significant difference. To explain these contradictory findings between median and mean survival findings, Kansal et al. suggested an artifact in the analysis due to heterogeneity in the included studies (sampling methods and regional context). They also raised the hypothesis that the presence of a negative or positive skew could be a statistical reflection of the survival profile. Indeed, a negative skew could reflect a young- to mid-life onset disease with a sufficiently long course and few premature deaths. A positive skew would be more likely associated with a disease characterized by a very short course, as the outliers are those with unusually long survival. Patients with nfvPPA had the longest mean survival between all the neurodegenerative diseases with a significant difference, compared to the PSP and CBD groups. Mean and median YLL estimated from survival (years) were, respectively, estimated to be 10.54 years and 8.97 years for nfvPPA, and 13.56 and 7.71 years for svPPA. The main limitations of this study were the heterogeneity of the data, the absence of lvPPA patients, the presence of uncertain associations between clinical and sociodemographic outcomes, and contradictory results in some of the studies included. Moreover, they did not examine the effects of confounding features and comorbid conditions. Causes of death were not reported either. Indeed, it would be helpful to know if death occurred before the final stages because of intercurrent medical conditions.

In this special issue, our group has provided recent insights into survival in the three PPAs. Indeed, significant differences in survival were found with svPPA showing the longest and nfvPPA showing more neurologically-related causes of death [70].

4.6. Limitations

Studies in the current review included were very heterogeneous, with none assessing longitudinal changes in cognition, language, BSPD, and functioning altogether. In fact, only six of them studied at least three of our outcomes of interest. Moreover, these clinical aspects were mostly assessed through different assessment tools (e.g., MMSE, ACE-R) and assessments by specialists such as speech–language pathologists or neuropsychologists were uncommon. Duration of follow-up was relatively short (only one or two years after diagnosis), and very few patients were followed until death. All these elements prevented us from extracting solid data about the long-term outcomes of PPA patients. Moreover, PPA being a rare type of dementia, sample sizes were relatively small, with a mean of 21 patients per variant among the studies. The lvPPA variant was also underrepresented, especially since five studies included only FTD variants (svPPA and nfvPPA). This underrepresentation of lvPPA patients could be explained by the fact that lvPPA is the most recently described variant. Indeed, no survival data were available for this variant and epidemiological information remains lacking.

The research method used in the present review also has its limitations. First, only two databases were explored, and restrictive inclusion criteria were used. Moreover, no assessment of the methodological quality of each study was performed. A systematic review and meta-analysis conducted according to the Preferred Reporting Items for Systematic Reviews and Meta-Analyses statement (PRISMA) [71] could contribute to additional and more significant outcomes. Indeed, clinical assessments of at least 745 PPA patients are currently available, with 172 lvPPA, 277 nfvPPA, 281 svPPA, and 15 unclassified PPAs. However, as some studies were conducted within the same research centres, it is possible that the actual number of participants censored was lower, therefore limiting the generalization of findings. Further studies with a longer follow-up period until death remains to be conducted. It could be particularly useful to provide a more exhaustive evaluation of the end stages of PPA, causes of death, and epidemiological data. Finally, in the near future, the focus should be on lvPPA patients whose data are actually less well represented in the literature, compared to FTLD variants.

5. Conclusions

This study sheds further light on our current understanding of the longitudinal changes in cognition, behaviours, and functional abilities in PPA variants but, most importantly, it provides useful information for patients and their families. In addition to confirming general tendencies for the evolution of PPA, our study highlights differences in the progression of the three variants with svPPA, showing more behavioural disturbances, nfvPPA progressing towards more language and functional deficits, and lvPPA displaying a worse decline in global cognition, especially memory. These findings are also relevant to prioritizing the clinical care offered to PPA patients and their caregivers by highlighting the challenges they are most likely to face. For example, education and support groups for patients and their caregivers were demonstrated as a worthy component of PPA patients' care and are indicated regardless of the variant [22,72]. However, referrals to groups aiming at managing behavioural disturbances are likely to be highly important for patients with svPPA and their caregivers, given the high prevalence of BPSD in this variant. The decline in the language is expected in all three variants, and therefore, referral to a speech–language pathologist can be useful. Indeed, previous studies highlighted the efficacy of speech and language interventions for PPA patients, as well as maintenance of the gains after the treatment period [11,19,73,74]. Even teletherapy proved to be beneficial in mild-to-moderate cases, therefore opening the door to new possibilities and better access to therapy [75]. Knowing that nfvPPA patients are most likely to have a faster language decline and to progress to mutism, implementation of compensatory communication tools such as assistive augmentative communication (AAC) devices should be a priority in the management of the disease and as early as possible so that the patient is still able to learn to use it. Indeed, the use of AAC devices can allow patients to maintain effective communication [12]. Moreover, clinical care of nfvPPA should include information to caregivers about the different options regarding home care given the faster decline in the level of functioning in this variant. Finally, given the likely memory impairment in lvPPA, patients and caregivers should be informed about ways to compensate for this deficit in their daily life.

This study also highlights important shortcomings in the literature and the need for more research on this subject, especially regarding the logopenic variant. Future studies should aim at better documenting cognitive and language functions, BPSD, and functioning in everyday life, throughout the disease, in order to improve management of PPA in clinical settings.

Author Contributions: All four authors contributed to the writing of the article: Conceptualization, J.d.l.S., M.L., M.T., and R.L.J.; writing—original draft preparation, J.d.l.S.; writing—review and editing, J.d.l.S., M.L., M.T., and R.L.J.; funding acquisition, R.L.J. All authors have read and agreed to the published version of the manuscript.

Funding: This research received funding from Chaire de Recherche sur les Aphasies Primaires Progressives—Fondation de la Famille Lemaire, grant number 0647.

Conflicts of Interest: The authors declare no conflict of interest.

References

1. Mesulam, M.-M. Primary progressive aphasia. *Ann. Neurol.* **2001**, *49*, 425–432. [CrossRef]
2. Mesulam, M.-M. Slowly progressive aphasia without generalized dementia. *Ann. Neurol.* **1982**, *11*, 592–598. [CrossRef] [PubMed]
3. Snowden, J.; Goulding, P.J.; David, N. Semantic dementia: A form of circumscribed cerebral atrophy. *Behav. Neurol.* **1989**, *2*, 167–182. [CrossRef]
4. Neary, D.; Snowden, J.; Gustafson, L.; Passant, U.; Stuss, D.; Black, S.; Freedman, M.; Kertesz, A.; Robert, P.H.; Albert, M.; et al. Frontotemporal lobar degeneration: A consensus on clinical diagnostic criteria. *Neurology* **1998**, *51*, 1546–1554. [CrossRef] [PubMed]
5. Gorno-Tempini, M.L.; Dronkers, N.F.; Rankin, K.P.; Ogar, M.J.; Phengrasamy, L.; Rosen, H.J.; Johnson, J.K.; Weiner, M.W.; Miller, B.L. Cognition and anatomy in three variants of primary progressive aphasia. *Ann. Neurol.* **2004**, *55*, 335–346. [CrossRef]
6. Gorno-Tempini, M.L.; Hillis, A.E.; Weintraub, S.; Kertesz, A.; Mendez, M.; Cappa, S.F.; Ogar, J.M.; Rohrer, J.D.; Black, S.; Boeve, B.F.; et al. Classification of primary progressive aphasia and its variants. *Neurology* **2011**, *76*, 1006–1014. [CrossRef]
7. Botha, H.; Josephs, K.A. Primary Progressive Aphasias and Apraxia of Speech. *Contin. Lifelong Learn. Neurol.* **2019**, *25*, 101–127. [CrossRef]
8. Bergeron, D.; Gorno-Tempini, M.L.; Rabinovici, G.D.; Santos-Santos, M.A.; Seeley, W.; Miller, B.L.; Pijnenburg, Y.; Keulen, M.A.; Groot, C.; Van Berckel, B.N.M.; et al. Prevalence of amyloid-β pathology in distinct variants of primary progressive aphasia. *Ann. Neurol.* **2018**, *84*, 729–740. [CrossRef]
9. Botha, H.; Duffy, J.R.; Whitwell, J.L.; Strand, E.A.; Machulda, M.M.; Schwarz, C.; Reid, R.I.; Spychalla, A.J.; Senjem, M.L.; Jones, D.T.; et al. Classification and clinicoradiologic features of primary progressive aphasia (PPA) and apraxia of speech. *Cortex* **2015**, *69*, 220–236. [CrossRef]
10. Perry, D.C.; Datta, S.; Miller, Z.; Rankin, K.P.; Gorno-Tempini, M.L.; Kramer, J.H.; Rosen, H.J.; Seeley, W.W.; Miller, B.L. Factors that predict diagnostic stability in neurodegenerative dementia. *J. Neurol.* **2019**, *266*, 1998–2009. [CrossRef]
11. Volkmer, A.; Rogalski, E.; Henry, M.; Taylor-Rubin, C.; Ruggero, L.; Khayum, R.; Kindell, J.; Gorno-Tempini, M.L.; Warren, J.D.; Rohrer, J.D. Speech and language therapy approaches to managing primary progressive aphasia. *Pract. Neurol.* **2020**, *20*, 154–161. [CrossRef]

67. Ulugut Erkoyun, H.; Groot, C.; Heilbron, R.; Nelissen, A.; Van Rossum, J.; Jutten, R.; Koene, T.; Van Der Flier, W.M.; Wattjes, M.P.; Scheltens, P.; et al. A clinical-radiological framework of the right temporal variant of frontotemporal dementia. *Brain* **2020**, *143*, 2831–2843. [CrossRef] [PubMed]

68. Sebastian, R.; Thompson, C.B.; Wang, N.-Y.; Wright, A.; Meyer, A.; Friedman, R.B.; Hillis, A.E.; Tippett, D.C. Patterns of decline in naming and semantic knowledge in primary progressive aphasia. *Aphasiology* **2017**, *32*, 1010–1030. [CrossRef] [PubMed]

69. Rohrer, J.; Rossor, M.; Warren, J.D. Alzheimer's pathology in primary progressive aphasia. *Neurobiol. Aging* **2012**, *33*, 744–752. [CrossRef]

70. Tastevin, M.; Lavoie, M.; de la Sablonnière, J.; Carrier-Auclair, J.; Laforce, R., Jr. Survival in the Three Common Variants of Primary Progressive Aphasia: A Retrospective Study in a Tertiary Memory Clinic. *Brain Sci.* **2021**, *11*, 1113. [CrossRef]

71. Moher, D.; Liberati, A.; Tetzlaff, J.; Altman, D.G.; Prisma Group. Preferred reporting items for systematic reviews and meta-analyses: The PRISMA statement. *BMJ* **2009**, *339*, b2535. [CrossRef]

72. Morhardt, D.J.; O'Hara, M.C.; Zachrich, K.; Wieneke, C.; Rogalski, E.J. Development of a Psycho-Educational Support Program for Individuals with Primary Progressive Aphasia and their Care-Partners. *Dementia* **2019**, *18*, 1310–1327. [CrossRef]

73. Henry, M.L.; Hubbard, H.I.; Grasso, S.M.; Dial, H.R.; Beeson, P.M.; Miller, B.L.; Gorno-Tempini, M.L. Treatment for Word Retrieval in Semantic and Logopenic Variants of Primary Progressive Aphasia: Immediate and Long-Term Outcomes. *J. Speech Lang. Hear. Res.* **2019**, *62*, 2723–2749. [CrossRef] [PubMed]

74. Pagnoni, I.; Gobbi, E.; Premi, E.; Borroni, B.; Binetti, G.; Cotelli, M.; Manenti, R. Language training for oral and written naming impairment in primary progressive aphasia: A review. *Transl. Neurodegener.* **2021**, *10*, 24. [CrossRef] [PubMed]

75. Dial, H.R.; Hinshelwood, H.A.; Grasso, S.M.; Hubbard, H.I.; Gorno-Tempini, M.-L.; Henry, M.L. Investigating the utility of teletherapy in individuals with primary progressive aphasia. *Clin. Interv. Aging* **2019**, *14*, 453–471. [CrossRef] [PubMed]

Article

Primary Progressive Aphasia: Use of Graphical Markers for an Early and Differential Diagnosis

Alexandra Plonka [1,2,3,*], Aurélie Mouton [1,2,4], Joël Macoir [5,6], Thi-Mai Tran [7], Alexandre Derremaux [2], Philippe Robert [1,2,4], Valeria Manera [2] and Auriane Gros [1,2,4]

[1] Département d'Orthophonie de Nice, Faculté de Médecine, Université Côte d'Azur, 06000 Nice, France; mouton.a2@chu-nice.fr (A.M.); philippe.robert@unice.fr (P.R.); auriane.gros@univ-cotedazur.fr (A.G.)

[2] Laboratoire CoBTeK (Cognition Behaviour Technology), Université Côte d'Azur, 06000 Nice, France; derreumaux.a2@chu-nice.fr (A.D.); valeria.manera@univ-cotedazur.fr (V.M.)

[3] Institut NeuroMod, Université Côte d'Azur, 06902 Sophia-Antipolis, France

[4] Service Clinique Gériatrique du Cerveau et du Mouvement, CMRR, Centre Hospitalier Universitaire, 06000 Nice, France

[5] Département de Réadaptation, Faculté de Médecine, Université Laval, Québec, QC G1V 0A6, Canada; joel.macoir@rea.ulaval.ca

[6] Centre de Recherche CERVO (CERVO Brain Research Centre), Québec, QC G1J 2G3, Canada

[7] Laboratoire STL, UMR 8163, Département d'Orthophonie, UFR3S, Université de Lille, 59000 Lille, France; thi-mai.tran@univ-lille.fr

* Correspondence: alexandra.plonka@outlook.com

Citation: Plonka, A.; Mouton, A.; Macoir, J.; Tran, T.-M.; Derremaux, A.; Robert, P.; Manera, V.; Gros, A. Primary Progressive Aphasia: Use of Graphical Markers for an Early and Differential Diagnosis. *Brain Sci.* **2021**, *11*, 1198. https://doi.org/10.3390/brainsci11091198

Academic Editors: Jordi A. Matias-Guiu, Robert Jr Laforce and Rene L. Utianski

Received: 26 July 2021
Accepted: 9 September 2021
Published: 11 September 2021

Publisher's Note: MDPI stays neutral with regard to jurisdictional claims in published maps and institutional affiliations.

Abstract: Primary progressive aphasia (PPA) brings together neurodegenerative pathologies whose main characteristic is to start with a progressive language disorder. PPA diagnosis is often delayed in non-specialised clinical settings. With the technologies' development, new writing parameters can be extracted, such as the writing pressure on a touch pad. Despite some studies having highlighted differences between patients with typical Alzheimer's disease (AD) and healthy controls, writing parameters in PPAs are understudied. The objective was to verify if the writing pressure in different linguistic and non-linguistic tasks can differentiate patients with PPA from patients with AD and healthy subjects. Patients with PPA (n = 32), patients with AD (n = 22) and healthy controls (n = 26) were included in this study. They performed a set of handwriting tasks on an iPad® digital tablet, including linguistic, cognitive non-linguistic, and non-cognitive non-linguistic tasks. Average and maximum writing pressures were extracted for each task. We found significant differences in writing pressure, between healthy controls and patients with PPA, and between patients with PPA and AD. However, the classification of performances was dependent on the nature of the tasks. These results suggest that measuring writing pressure in graphical tasks may improve the early diagnosis of PPA, and the differential diagnosis between PPA and AD.

Keywords: primary progressive aphasia; Alzheimer's disease; graphical markers; graphical parameters; writing pressure; differential diagnosis

1. Introduction

Primary progressive aphasia (PPA) assembles a heterogeneous syndromic group of neurodegenerative pathologies characterised by a foreground and initially isolated language impairment that can later extend to cognitive functions such as computation, praxis, memory or executive functions [1–3]. It is a focal form of atrophy with great neuropathological heterogeneity, ranging from tauopathy to amyloidopathy or TDP-43 inclusions [4]. The prevalence of this disease is estimated at 3 per 100,000 [4], with a starting age assessed between 50 and 65 years [5] and a life expectancy of 10 to 15 years [6].

1.1. Diagnosis and Classification

PPA is diagnosed when three criteria overlap: (1) language is mainly damaged; (2) daily living activities are impaired during the initial stages of illness; and (3) word production and comprehension are impaired due to a progressive aphasic disorder and there is an underlying neurodegenerative disease [7]. This last criterion is still debated, based on the fact that PPA evolution from isolated language alteration to global cognitive impairment with multiple neuropsychiatric symptoms can lead to a change in diagnosis [8]. Additionally, language impairment that commonly lasts for about 6 years can represent the only symptom for 10 to 14 years, and is quickly impaired all along the degenerative process before being added to psychiatric and neurologic symptoms [9,10].

In 2011, a broad-ranging International Consensus Group published recommendations for the diagnosis and classification of PPA, establishing three different subtypes of this disease depending of the affected brain regions and the type of aphasic disorder [7]: the logopenic subtype (lvPPA), the non-fluent/agrammatic subtype (nfavPPA) and the semantic subtype (svPPA). A fourth subtype came to complete this classification: a mixed form or non-classified form [11].

lvPPA is defined by impaired word retrieval and phonologic errors that alter language fluidity. Sentence and word repetition are difficult due to a phonological loop disorder that also affects the understanding of long sentences with illness evolution [7,12,13]. lvPPA is characterised by a left posterior parietal or Perisylvian hypometabolism and an atrophy in the left posterior parietal lobe [12,14,15]. Studies have shown that AD is the most common underlying pathology of lvPPA [16].

nfavPPA is characterised by the presence of agrammatism in speech production, with impairments in understanding syntactically complex sentences [3,17,18]. Language production is laborious due to apraxia of speech with phonetic errors, although word comprehension is preserved. nfavPPA is related to dysfunctions in the frontal lobe, in Broca's region, and the anterior parts of the insula [11,12,15,19]. Some studies have also exhibited parietal and temporal involvement [20]. Disorder of nfavPPA is most often frontotemporal lobar degeneration (FTLD) [16].

svPPA is characterised by the presence of a semantic language disorder with paraphasias in the expressive side and impairments in word comprehension in written or oral modalities, associated with a non-verbal semantic disorder [21,22]. svPPA's anatomical lesions are located in the anterior temporal cortex and the inferior and middle temporal cortex [12,23–25]. An infiltration of several connecting beams passing through the temporal lobe have also been reported by fibre-tracking method (DTI) on a small sample (n = 5) [26]. As for nfavPPA, FTLD-type disorder changes are the most common in svPPA [16].

Mixed PPA is characterised by a combination of symptoms of the three main PPA variants with frequent impairment of word comprehension, apraxia of speech or agrammatism [27].

1.2. Early Diagnosis

Early diagnosis of PPA is important in clinical practice because its phenotype is complex, constantly evolving, and is crucial because it increases the possibilities of appropriate clinical interventions. In addition, diagnosis is complex: it has been shown that there is a delay of approximately 4 years between the onset of troubles and PPA diagnosis [6,28,29]. Moreover, the three PPA variants differ in terms of progression over time. lvPPA seems to follow the pattern of Alzheimer's disease (AD) [30], which evolves to a generalised cognitive impairment, whereas other PPA types can be related to different diseases such as behavioural variants of FTLD, corticobasal degeneration or progressive supranuclear palsy [31].

Thus far, there has been no pharmacological treatment modifying or delaying PPA, but non-pharmacological interventions, such as speech therapy, have proven to be useful in compensating for and maintaining functional communications [32]. Early diagnosis is thus crucial to implement early and adapted interventions. Most of the scales available so far for

PPA diagnosis are based on language production and comprehension in oral and written modalities. The main parameters assessed are performance (correct responses and mistakes) and response times. Writing disorders are also considered, such as dysorthography and, more specifically, spelling impairment [33,34], but no study has used graphical parameters such as writing pressure so far.

The use of new technologies allows more ecological and reproducible tests in comparison to certain scales or paper–pencil tests [35,36]. Computerised assessment batteries can build upon standardised and validated pencil-and-paper tests [37].

1.3. Contribution of Graphical Markers

With language symptoms being the earliest and most prominent signs in the early stages of the disease, graphical writing markers may constitute ecological markers of great interest for the early diagnosis of PPA [38].

Several studies have shown that graphic parameters are affected early in people with moderate to severe Alzheimer's disease [38]. Studies have also shown that motor activity reveals language-related characteristics, due to the involvement of motor areas of the brain in writing, and that even mild disorders can be detected using motor parameters (reduction in written pressure) during language production tasks [39].

Handwriting requires the implementation of cognitive processes related to language as well as planning, coordination and motor execution. It has thus been shown that people with cognitive decline overall have a lower writing speed and pressure with a longer writing time, especially when analysing cursive loops [40]. Handwriting performance therefore exhibits significant changes, which it would be interesting to take into account within the framework of a classification of parameters characteristic of the neurodegenerative diseases such as AD, Parkinson's disease (PD) or PPAs [41].

The use of a digital tablet with a stylus makes it possible to objectify the kinematic parameters of writing (pressure, stroke, velocity, jerk, and writing task time); therefore, this would allow a low-cost dissemination of this technology, especially if included in existing screening batteries [42].

The aim of this study was to confirm the initial findings of Gros et al. on a larger sample of PPA [41], concerning the role of writing pressure in differentiating PPA and controls, and to verify if writing pressure is also relevant to distinguish patients with PPA and Alzheimer's disease.

2. Materials and Methods

2.1. Ethics

This study was approved by CPP Ile de France X (N° IDRCB: 2019-A00342-55 accepted on 11 September 2019). At the time of diagnosis, patients and relatives were informed of their inclusion in this study and could decline their participation or withdraw consent. Data were anonymised before the analyses.

2.2. Population

This was a prospective, multicentric study that included 5 French Neurology Departments (Nice, Angers, Nîmes, Saint-Brieuc, and La Rochelle). The patients were recruited from memory consultations in the various centres from June 2019 to February 2020. Eighty adults participated in this study, including patients with PPA (n = 32), patients with typical AD (n = 22) and healthy controls (HC) (n = 26) recruited in the memory centres. All the healthy controls were in good physical and mental health, reported no significant complaints related to cognition, and performed within the normal range on standardised neuropsychological tests. Only two patients (1 PPA, 1 HC) were left-handed. The demographic and clinical features of the three groups of participants are summarised in Table 1.

Table 1. Demographic features of the groups of participants.

	lvPPA	nfavPPA	svPPA	AD	HC	*p*-Value
N	20	6	6	22	26	
*Female, n (%) *	8 (40%)	3 (50%)	3 (50%)	9 (40.9%)	18 (69%)	0.081
Age range (y)	55–85	58–85	70–75	57–87	48–80	
*Mean age **	73.1	69.5	71.3	73.6	65.7	0.003
*SD age **	8.2	8.9	3.1	8.9	8.6	
*Mean Education (y) **	10.1	11.5	9.8	10.1	11.1	0.738
*Education SD **	3.8	3.7	5	4.8	5.4	
*Mean MMSE score **	23.6	20.5	20.7	21.5	28.5	<0.001
*MMSE SD **	5.4	4.2	5	4.9	1.7	
*Mean DTLA score ***	74.5	44.6	58.3	74.8	95.9	<0.001
*DTLA SD ***	16.6	10.9	19.9	17.1	5.6	

* χ^2; ** *ANOVA*; *** *Kruskal–Wallis*, *p*-values refer to the overall comparisons between the three diagnostic groups (PPA, AD and HC).

To be included in the study, the patients had to: be aged 40 years or more, have been diagnosed with PPA or AD according to the DSM-5TM criteria [42], have consulted in one of the investigation centres for cognitive, behavioural and/or motor difficulties, be able to read, write and speak French, benefit from social security coverage, and have no objection for inclusion on the study after reading the information note. The exclusion criteria for the patients and the healthy controls were the presence of a protective measure (guardianship or curatorship), a history of cerebrovascular disease, a history of psychiatric disorder according to the DSM-IVTR criteria [43], any neurological condition (except PPA and AD), traumatic brain injury, untreated medical or metabolic condition (e.g., diabetes, hypothyroidism) uncorrected hearing and vision problems, or prescribed medication with central nervous system sides effects likely to interfere with the carrying out of the tests.

Clinical data were reported retrospectively by the investigators and included: the etiological diagnosis of PPA, PPA variant according to Gorno-Tempini et al. criteria [7], the etiological diagnosis of AD according to the DSM-5TM criteria [42], the results of the various paraclinical examinations (cerebral MRI, PET-Scan, DAT scan, lumbar puncture), the current treatments, including the use of anticholinesterases or Memantine, the global level of cognitive functioning with the Mini Mental State Examination (MMSE), and the status of memory and language capabilities as well as their severity level.

2.3. Procedure

During the first visit, patients received explanations about the study and were given an information sheet. The investigator checked the inclusion criteria and signed a no-objection form. Various elements of the anamnesis were collected: age, gender, level of education, laterality, duration of the disease and familiarity or not with the touchpad devices.

When a patient was included in the study, the practitioner administered the Detection Test of Language impairments in Adult (DTLA) and the tasks of graphic markers on an iPad® tablet [43]. The DTLA test was chosen because of its accuracy for language disorders associated with neurodegenerative diseases. It is a standardised, rapid test, scored on 100 points, validated, and standardised in four French-speaking countries, as well as standardised according to 2 age groups and 2 levels of study. The DTLA test is composed of 9 subtests exploring the language functions most affected in neurodegenerative diseases, and its validation study showed that it has a good convergent validity, a good discriminant validity with healthy controls and a good test–retest fidelity.

2.4. Material and Variables

Graphical markers were collected on the written tasks of the DTLA with an Apple iPad ® 2018 touchpad (model MR7F2NF/A) and an Apple Pencil ® stylus model A1603. The stylus sample rate was 60 Hz, the screen accuracy was 1 pixel, and its resolution was 2048 × 1536. The application retrieved the position and tap pressure provided by the

Apple stylus through the Safari browser. Pressure was measured as a percentage of the maximum pressure allowed by the stylus. These values were measured during plots, and updated every 17 ms.

The following ten written tasks were analysed: four linguistic tasks, consisting of writing words to dictation, writing nonsense words to dictation, writing a spontaneous sentence, all part of the DTLA, and writing letter 'l' loops. Four cognitive non-linguistic tasks, consisting of writing vertical and horizontal lines, diagonals, and a spiral, and two non-cognitive non-linguistic tasks, consisting of writing dots and filling loops were performed. For the cognitive non-linguistic tasks of writing diagonals, the participants had 30 s to go back and forth as fast as possible between two squares presented on the screen. For the non-cognitive non-linguistic tasks, they had to fill the screen with dots and loops (Figure 1). For each task, we extracted the average (avgP) and the maximum (maxP) writing pressure, representing the pressure of the stylus on the screen (ranging from 0 to 1).

Linguistic Tasks	Cognitive-Non-Linguistic Tasks	Non-Cognitive-Non-Linguistic Tasks

Figure 1. Graphical marker tasks. Linguistic tasks: words, nonsense words, sentence, letter 'l' loops. Cognitive non-linguistic tasks: vertical and horizontal lines, spiral, diagonals. Non-cognitive non-linguistic tasks: dots, filling loops. Writing pressure was collected on an iPad® tablet. Red colour indicates the maximum pressure.

2.5. Statistical Analyses

Descriptive statistics were used to present demographic and clinical characteristics. Qualitative variables (sex) were presented using the frequency and percentage, and quantitative variables (age, years of education, MMSE score, and DTLA score) were presented using the mean and standard deviation (SD). The effects of the diagnostic group (PPA, AD and healthy controls) on quantitative demographic variables were tested using one-way ANOVAs for normally distributed variables (followed by LSD-corrected post hoc tests) and Kruskal–Wallis for non-normally distributed variables (followed by Bonferroni-corrected post hoc tests). The diagnostic groups differed in terms of mean age; therefore, we performed ANCOVAs on the average and maximum writing pressure using the diagnostic group (PPA, AD and healthy controls) as between-subject factor, and the age as a covariate (followed by LSD-corrected post hoc tests).

Qualitative variables (such as sex) were compared using the χ^2 test. All statistical analyses were performed using IMB SPSS Statistics V20.0 software.

3. Results

3.1. Demographic and Clinical Information

Characteristics and clinical information of each group are reported in Table 1. No significant differences in gender ($\chi^2_{(2)} = 5.03$, $p = 0.081$) and the number of years of education ($F_{(2,77)} = 0.31$, $p = 0.738$) were found across the three groups. Age varied significantly across the groups ($F_{(2,77)} = 6.34$, $p = 0.003$). Specifically, post hoc LSD tests showed that participants in the control group were significantly younger than participants with PPA ($p = 0.005$) and AD ($p = 0.002$), whereas no difference between PPA and AD groups was found ($p = 0.521$). As expected, MMSE scores varied significantly across groups ($F_{(2,51)} = 8.66$, $p = 0.001$), with participants in the control group showing significantly higher MMSE scores than participants in the PPA ($p = 0.001$) and the AD ($p < 0.001$) groups. No difference between PPA and AD groups was found ($p = 0.493$). A significant difference in the results of the DTLA scale was found ($H_{(2)} = 46.20$, $p < 0.001$). Bonferroni-corrected post hoc tests revealed that participants in the control group had significantly higher DTLA scores than participants in the PPA ($p < 0.001$) and the AD ($p < 0.001$) groups. The difference between PPA and AD groups did not reach statistical significance ($p = 0.838$).

3.2. Graphical Markers

3.2.1. Average Pressure (avgP)

Descriptive analyses (mean and standard deviation) for the average pressure in each task and for differences between linguistic and non-linguistic tasks are reported in Table 2.

The ANCOVA with Group as the between-subject factor and Age as a covariate revealed a significant effect of Group on avgP in the horizontal lines (cognitive non-linguistic) task ($F_{(2,41)} = 3.26$, $p = 0.049$). Specifically, paired post hoc comparisons (LSD-corrected) revealed that avgP was significantly higher in AD compared to controls ($p = 0.035$), and almost significantly higher in AD compared to PPA ($p = 0.057$). No significant effect of Group was found for the other tasks.

Concerning the differences between linguistic and non-linguistic tasks, a significant effect of Group was found on the difference between words and horizontal lines ($F_{(2,40)} = 3.94$, $p = 0.027$); specifically, subjects with AD showed a higher avgP in the horizontal lines compared to the words task, whereas the opposite was true for controls ($p = 0.016$) and PPA subjects ($p = 0.049$). The same pattern was also found for the difference between non-words and horizontal lines ($F_{(2,40)} = 4.24$, $p = 0,021$)—subjects with AD showed a higher avgP in the horizontal lines compared to the non-words task, whereas the opposite was true for controls ($p = 0.016$) and PPA subjects ($p = 0.031$)—and for the difference between horizontal lines and sentence tasks ($F_{(2,40)} = 3.99$, $p = 0,026$), with subjects with AD showing a higher avgP in the horizontal lines compared to the sentence task, whereas the opposite was true for controls ($p = 0.032$) and PPA subjects ($p = 0.021$). Finally, a significant effect of Group was found on the difference between letter 'l' loops (linguistic) task and (cognitive non-linguistic) diagonals task ($F_{(2,74)}$ = 3.38, p = 0,039), with subjects with PPA showing a higher avgP in the diagonals compared to the cursive loops task, whereas the opposite was true for controls (p = 0,026) and AD subjects (p = 0,046). No other significant difference was found.

Table 2. Average writing pressure in participants with PPA, AD and Healthy Controls.

	Task	Diagnosis	Mean	Standard Deviation
Linguistic Tasks	Words	PPA	0.20	0.09
		AD	0.22	0.13
		Controls	0.20	0.08
	Nonsense words	PPA	0.22	0.10
		AD	0.23	0.14
		Controls	0.21	0.09
	Sentence	PPA	0.23	0.11
		AD	0.26	0.13
		Controls	0.22	0.09
	Letter 'l' loops	PPA	0.25	0.12
		AD	0.28	0.15
		Controls	0.26	0.10
Cognitive Non-Linguistic Tasks	Diagonal	PPA	0.28	0.14
		AD	0.26	0.13
		Controls	0.24	0.10
	Vertical	PPA	0.28	0.08
		AD	0.30	0.15
		Controls	0.22	0.12
	Horizontal	PPA	0.21	0.55
		AD	0.31	0.21
		Controls	0.18	0.08
	Spiral	PPA	0.25	0.12
		AD	0.26	0.11
		Controls	0.25	0.09
Non-Cognitive Non-Linguistic Tasks	Dots	PPA	0.17	0.07
		AD	0.19	0.09
		Controls	0.13	0.04
	Filling Loops	PPA	0.28	0.11
		AD	0.31	0.15
		Controls	0.27	0.09

3.2.2. Maximum Pressure (maxP)

Descriptive analyses (mean and standard deviation) for the average pressure in each task and for differences between linguistic and non-linguistic tasks are reported in Table 3.

The ANCOVA with Group as a between-subject factor and Age as a covariate revealed a significant effect of Group on maxP for the sentences (linguistic) task ($F_{(2,74)}$ = 3.65, p = 0.031), with AD subjects showing a significantly higher maxP compared to the controls (p = 0.009). A significant effect of Group was also found for the horizontal lines (cognitive non-linguistic) task ($F_{(2,41)}$ = 3.24, p = 0,049)—AD subjects showed a significantly higher maxP compared to the controls (p = 0.021)—and for the dots (non-cognitive non-linguistic) task ($F_{(2,74)}$ = 4.12, p = 0,020), with subjects with PPA (p = 0.007) and AD (p = 0.032) showing a higher maxP compared to the controls. No significant effect of Group was found for the other tasks.

Concerning the differences between linguistic and non-linguistic tasks, a significant effect of group was found on the difference between letter 'l' loops and dots ($F_{(2,75)}$ = 5.27, p = 0.007). Specifically, all subjects showed a higher maxP in the dots compared to the cursive loops task, but the difference was higher for PPA (p = 0.002) and AD subjects (p = 0.027) compared to the controls. Furthermore, an almost-significant effect of Group was found on the difference between letter 'l' loops and horizontal lines ($F_{(2,42)}$ = 3.03, p = 0.059) with controls showing a higher maxP in the letter 'l' loops vs. the horizontal lines task, whereas the opposite was true for subjects with AD (p = 0.028). No other significant difference was found.

Table 3. Maximum writing pressure in participants with PPA, AD and Healthy Controls.

	Task	Diagnosis	Mean	Standard Deviation
		PPA	0.61	0.30
	Words	AD	0.66	0.26
		Controls	0.55	0.23
		PPA	0.58	0.30
	Nonsense words	AD	0.61	0.30
Linguistic Tasks		Controls	0.54	0.22
		PPA	0.65	0.33
	Sentence	AD	0.78	0.27
		Controls	0.53	0.25
		PPA	0.45	0.23
	Letter 'l' loops	AD	0.48	0.22
		Controls	0.44	0.23
		PPA	0.54	0.27
	Diagonal	AD	0.49	0.25
		Controls	0.42	0.21
		PPA	0.55	0.19
	Vertical	AD	0.57	0.23
Cognitive Non-Linguistic Tasks		Controls	0.43	0.21
		PPA	0.45	0.21
	Horizontal	AD	0.58	0.30
		Controls	0.32	0.16
		PPA	0.44	0.26
	Spiral	AD	0.47	0.21
		Controls	0.40	0.19
		PPA	0.73	0.27
	Dots	AD	0.71	0.25
Non-Cognitive Non-Linguistic Tasks		Controls	0.53	0.24
		PPA	0.52	0.20
	Filling Loops	AD	0.54	0.22
		Controls	0.51	0.20

3.2.3. Summary of the Main Differences between PPA and Healthy Controls

Considering post hoc corrected comparisons, the most relevant tasks to distinguish PPA patients from healthy controls seemed to be the dots (non-cognitive non-linguistic) task and the letter 'l' loops (linguistic) task. Specifically, the maxP (p = 0,007) in the dots task was higher in PPA compared to healthy controls. Furthermore, the difference in maxP in the dots compared to the letter 'l' loops task was higher for PPA than for controls (p = 0.002). Finally, subjects with PPA had a higher avgP in the diagonals compared to the letter 'l' loops task, whereas the opposite was true for controls (p = 0.026) (Figure 2).

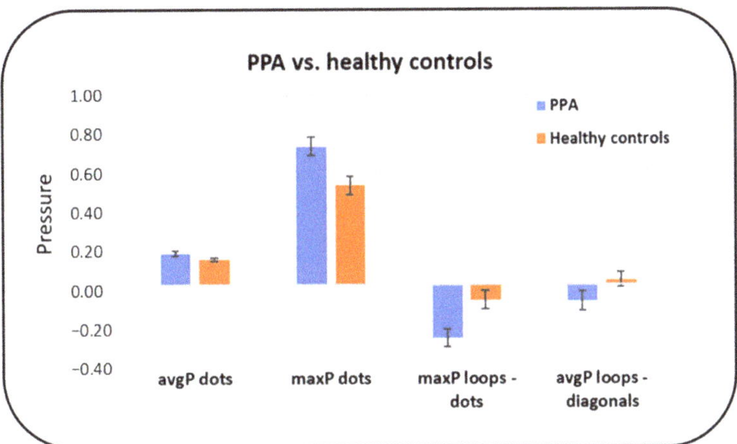

Figure 2. Differences in average and maximum writing pressure between patients with PPA and healthy controls.

3.2.4. Summary of the Main Differences between PPA and AD

Considering post hoc corrected comparisons, the most relevant feature distinguishing between PPA and AD patients was the avgP, whereas no significant differences were found for the maxP. In terms of tasks, the most relevant seemed to be the horizontal lines and diagonal lines (cognitive non-linguistic) tasks and the linguistic tasks. Indeed, differences in avgP were found for the horizontal lines task (AD>PPA, $p = 0.057$) and for the difference between horizontal lines and three linguistic tasks (words, non-words and sentence, $p = 0.049, 0.031$ and 0.021, respectively). Specifically, avgP in AD was higher in the cognitive non-linguistic tasks compared to the linguistic tasks, whereas avgP in PPA was higher in the linguistic tasks compared to the cognitive non-linguistic task. Finally, subjects with PPA showed a higher avgP in the diagonals task compared to the letter 'l' loops (linguistic) task, whereas the opposite was true for AD subjects ($p = 0,046$) (Figure 3).

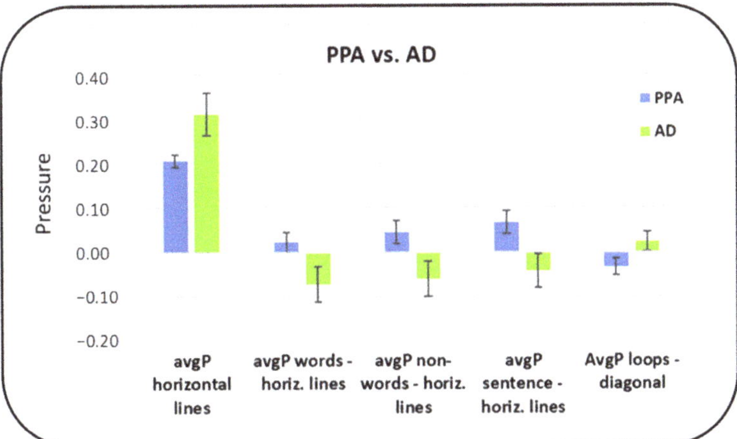

Figure 3. Differences in average and maximum writing pressure between patients with PPA and patients with AD.

4. Discussion

In the present study, we investigated the usefulness of graphical parameters collected in a handwriting protocol to differentiate patients with PPA from healthy controls, and patients with PPA from patients with AD. Significant differences in the average pressure and maximum pressure between PPA participants and healthy controls were found in the non-linguistic non-cognitive 'dots' task, and in the pressure difference between linguistic and non-linguistic 'letter l loops' and 'dots' tasks. These results show that PPA patients have a higher difference in the maximum pressure between a linguistic ('letter l loops') and a non-linguistic non-cognitive task ('dots) than healthy controls. A previous study already showed that motor activity reveals language-related characteristics, due to the involvement of motor areas of the brain in writing [39]. This suggests that motor performance involved in linguistic and non-linguistic tasks may change in the presence of language disorders.

Other studies have shown an overall lower writing pressure in people with cognitive decline associated with AD compared to healthy people [40], with a lower pressure in most cognitively deteriorated groups [44]. Our results suggest the opposite with PPA patients in whom writing maximum pressure was significatively higher compared to healthy controls in the non-cognitive 'dots' task. Two major processes enter in handwriting: language processes and motor processes. Thus, writing could experience variations in different tasks depending on which process is reached [45].

Differences in pressure between a non-linguistic task and a linguistic task may suggest a decrease in the activity of the motor cortex during the graphic act, associated with a linguistic task for PPA patients (with a smaller difference between both). These results may be explained by the need for recruiting more cognitive resources during a linguistic task than during a non-linguistic task for PPA participants. Indeed, non-linguistic areas of the brain are usually more preserved in PPA than linguistic areas. This interpretation must be confirmed by an EEG exploration during writing in linguistic and non-linguistic tasks. These results are in line with other studies that show a relationship between language and gesture processing and the partial overlap of their neural representations. Indeed, a study demonstrated that PPA patients showed significant deficits on gesture discrimination tasks clustered with linguistic tasks as word and nonsense-word repetition, and writing-to-dictation [46].

The last aim of this study was to verify if graphical parameters could differentiate participants with PPA from participants with AD. Several studies have analysed graphical markers in patients with AD, but none in PPAs. Indeed, studies on PPAs focused only on the content of language in writing, and not on the graphic parameters. Thus, studies have shown letter insertion errors in patients with PPA, whereas they were absent in AD and mild cognitive impairment (MCI) patients, and that patients with PPA use more verbs than patients with AD [47].

Although the symptoms of AD are more cognitive than motor, it has been shown that motor dysfunction quantified by kinematic handwriting analysis is significantly correlated with MMSE scores in AD [48], and that pressure is lower in more cognitively deteriorated groups [44]. Graphic parameters and variability in the performance of patients with AD have been explained by a degradation of the motor programming, resembling that of Huntington's rather than Parkinson's disease patients, and may reflect frontal rather than basal ganglia dysfunction [49]. Finally, these studies suggest that MCI is also characterised by motor dysfunction and that writing with accuracy constraints may help identify those at risk of AD [50]. According to these studies, these deficits in graphical parameters seem to be more related to a motor dysfunction than a language impairment. Indeed, it has already been shown that in the mild phase of AD, lexico-semantic problems in the speaking process are possible but not predominant [51]. Thus, graphical markers in patients with AD seem more related to a deterioration in fine motor control and coordination [52,53].

Indeed, graphical markers seem to reflect the type of specific disorders in different pathologies and permit better comprehension of the nature of these deficits. In the same way, we have recently demonstrated a reduction in pressure, particularly in graphical activities, which have a spatial component in posterior cortical atrophy [54]. This result of a writing pressure change depending on the graphical task performed is in line with the results of a previous study on AD, and can be explained by the difference type of impairment between these pathologies [55].

Inconsistently with the literature on writing in patients with AD, our results show a difference between patients with PPA and patients with AD, with a predominant impairment in linguistic tasks in AD. Indeed, significant differences between the two groups were found for the cognitive non-linguistic horizontal lines task and for the difference between horizontal lines and three linguistic tasks. The average pressure in AD was higher in the cognitive non-linguistic tasks compared to the linguistic tasks, whereas the average pressure in patients with PPA was higher in the linguistic tasks compared to the cognitive non-linguistic tasks.

Contrary to the literature, these results suggest that graphical markers are not only a sign of motor and coordination disorders, but also a sign of cognitive and, more specifically, language disorders. Indeed, our results may suggest that patients with AD, despite an overall cognitive impairment, have a higher cognitive load than patients with PPA in linguistic tasks. In the same way, patients with PPA seem to have a high cognitive load for linguistic tasks but also in cognitive tasks (dysexecutive impairment). These results are in line with other studies that show early dysexecutive symptoms in patients with PPA [56] and a severe language impairment in patients with AD [57].

In conclusion, graphical markers may allow the performance of an early and differential diagnosis of patients with PPA and patients with AD. Writing pressure comparisons between linguistic and cognitive non-linguistic tasks reveal a difference in pressure between patients with PPA and healthy controls and patients with PPA and patients with AD. Indeed, in patients with AD, although the cognitive impairment is global, language impairment appears as an important diagnosis marker, such as in patients with PPA.

Other graphical kinematic parameters such as writing velocity could also be of interest for the classification of different subtypes of PPAs, because of the different anatomical pathways of degeneration. Thus, it has been shown that people with cognitive decline have a lower writing speed and pressure overall, with longer writing times [40]. However, to confirm these first results, a larger and more balanced PPA sample seems necessary.

Finally, this study highlights two main elements.

First, and on the scientific side, studying patients suffering from primary progressive aphasia, a clinical syndrome characterised by comparatively isolated language deficits, may provide direct evidence for anatomical and functional association between language deficits and gesture graphic particularity.

Second, on the clinical side, this study has shown the benefits of associating graphical markers to a rapid screening battery such as DTLA for the earlier and differential diagnosis of PPAs.

Author Contributions: Conceptualization, A.G., V.M. and A.P.; methodology, A.G.; validation, A.G., J.M. and P.R.; formal analysis, V.M. and A.D.; investigation, A.P. and A.G.; resources, A.G. and A.P.; data curation, A.G.; writing—original draft preparation, A.P.; writing—review and editing, A.G., V.M., J.M., and T.-M.T.; supervision, P.R. and A.G.; project administration, A.G., P.R. and A.M.; funding acquisition, A.G. All authors have read and agreed to the published version of the manuscript.

Funding: This research received no external funding.

Institutional Review Board Statement: The study was conducted according to the guidelines of the Declaration of Helsinki, and approved by the Ethics Committee of CPP Ile de France X (N° IDRCB: 2019-A00342-55 accepted on 11 September 2019). At the time of diagnosis, patients and relatives were informed of their inclusion in this study and could decline their participation or withdraw consent. Data were anonymised before the analyses.

Informed Consent Statement: Informed consent was obtained from all subjects involved in the study.

Data Availability Statement: The data reported are part of an ongoing registration program. Deidentified participant data are not available for legal and ethical reasons. Anonymised data will be made available for research purposes, upon request and specifical approval of the database advisory board and ethical committee.

Acknowledgments: This work was supported by University Côte d'Azur and NeuroMod Institute and by a grant from the Association Innovation Alzheimer. Thanks to all patients that participated in this study and to all participants.

Conflicts of Interest: The authors declare no conflict of interest. The funders had no role in the design of the study; in the collection, analyses, or interpretation of data; in the writing of the manuscript, or in the decision to publish the results.

References

1. Mesulam, M.M. Primary progressive aphasia. *Ann. Neurol.* **2001**, *49*, 425–432. [CrossRef]
2. Mesulam, M.-M. Primary Progressive Aphasia—A Language-Based Dementia. *N. Engl. J. Med.* **2003**, *349*, 1535–1542. [CrossRef] [PubMed]
3. Grossman, M.; Ash, S. Primary Progressive Aphasia: A Review. *Neurocase* **2004**, *10*, 3–18. [CrossRef]
4. Marshall, C.R.; Hardy, C.J.D.; Volkmer, A.; Russell, L.L.; Bond, R.L.; Fletcher, P.D.; Clark, C.N.; Mummery, C.J.; Schott, J.M.; Rossor, M.N.; et al. Primary progressive aphasia: A clinical approach. *J. Neurol.* **2018**, *265*, 1474–1490. [CrossRef] [PubMed]
5. Ratnavalli, E.; Brayne, C.; Dawson, K.; Hodges, J.R. The prevalence of frontotemporal dementia. *Neurology* **2002**, *58*, 1615–1621. [CrossRef] [PubMed]
6. Le Rhun, E.; Richard, F.; Pasquier, F. Natural history of primary progressive aphasia. *Neurology* **2005**, *65*, 887–891. [CrossRef]
7. Gorno-Tempini, M.L.; Hillis, A.E.; Weintraub, S.; Kertesz, A.; Mendez, M.; Cappa, S.F.; Ogar, J.M.; Rohrer, J.D.; Black, S.; Boeve, B.F.; et al. Classification of primary progressive aphasia and its variants. *Neurology* **2011**, *76*, 1006–1014. [CrossRef] [PubMed]
8. Kertesz, A.; Morlog, D.; Light, M.; Blair, M.; Davidson, W.; Jesso, S.; Brashear, R. Galantamine in Frontotemporal Dementia and Primary Progressive Aphasia. *Dement. Geriatr. Cogn. Disord.* **2008**, *25*, 178–185. [CrossRef]
9. Weintraub, S.; Rubin, N.P.; Mesulam, M.M. Primary progressive aphasia. Longitudinal course, neuropsychological profile, and language features. *Arch. Neurol.* **1990**, *47*, 1329–1335. [CrossRef]
10. Mesulam, M. Primary progressive aphasia: A dementia of the language network. *Dement. Neuropsychol.* **2013**, *7*, 2–9. [CrossRef]
11. Josephs, K.A. Clinicopathological and imaging correlates of progressive aphasia and apraxia of speech. *Brain* **2006**, *129*, 1385–1398. [CrossRef]
12. Gorno-Tempini, M.L.; Dronkers, N.F.; Rankin, K.P.; Ogar, J.M.; Phengrasamy, L.; Rosen, H.J.; Johnson, J.K.; Weiner, M.W.; Miller, B.L. Cognition and anatomy in three variants of primary progressive aphasia. *Ann. Neurol.* **2004**, *55*, 335–346. [CrossRef]
13. Gorno-Tempini, M.L.; Brambati, S.M.; Ginex, V.; Ogar, J.; Dronkers, N.F.; Marcone, A.; Perani, D.; Garibotto, V.; Cappa, S.F.; Miller, B.L. The logopenic/phonological variant of primary progressive aphasia. *Neurology* **2008**, *71*, 1227–1234. [CrossRef] [PubMed]
14. Wilson, S.M.; Henry, M.L.; Besbris, M.; Ogar, J.M.; Dronkers, N.F.; Jarrold, W.; Miller, B.L.; Gorno-Tempini, M.L. Connected speech production in three variants of primary progressive aphasia. *Brain* **2010**, *133*, 2069–2088. [CrossRef]
15. Rohrer, J.D.; Warren, J.D.; Modat, M.; Ridgway, G.R.; Douiri, A.; Rossor, M.N.; Ourselin, S.; Fox, N.C. Patterns of cortical thinning in the language variants of frontotemporal lobar degeneration. *Neurology* **2009**, *72*, 1562–1569. [CrossRef]
16. Montembeault, M.; Brambati, S.M.; Gorno-Tempini, M.L.; Migliaccio, R. Clinical, Anatomical, and Pathological Features in the Three Variants of Primary Progressive Aphasia: A Review. *Front. Neurol.* **2018**, *9*, 692. [CrossRef] [PubMed]
17. Grossman, M.; Mickanin, J.; Onishi, K.; Hughes, E.; D'Esposito, M.; Ding, X.-S.; Alavi, A.; Reivich, M. Progressive Nonfluent Aphasia: Language, Cognitive, and PET Measures Contrasted with Probable Alzheimer's Disease. *J. Cogn. Neurosci.* **1996**, *8*, 135–154. [CrossRef] [PubMed]
18. Knibb, J.A.; Woollams, A.M.; Hodges, J.R.; Patterson, K. Making sense of progressive non-fluent aphasia: An analysis of conversational speech. *Brain* **2009**, *132*, 2734–2746. [CrossRef]
19. Rohrer, J.D.; Warren, J.D. Phenomenology and anatomy of abnormal behaviours in primary progressive aphasia. *J. Neurol. Sci.* **2010**, *293*, 35–38. [CrossRef] [PubMed]
20. Rohrer, J.D.; Ridgway, G.R.; Crutch, S.J.; Hailstone, J.; Goll, J.C.; Clarkson, M.J.; Mead, S.; Beck, J.; Mummery, C.; Ourselin, S.; et al. Progressive logopenic/phonological aphasia: Erosion of the language network. *NeuroImage* **2010**, *49*, 984–993. [CrossRef]
21. Hodges, J.R.; Patterson, K.; Oxbury, S.; Funnell, E. Semantic dementia. Progressive fluent aphasia with temporal lobe atrophy. *Brain J. Neurol.* **1992**, *115 Pt 6*, 1783–1806. [CrossRef]
22. Adlam, A.-L.R.; Patterson, K.; Rogers, T.T.; Nestor, P.J.; Salmond, C.H.; Acosta-Cabronero, J.; Hodges, J.R. Semantic dementia and fluent primary progressive aphasia: Two sides of the same coin? *Brain* **2006**, *129*, 3066–3080. [CrossRef]
23. Mummery, C.J.; Patterson, K.; Price, C.J.; Ashburner, J.; Frackowiak, R.S.; Hodges, J.R. A voxel-based morphometry study of semantic dementia: Relationship between temporal lobe atrophy and semantic memory. *Ann. Neurol.* **2000**, *47*, 36–45. [CrossRef]
24. Rosen, H.J.; Perry, R.J.; Murphy, J.; Kramer, J.H.; Mychack, P.; Schuff, N.; Weiner, M.; Levenson, R.W.; Miller, B.L. Emotion comprehension in the temporal variant of frontotemporal dementia. *Brain J. Neurol.* **2002**, *125*, 2286–2295. [CrossRef]

25. Garrard, P.; Hodges, J.R. Semantic dementia: Clinical, radiological and pathological perspectives. *J. Neurol.* **2000**, *247*, 409–422. [CrossRef]
26. Agosta, F.; Henry, R.G.; Migliaccio, R.; Neuhaus, J.; Miller, B.L.; Dronkers, N.F.; Brambati, S.M.; Filippi, M.; Ogar, J.M.; Wilson, S.M.; et al. Language networks in semantic dementia. *Brain* **2010**, *133*, 286–299. [CrossRef] [PubMed]
27. Vandenberghe, R. Classification of the primary progressive aphasias: Principles and review of progress since 2011. *Alzheimers Res. Ther.* **2016**, *8*, 16. [CrossRef] [PubMed]
28. Westbury, C.; Bub, D. Primary Progressive Aphasia: A Review of 112 Cases. *Brain Lang.* **1997**, *60*, 381–406. [CrossRef] [PubMed]
29. Mouton, A.; Plonka, A.; Fabre, R.; Tran, M.; Robert, P.; Macoir, J.; Manera, V.; Gros, A. The Course of Primary Progressive Aphasia Diagnosis: A Cross-Sectional Study. Available online: https://www.researchsquare.com/article/rs-440319/v1 (accessed on 22 April 2021).
30. Dubois, B.; Feldman, H.H.; Jacova, C.; Cummings, J.L.; Dekosky, S.T.; Barberger-Gateau, P.; Delacourte, A.; Frisoni, G.; Fox, N.C.; Galasko, D.; et al. Revising the definition of Alzheimer's disease: A new lexicon. *Lancet Neurol.* **2010**, *9*, 1118–1127. [CrossRef]
31. Harciarek, M.; Sitek, E.J.; Kertesz, A. The patterns of progression in primary progressive aphasia—Implications for assessment and management. *Aphasiology* **2014**, *28*, 964–980. [CrossRef]
32. Routhier, S.; Gravel-Laflamme, K.; Macoir, J. Non-pharmacological therapies for language deficits in the agrammatic and logopenic variants of primary progressive aphasia: A literature review. *Gériatrie Psychol. Neuropsychiatr. Viellissement* **2013**, *11*, 87–97. [CrossRef]
33. Nagai, C.; Iwata, M. Writing disorders in primary progressive aphasia. *Rinsho Shinkeigaku* **2003**, *43*, 84–92.
34. Graham, N.L. Dysgraphia in primary progressive aphasia: Characterisation of impairments and therapy options. *Aphasiology* **2014**, *28*, 1092–1111. [CrossRef]
35. Gomez-Vilda, P.; Perez-Broncano, O.; Martinez-Olalla, R.; Rodellar-Biarge, V.; Lopez de Ipina Pena, K.; Ecay, M.; Martinez-Lage, P. Biomechanical characterization of phonation in Alzheimer's Disease. In *Proceedings of the 3rd IEEE International Work-Conference on Bioinspired Intelligence*; IEEE: Liberia, Costa Rica, 2014; pp. 14–20. Available online: http://ieeexplore.ieee.org/document/6913931/ (accessed on 2 October 2014).
36. Brown, L.J.E.; Adlam, T.; Hwang, F.; Khadra, H.; Maclean, L.M.; Rudd, B.; Smith, T.; Timon, C.; Williams, E.A.; Astell, A.J. Computer-based tools for assessing micro-longitudinal patterns of cognitive function in older adults. *AGE* **2016**, *38*, 335–350. [CrossRef]
37. Wild, K.; Howieson, D.; Webbe, F.; Seelye, A.; Kaye, J. Status of computerized cognitive testing in aging: A systematic review. *Alzheimers Dement.* **2008**, *4*, 428–437. [CrossRef] [PubMed]
38. Afonso, O.; Álvarez, C.J.; Martínez, C.; Cuetos, F. Writing difficulties in Alzheimer's disease and mild cognitive impairment. *Read. Writ.* **2019**, *32*, 217–233. [CrossRef]
39. Nazir, T.A.; Hrycyk, L.; Moreau, Q.; Frak, V.; Cheylus, A.; Ott, L.; Lindemann, O.; Fischer, M.H.; Paulignan, Y.; Delevoye-Turrell, Y. A simple technique to study embodied language processes: The grip force sensor. *Behav. Res. Methods* **2017**, *49*, 61–73. [CrossRef] [PubMed]
40. Kahindo, C.; El-Yacoubi, M.A.; Garcia-Salicetti, S.; Rigaud, A.-S.; Cristancho-Lacroix, V. Characterizing Early-Stage Alzheimer Through Spatiotemporal Dynamics of Handwriting. *IEEE Signal Process. Lett.* **2018**, *25*, 1136–1140. [CrossRef]
41. Gros, A.; Plonka, A.; Manera, V. Graphic markers: Towards an early diagnosis of primary progressive aphasia. *Alzheimers Dement.* **2019**, *15*, P351–P352. [CrossRef]
42. Macoir, J.; Fossard, M.; Lefebvre, L.; Monetta, L.; Renard, A.; Tran, T.M.; Wilson, M.A. Detection Test for Language Impairments in Adults and the Aged—A New Screening Test for Language Impairment Associated With Neurodegenerative Diseases: Validation and Normative Data. *Am. J. Alzheimers Dis. Dementiasr* **2017**, *32*, 382–392. [CrossRef]
43. Macoir, J.; Fossard, M.; Lefebvre, L.; Monetta, L.; Renard, A.; Tran, T.M.; Wilson, M. DTLA-Détection des Troubles du Langage Chez L'adulte et la Personne Agée. 2017. Available online: https://www.researchgate.net/publication/317905040_DTLA_-_Detection_des_troubles_du_langage_chez_l%27adulte_et_la_personne_agee (accessed on 2 October 2014). [CrossRef]
44. Werner, P.; Rosenblum, S.; Bar-On, G.; Heinik, J.; Korczyn, A. Handwriting process variables discriminating mild Alzheimer's disease and mild cognitive impairment. *J. Gerontol. B. Psychol. Sci. Soc. Sci.* **2006**, *61*, 228–236. [CrossRef]
45. Van Galen, G.P. Handwriting: Issues for a psychomotor theory. *Hum. Mov. Sci.* **1991**, *10*, 165–191. [CrossRef]
46. Nelissen, N.; Pazzaglia, M.; Vandenbulcke, M.; Sunaert, S.; Fannes, K.; Dupont, P.; Aglioti, S.M.; Vandenberghe, R. Gesture Discrimination in Primary Progressive Aphasia: The Intersection between Gesture and Language Processing Pathways. *J. Neurosci.* **2010**, *30*, 6334–6341. [CrossRef] [PubMed]
47. Sitek, E.J.; Barczak, A.; Kluj-Kozłowska, K.; Kozłowski, M.; Barcikowska, M.; Sławek, J. Is descriptive writing useful in the differential diagnosis of logopenic variant of primary progressive aphasia, Alzheimer's disease and mild cognitive impairment? *Neurol. Neurochir. Pol.* **2015**, *49*, 239–244. [CrossRef] [PubMed]
48. Schröter, A.; Mergl, R.; Bürger, K.; Hampel, H.; Möller, H.-J.; Hegerl, U. Kinematic Analysis of Handwriting Movements in Patients with Alzheimer's Disease, Mild Cognitive Impairment, Depression and Healthy Subjects. *Dement. Geriatr. Cogn. Disord.* **2003**, *15*, 132–142. [CrossRef]
49. Slavin, M.J.; Phillips, J.G.; Bradshaw, J.L.; Hall, K.A.; Presnell, I. Consistency of handwriting movements in dementia of the Alzheimer's type: A comparison with Huntington's and Parkinson's diseases. *J. Int. Neuropsychol. Soc.* **1999**, *5*, 20–25. [CrossRef]

50. Yu, N.-Y.; Chang, S.-H. Kinematic Analyses of Graphomotor Functions in Individuals with Alzheimer's Disease and Amnestic Mild Cognitive Impairment. *J. Med. Biol. Eng.* **2016**, *36*, 334–343. [CrossRef]
51. Szatloczki, G.; Hoffmann, I.; Vincze, V.; Kalman, J.; Pakaski, M. Speaking in Alzheimer's Disease, is That an Early Sign? Importance of Changes in Language Abilities in Alzheimer's Disease. *Front. Aging Neurosci.* **2015**, *7*. [CrossRef]
52. Platel, H.; Lambert, J.; Eustache, F.; Cadet, B.; Dary, M.; Viader, F.; Lechevalier, B. Characteristics and evolution of writing impairmant in Alzheimer's disease. *Neuropsychologia* **1993**, *31*, 1147–1158. [CrossRef]
53. Yan, J.H.; Rountree, S.; Massman, P.; Doody, R.S.; Li, H. Alzheimer's disease and mild cognitive impairment deteriorate fine movement control. *J. Psychiatr. Res.* **2008**, *42*, 1203–1212. [CrossRef]
54. Videt-Dussert, A.; Plonka, A.; Derreumaux, A.; Manera, V.; Leone, E.; Gros, A. Handwriting graphical parameters analysis in Posterior Cortical Atrophy: A case report. *Clin. Neurol. Neurosurg.* **2021**, *208*, 106876. [CrossRef] [PubMed]
55. Impedovo, D.; Pirlo, G. Dynamic Handwriting Analysis for the Assessment of Neurodegenerative Diseases: A Pattern Recognition Perspective. *IEEE Rev. Biomed. Eng.* **2019**, *12*, 209–220. [CrossRef] [PubMed]
56. Macoir, J.; Lavoie, M.; Laforce, R.; Brambati, S.M.; Wilson, M.A. Dysexecutive Symptoms in Primary Progressive Aphasia: Beyond Diagnostic Criteria. *J. Geriatr. Psychiatry Neurol.* **2017**, *30*, 151–161. [CrossRef] [PubMed]
57. Fraser, K.C.; Meltzer, J.A.; Rudzicz, F. Linguistic Features Identify Alzheimer's Disease in Narrative Speech. *J. Alzheimers Dis.* **2015**, *49*, 407–422. [CrossRef] [PubMed]

Article

Survival in the Three Common Variants of Primary Progressive Aphasia: A Retrospective Study in a Tertiary Memory Clinic

Maud Tastevin *, Monica Lavoie, Justine de la Sablonnière, Julie Carrier-Auclair and Robert Laforce Jr.

Clinique Interdisciplinaire de Mémoire, Département des Sciences Neurologiques du CHU de Québec, Faculté de Médecine, Université Laval, Québec, QC G1V 0A6, Canada; monica.lavoie.1@ulaval.ca (M.L.); justine.de-la-sablonniere.1@ulaval.ca (J.d.l.S.); julie.carrier-auclair.1@ulaval.ca (J.C.-A.); robert.laforce@fmed.ulaval.ca (R.L.J.)
* Correspondence: maud.tastevin.1@ulaval.ca

Abstract: Knowledge on the natural history of the three main variants of primary progressive aphasia (PPA) is lacking, particularly regarding mortality. Moreover, advanced stages and end of life issues are rarely discussed with caregivers and families at diagnosis, which can cause more psychological distress. We analyzed data from 83 deceased patients with a diagnosis of PPA. We studied survival in patients with a diagnosis of logopenic variant (lvPPA), semantic variant (svPPA), or non-fluent variant (nfvPPA) and examined causes of death. From medical records, we retrospectively collected data for each patient at several time points spanning five years before the first visit to death. When possible, interviews were performed with proxies of patients to complete missing data. Results showed that survival from symptom onset and diagnosis was significantly longer in svPPA than in lvPPA ($p = 0.002$) and nfvPPA ($p < 0.001$). No relevant confounders were associated with survival. Mean survival from symptom onset was 7.6 years for lvPPA, 7.1 years for nfvPPA, and 12 years for svPPA. The most common causes of death were natural cardio-pulmonary arrest and pneumonia. Aspiration pneumonia represented 23% of deaths in nfvPPA. In conclusion, this pilot study found significant differences in survival between the three variants of PPA with svPPA showing the longest and nfvPPA showing more neurologically-related causes of death.

Keywords: primary progressive aphasia; natural history; mortality; survival; memory clinic

Citation: Tastevin, M.; Lavoie, M.; de la Sablonnière, J.; Carrier-Auclair, J.; Laforce, R., Jr. Survival in the Three Common Variants of Primary Progressive Aphasia: A Retrospective Study in a Tertiary Memory Clinic. *Brain Sci.* **2021**, *11*, 1113. https://doi.org/10.3390/brainsci11091113

Academic Editor: Yang Zhang

Received: 4 August 2021
Accepted: 22 August 2021
Published: 24 August 2021

Publisher's Note: MDPI stays neutral with regard to jurisdictional claims in published maps and institutional affiliations.

1. Introduction

Primary progressive aphasias (PPAs) are a group of neurodegenerative diseases characterized by a predominant and progressive deterioration of language, with relative preservation of other cognitive functions over at least two years after the onset of the disease [1]. Since 2011, PPAs have been classified into three variants based on their clinical manifestations: the semantic variant (svPPA), the non-fluent/agrammatic variant (nfvPPA), and the logopenic variant (lvPPA) [2].

Demographic and epidemiological data regarding PPAs are lacking and most estimations are based on Frontotemporal Lobar Degeneration (FTLD) studies. Indeed, PPAs represent 20–40% of FTLD cases [3] with an estimated prevalence between 3.6 and 8.1/100,000 inhabitants [4–6]. A recent study suggested a prevalence of 3.1/100,000 (95% confidence interval [2.96–3.23]) from a French database including 2035 PPAs patients followed in tertiary centers [7].

While nfvPPA and svPPA are commonly considered as clinical presentations of FTLD with a predominant FTLD-tau pathology for nfvPPA (64% of cases) and FTLD-TDP-43 for svPPA (80% of cases), it is estimated that 86% of lvPPA are associated with Alzheimer's pathology [8].

Like all neurodegenerative diseases, the impact of PPA on the functional, socio-economic, and quality of life aspects is significant [9,10]. In addition, at diagnostic announcement, advanced stages and end of life issues are rarely discussed with families and caregivers. Moreover, no pertinent guidelines are available for assisting the medical team in prognosis announcement, which can have a significant psychological impact for patients and their families [11,12].

Although some authors have studied the natural history of FTLD variants, survival analyses remain scarce in the literature especially for PPAs. In the most recent mixed effects meta-analysis of survival in FTLD published in 2016 by Kansal et al. [13], the mean and median survival in svPPA variant were respectively 7.45 and 12.22 years. For nfvPPA, mean and median survival were 7.69 and 8.11 years, respectively [13]. To date, no study included the three PPA variants in survival analyses and epidemiological and survival data are still unknown for lvPPA. Therefore, it is essential to improve our knowledge of the natural history of PPAs, so that patients and their families can be informed about the onset and management of the different clinical manifestations as well as be properly prepared for end of life issues. The aim to the present pilot study was to analyze and compare survival data between the three PPAs variants and to describe causes of death.

2. Materials and Methods

2.1. Subjects

We conducted a retrospective study including deceased patients with a diagnosis of PPA that had been followed at our tertiary memory clinic over the past twenty years (n = 83). Initial diagnostic evaluation was obtained by a neurologist, a geriatrician or a neuropsychiatrist with a strong clinical experience in cognitive disorders. Diagnosis was based on an extensive clinical and paraclinical evaluation including speech and neuropsychological examination, structural (MRI–dementia protocol) and molecular neuroimaging (FDG-PET), or cerebrospinal fluid AD biomarkers (aB1-42, total-tau, phospho-tau). For all patients recruited after 2011, diagnosis was established according to the clinical criteria by Gorno-Tempini et al. [2]. According to these criteria, svPPA is associated with impaired confrontation naming and single-word comprehension. Additionally, patients may show impaired object knowledge as well as surface dyslexia or dysgraphia. In this variant, repetition and speech production are usually spared. Secondly, patients with nfvPPA must show apraxia of speech and/or agrammatism in speech production. Comprehension of syntactically complex sentences can also be impaired. Spared single-word comprehension and object knowledge is expected. Finally, lvPPA is defined by impaired single-word retrieval in naming and spontaneous speech as well as impaired sentence repetition, particularly for long sentences. Patients may also produce phonologic errors. In this variant, single-word comprehension, object knowledge, grammar and speech production are usually spared. Patients recruited before 2011 were screened and reclassified a posteriori using these diagnostic criteria. Only patients for whom the date of death was not available were excluded from this study, otherwise this study recruited all deceased PPA patients.

2.2. Data Collection

For each patient, data were collected from medical records at several time periods five years before the first visit until death that is at five years, two years and one year before the first visit, the day of the first visit, as well as one year, two years, five years and 10 years after the first visit. Our follow-up strategy is illustrated in Figure 1.

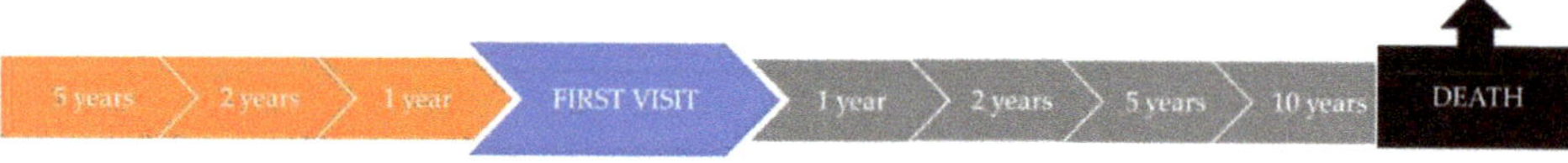

Figure 1. Follow-up strategy for data extraction.

Three of the authors (psychiatrist, neurologist, and a speech therapist) independently screened and extracted the data for all 83 patients. Subsequently, each diagnosis and data were reviewed, and disagreements were resolved by a consensus discussion between the authors.

Data were collected using a standard electronic form to ensure consistency of the appraisals and diagnosis for each patient.

For the present study, the following data were extracted from the database:

1. Socio-demographic data: gender, years of education;
2. Clinical data: diagnosis, MMSE at onset, age at symptoms onset, age at first visit, age at diagnosis, age of death, cause of death, duration from onset to diagnosis, duration from onset to first visit, duration from first visit to diagnosis, disease duration from diagnosis, and disease duration from onset.

When possible, data were validated and corroborated using a semi-structured telephone interview with caregivers. For the present study, 19 interviews were available.

The presence of comorbidities (i.e., cardiovascular risk factors and pulmonary risk factors) as well as intake of medication for cardiovascular conditions (hypertension, diabetes, dyslipidemia) and psychological disorders (anxiety, depression, psychosis, hallucinations, etc.) were extracted and all three groups were similar. Of note, some comorbidity data were not available, and we therefore only focused on those related to cardiovascular and pulmonary data.

2.3. Statistical Analyses

Statistical analyses were performed using the R Core Team software [14]. Mean and standard deviations were used to present patient characteristics whereas we used proportions to describe causes of death. A Khi-2 test was conducted for qualitative data. A one-way analysis of variance ANOVA was conducted to compare mean values between the three PPAs variants. A Kaplan–Meier analysis was performed to analyze survival and completed by log rank tests to examine survival curves across diagnostic groups. Cox's proportional hazard regression were performed to evaluate the effect of possible confounders. All statistical analyses were performed two-sided, and a p value of <0.05 was considered as significant.

3. Results

3.1. Patient Characteristics

Sociodemographic and clinical characteristics of the patients are described in Table 1. Eighty-three patients were included in the analyses: 35 lvPPA, 18 svPPA, and 30 nfvPPA. The proportion of male and female was equivalent in each group and there was no statistically significant difference between the three PPA variants. MMSE scores at first visit and education level were also not significantly different across the variants. However, svPPA patients with more years of education (>12 years) were overrepresented compared to nfvPPA and lvPPA ($p = 0.012$).

Table 1. Patient characteristics.

	lvPPA ($n = 35$)	**nfvPPA ($n = 30$)**	**svPPA ($n = 18$)**	p
Male/Female	19/16	15/15	9/9	n.s
Education years (mean ± SD)	11.7 ±4.4	11.03 ± 4.2	13.17 ± 4.1	n.s
Education level (>12 years)	15	9	12	0.026
MMSE 1st visit (mean ± SD)	21.03 ± 5.9	21.67 ± −6.8	22.35 ± 7.4	n.s
Age of onset (mean ± SD)	69.05 ± 10.8	70.13 ± 6.9	64.38 ± 7.8	n.s
Age at diagnosis (mean ± SD)	71.65 ± 10.22	73.1 ± 6.79	68.68 ± 8.5	n.s
Age of death (mean ± SD)	75.90 ± 9.5	76.8 ± 6.3	74.11 ± 8.3	n.s

n.s = not significant. SD = standard deviation.

Mean age of onset was 69.05 ± 10.8 years for lvPPA and 70.13 ± 6.9 for nfvPPA. The mean age of onset was earlier for svPPA (64.38 ± 7.8 years), but no statistically significant difference was found compared to lvPPA ($p = 0.17$) and nfvPPA ($p = 0.08$). No significant differences were found regarding age of death which occurred after 74 years-old for each subgroup. Mean age at diagnosis did not significantly differ between the three PPA variants.

3.2. Disease Duration

Several interval times were analyzed to describe and compare disease duration. Duration from onset was significantly longer for svPPA ($p = 0.001$ versus nfvPPA and lvPPA). Disease duration since diagnosis was also longer for svPPA but a statistically significant difference was only observed when compared with nfvPPA ($p = 0.01$). Diagnosis latency, which corresponds to the mean time needed to provide a diagnosis, was significantly longer for svPPA versus other variants. A mean of 13 ± 18 months was necessary to establish a diagnosis of svPPA after the first visit (see Figure 2) and diagnostic latency from first symptoms was 4.47 ± 2.03 years (see Figure 3). No differences were found between the variants in the interval between first symptoms and first medical visit.

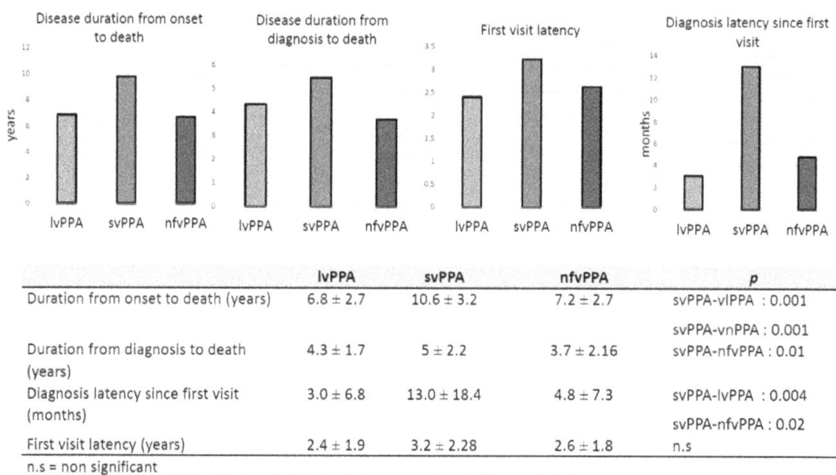

	lvPPA	svPPA	nfvPPA	*p*
Duration from onset to death (years)	6.8 ± 2.7	10.6 ± 3.2	7.2 ± 2.7	svPPA-vlPPA : 0.001
				svPPA-vnPPA : 0.001
Duration from diagnosis to death (years)	4.3 ± 1.7	5 ± 2.2	3.7 ± 2.16	svPPA-nfvPPA : 0.01
Diagnosis latency since first visit (months)	3.0 ± 6.8	13.0 ± 18.4	4.8 ± 7.3	svPPA-lvPPA : 0.004
				svPPA-nfvPPA : 0.02
First visit latency (years)	2.4 ± 1.9	3.2 ± 2.28	2.6 ± 1.8	n.s
n.s = non significant				

Figure 2. Disease duration.

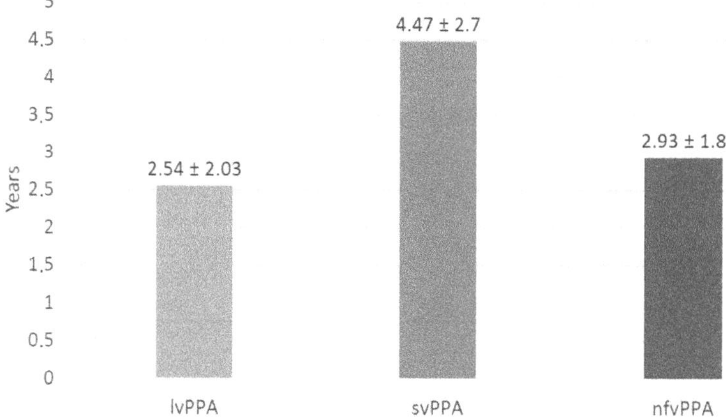

Figure 3. Diagnosis latency since onset.

3.3. Survival Analyses

Mean survival time is summarized in Table 2. Mean and estimated median survival time in patients with svPPA were respectively 12 years and 10 years from onset to death and 7.3 years and 7.5 years from diagnosis to death. Mean survival time in nfvPPA and lvPPA were respectively 7.1 and 7.6 years from onset to death, and 5.6 and 4.7 years from onset to diagnosis. Estimated median survival time was the same for both variants with six years from onset and five years from diagnosis. Causes of death were only available for 34 patients. The most common causes were natural cardio-pulmonary arrest (26.4%), followed by pneumonia (23.52%), cachexia (14.7%), and bedsores infections (11.7%). Major adverse cardiovascular events including stroke, cardiac infraction and systemic embolism represented 11.7% of causes. Half of the pneumonia were of the aspiration subtype and this accounted for 23% of deaths in nfvPPA.

Table 2. Mean and estimated median survival time (years).

	Mean Survival		Estimated Median Survival	
	Since Onset	**Since Diagnosis**	**Since Onset**	**Since Diagnosis**
lvPPA ($n = 35$)	7.6 CI95%: 6.2–8.9	5.6 CI95%: 4.5–6.5	6 CI95%: 5.0–10.0	5 CI95%: 0.0–10
nfvPPA ($n = 30$)	7.1 CI95%: 5.9–8.3	4.7 CI95%: 3.7–7.6	6 CI95%: 6.00–7.00	5 CI95%: 0.0–10.0
svPPA ($n = 18$)	12 CI95%: 9.5–14.4	7.3 CI95%: 6.0–8.6	10 CI95%: 0.0–20.0)	7.5 CI95%: 5.0–10.0

CI = Confidence interval.

Survival time since onset and diagnosis were significantly longer in svPPA than in lvPPA ($p = 0.02$ and $p = 0.04$, respectively) and nfvPPA ($p < 0.0001$ and $p = 0.004$, respectively) (see Figure 4). A poor association was observed between age of onset and survival from onset in lvPPA with a hazard ratio (HR) next to 1 (HR = 1.008, $p = 0.019$), not pertinent for interpretation (see Table 3). No relevant confounders were associated with survival since diagnosis (see Table 4).

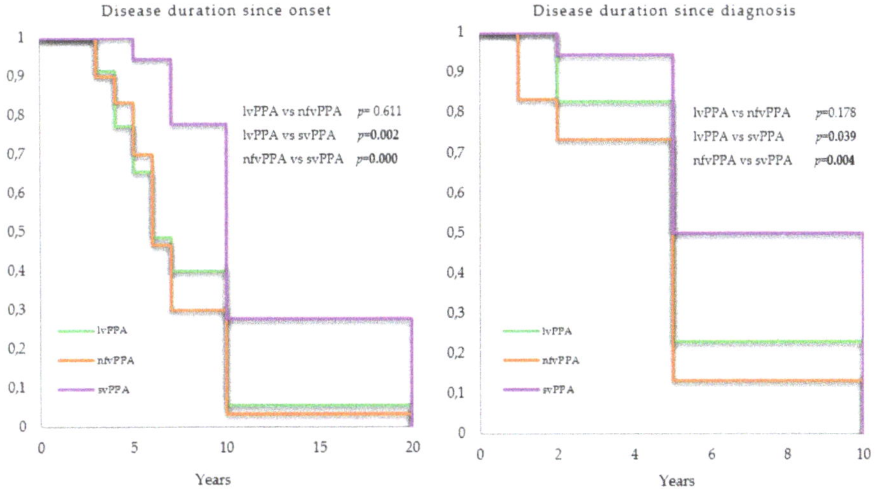

Figure 4. Kaplan–Meier survival curves.

Table 3. Hazard ratios since onset.

	lvPPA		nfvPPA		svPPA	
	HR (95%CI)	*p*-Value	HR (95%CI)	*p*-Value	HR (95%CI)	*p*-Value
MMSE 1st visit	0.96 (0.89–1.03)	0.30	0.96 (0.89–1.02)	0.21	0.99 (0.89–1.10)	0.90
Age of onset	1.05 (1.00–1.08)	0.02	1.03 (0.95–1.11)	0.52	0.96 (0.86–1.07)	0.43
Years of education	1.02 (0.94–1.11)	0.56	1.02 (0.88–1.17)	0.84	0.98 (0.82–1.16)	0.81
Gender	1.36 (0.56–3.32)	0.50	0.82 (0.33–2.08)	0.68	0.46 (0.13–1.64)	0.2

HR = Hazard ratio; CI = Confidence interval.

Table 4. Hazard ratios since diagnosis.

	lvPPA		nfvPPA		svPPA	
	HR (95%CI)	*p*-Value	HR (95%CI)	*p*-Value	HR (95%CI)	*p*-Value
MMSE 1st visit	0.98 (0.92–1.06)	0.66	0.98 (0.87–1.10)	0.39	0.99 (0.89–1.10)	0.91
Age of onset	1.01 (0.98–1.06)	0.48	1.02 (0.95–1.11)	0.52	1.00 (0.91–1.09)	0.99
Years of education	1.02 (0.94–1.10)	0.57	0.98 (0.87–1.10)	0.74	0.99 (0.82–1.20)	0.98
Gender	0.80 (0.36–1.81)	0.60	1.19 (0.49–2.84)	0.70	1.12 (0.28–4.52)	0.87

HR = Hazard ratio; CI = Confidence interval.

4. Discussion

The aim of this work was to study survival in a group of deceased PPA patients that have been followed at our memory clinic. To date, only five studies on survival in nfvPPA and svPPA were published [15–19]. Among these studies, only three compared the two variants. Moreover, publication date was before the most recent classification criteria by Gorno-Tempini and colleagues. Since 2011, no PPA survival analysis has been done and therefore no study included patients with lvPPA. To our knowledge, this is the first survival study including patients with the three PPA variants.

Despite the small sample size, PPA groups were representative of current practice, where approximately 42% of patients present with lvPPA, 36% with nfvPPA, and 22% with svPPA. Magnin et al. in 2016 [7], compared the demographical data across 2035 PPA subjects. They proposed three samples of patients. Sample 1 included all participants (n = 2035) and Sample 2 (n = 67) was a subgroup from Sample 1 with CSF biormarkers available. Sample 2 was divided between two subgroups: nfvPPA/ lvPPA/unclassifiable PPA and svPPA. Sample 3 (n = 97) was divided between lvPPA, svPPA, and nfvPPA, and the CSF biomarkers were available for all of them. In Sample 1, the proportion of svPPA patients represented 28.1% of the entire sample. Moreover, the repartition was comparable across each sample with less svPPA patients [7]. Our sample also showed a smaller proportion of svPPA patients compared to the two other variants. Moreover, the authors estimated that nfvPPA/lvPPA/unclassifiable PPA were more frequent than svPPA (2.2 versus 0.8/100,000 inhabitants; p < 0.00001) [7].

In this study, patients with svPPA were younger than lvPPA and nfvPPA patients with a mean age at onset of 64 years-old, which is similar to that found in the three survival studies cited earlier. However, no statistically significant differences across the PPA groups were found, which was also the case in the studies by Hodges et al. (2003), Nunnemann et al. (2011), and Roberson et al. (2005) [15,17,18]. On the other hand, Kertesz et al. (2007) showed a statistically significant younger age of onset in svPPA patients compared to nfvPPA patients [16]. The same profile was highlighted concerning age at diagnosis. Age at diagnosis was also younger in the eight natural studies of PPA including the three variants (between 60 and 70 years) and no differences were observed across groups [19–26]. Absence of differences could be explained by small sample sizes. Indeed, no data were available for sample sizes over 150 subjects. In Sample 1 (n = 2035) of the multicentric epidemiologic study published by Magnin et al. (2016) [7], the nfvPPA/lvPPA/unlassifiable PPA group was significantly older at disease onset and at diagnosis than the svPPA group ($p < 0.00001$). There was no significant differences in level of education or gender. Compared to typical AD, in svPPA patients, male predominance occurred after the age of 80, the level of education was higher, and the age of onset was younger (71.6 versus 78.61 years old), all differences being statistically significant with a $p < 0.0001$. In Sample 3 (n = 97), svPPA patients were also younger at disease onset than lvPPA patients (59.48 years-old versus 63.72 years-old) and no differences were observed on gender and age of education. After CSF biomarkers stratification, no significant difference was observed for age at onset, gender, or level of education between the "PPA-AD" group and the "PPA–not "AD" group [7].

In this study, no significant differences were observed across MMSE scores. In a study by Ulugut et al. published in 2021 [21], MMSE score also did not differ significantly among the PPA variants. However, lvPPA patients performed worse on executive and visuospatial specific testing and svPPA performed worse on the verbal memory test at baseline. These findings must, however, be interpreted with caution as cognitive assessment can be strongly impacted by language abilities especially on global scores (MoCA and MMSE). In clinical practice, it is frequent to observe a clinical dissociation between performance on cognitive tests and level of functioning on collateral history. None of the PPA patients were tested using specific assessment and neuropsychological tools that take into account language impairments [27].

In the present study, diagnostic latencies since onset and since the first visit were significantly longer for svPPA versus the other variants. Especially, the diagnostic latency since onset was 4.47 years in svPPA versus 2.54 years in lvPPA and 2.93 years in nfvPPA. Although latencies were not explored in PPA survival studies, several other observational studies have found similar results. Indeed, in the analysis of Sample 3, Magnin et al. (2016) found that the delay between first symptoms and PPA diagnosis was longer in svPPA patients (4.48 years) compared to lvPPA (3.02 years) and nfvPPA (2.26 years) [7]. In the study published by Van Langenhove et al. (2016), symptoms duration at baseline was 4.4 years for svPPA, compared to 2.3 and 3.5 years in nfvPPA and lvPPA, respectively [22]. Hseish et al. (2012) also found that svPPA had a mean disease duration of 4.2 years at time of diagnosis, compared with 2.3 and 3.9 years for nfvPPA and lvPPA [26]. These results could be explained by a more challenging recognition of the disease by the family and the physicians. Indeed, loss of semantic knowledge can be masked by a more fluent profile and patients are, therefore, more able to develop compensation mechanisms at the onset of disease. Moreover, it can also be masked by the first symptoms being characterized as a predominant memory and behavioral presentation [21].

Our survival results were compatible with findings from a meta-analysis published by Kansal et al. in 2016, with a median survival estimated at 10 years for svPPA (Kansal et al. showed a median svPPA survival of 12 years). In this study, we also found a significantly longer survival time in svPPA patients. However, Kansal et al. results revealed a significant difference only in median survival, whereas mean survival between svPPA and nfvPPA patients did not reach statistical significance [13]. The three survival studies comparing svPPA and nfvPPA did not show differences on survival time [15–17]. Moreover, contrary to Kansal et al., we found higher mean survival times than median survival times for svPPA and nfvPPA. The authors underlined this asymmetry between median and mean survival findings in svPPA results. In their discussion, Kansal et al. suggested an artefact in the analysis due to the heterogeneity the included studies (sampling methods and regional context) but they also proposed an interesting hypothesis where presence of a negative or positive skew could be a statistical reflection of the survival profile. A negative skew could concern a disease with young- or mid-life onset, a sufficiently long course, with few premature deaths. In contrast, a positive skew was more likely when a disease was characterized by a very short course, as the outliers are those with unusually long survival. It is important to consider that the meta-analysis was first limited by the small number of studies included, with only three comparing nfvPPA and svPPA. In the study by Nunnemann et al. [17], median survival in the svPPA group could not be defined because less than half of the patients had died at the end of the observational period. Moreover, in the three studies, the sample size of died patients was extremely limited with only nine svPPA and eight nfvPPA for Hodges et al. (2003) [15], three svPPA and seven nfvPPA for Nunnemann et al. (2011) [17], twelve svPPA and seven nfvPPA for Roberson et al. (2005) [18].

According to the WHO definition, years of life lost (YLL) is the age at which deaths occur by giving greater weight to deaths at younger age and lower weight to deaths at older age. The years of life lost (percentage of total) indicator measures the YLL due to a cause as a proportion of the total YLL lost in the population due to premature mortality. We did not include YLL in our survival analyses. It could have been interesting especially for patients and their family to know which variant is associated with premature mortality. In the meta-analysis of Kansal et al. (2016), mean YLL was significantly higher in svPPA compared to nfvPPA (13.56 years versus 10.54 years). These results could also be overestimated for svPPA because only premature deaths were analyzed for svPPA in the study of Nunnemann et al. (2011), given that less than half of the patients had died at the end of the observational period.

Finally, causes of death in PPA are poorly described in the current literature. In this study, we were only able to obtain causes for 34 patients, mostly by interviews with the caregivers. Our results were similar to the ones of Nunnemann et al. (2011) [17] with a predominance of respiratory system disorder including aspiration pneumonia, circulatory system disorder and cachexia. It has been our experience that a small proportion of PPA patients develop Progressive Supranuclear Palsy and/or Corticobasal Syndrome. These patients tend to show a poorer evolution over time. However, this study did not specifically address this or intend to compare AD with PSP and/ or CBS.

The main limitation of this pilot work was undoubtedly the small sample size, however, of higher or comparable size to that of studies already published. Our study was also the first to include lvPPA patients. Future multicentric studies should be conducted on the three PPA variants to obtain results for larger sample sizes. Moreover, in this study, few data were available regarding causes of death. Indeed, most patients died when in long term care homes and therefore were not admitted in the hospital at the time of death. Moreover, many came from all across the province of Quebec and their full medical record was not available to us. Finally, most diagnoses were not confirmed by pathological autopsy and misdiagnosis cannot be excluded for patients diagnosed prior to the 2011 criteria. However, our data extraction strategy was independently realized by a pluridisciplinary team experienced in the classification of PPAs, and secondarily completed by a consensus discussion to control for management of clinical misdiagnosis.

5. Conclusions

This is the first study to analyze survival data across the three variants of PPA according to Gorno-Tempini's criteria [2]. Our preliminary results tend to highlight a specific profile for svPPA patients in comparison to the two other variants. Patients suffering from svPPA seem to get a confirmed diagnosis later than the others but svPPA remains characterized by the best survival time. More survival analyses, integrating the most recent diagnostic criteria, all phenotypes and multicentric databases will be necessary to confirm our findings. Moreover, more complete survival analyses including median and mean survival time, YLL, comorbidities and treatments could offer better prevention information for patients and their caregivers by taking account premature death and their causes. To date, these preliminary findings already provide precious data to better prepare them to the progression of the disease and end stages of life.

Author Contributions: Conceptualization: R.L.J., M.T. and M.L.; methodology: R.L.J., M.T. and M.L.; software: M.T.; validation: M.T., M.L. and J.d.l.S.; formal analysis: M.T.; investigation: M.T., M.L., J.d.l.S. and J.C.-A.; resources: M.T.; data curation: M.T., M.L., J.d.l.S. and J.C.-A.; writing—original draft preparation: M.T.; writing—review and editing: M.T., M.L. and R.L.J.; visualization: R.L.J. and M.L.; supervision: R.L.J.; project administration: M.T., R.L.J. and M.L.; funding acquisition: R.L.J. All authors have read and agreed to the published version of the manuscript.

Funding: This research received external funding from the La Chaire de Recherche sur les Aphasies Primaires Progressives–Fondation de la famille Lemaire (https://app-ffl.ulaval.ca/), accessed on 15 January 2020.

Institutional Review Board Statement: The study was conducted according to the guidelines of the Declaration of Helsinki, and approved by Ethics Committee of CHU de Quebec (protocol code: 2021-5406, date of approval: 9 September 2020).

Informed Consent Statement: Written Informed consent was obtained from all caregivers involved in the study. The consent statement was not applicable because all patients included were dead.

Data Availability Statement: The data that support the findings of this study are available from the corresponding author upon reasonable request.

Conflicts of Interest: The authors declare no conflict of interest.

References

1. Mesulam, M.-M. Primary Progressive Aphasia. *Ann. Neurol.* **2001**, *49*, 425–432. [CrossRef]
2. Gorno-Tempini, M.L.; Hillis, A.E.; Weintraub, S.; Kertesz, A.; Mendez, M.; Cappa, S.F.; Ogar, J.M.; Rohrer, J.D.; Black, S.; Boeve, B.F.; et al. Classification of Primary Progressive Aphasia and Its Variants. *Neurology* **2011**, *76*, 1006–1014. [CrossRef]
3. Grossman, M. Primary Progressive Aphasia: Clinicopathological Correlations. *Nat. Rev. Neurol.* **2010**, *6*, 88–97. [CrossRef]
4. Ratnavalli, E.; Brayne, C.; Dawson, K.; Hodges, J.R. The Prevalence of Frontotemporal Dementia. *Neurology* **2002**, *58*, 1615–1621. [CrossRef]
5. Onyike, C.U.; Diehl-Schmid, J. The Epidemiology of Frontotemporal Dementia. *Int. Rev. Psychiatry* **2013**, *25*, 130–137. [CrossRef] [PubMed]
6. Gilberti, N.; Turla, M.; Alberici, A.; Bertasi, V.; Civelli, P.; Archetti, S.; Padovani, A.; Borroni, B. Prevalence of Frontotemporal Lobar Degeneration in an Isolated Population: The Vallecamonica Study. *Neurol. Sci.* **2012**, *33*, 899–904. [CrossRef] [PubMed]
7. Magnin, E.; Démonet, J.-F.; Wallon, D.; Dumurgier, J.; Troussière, A.-C.; Jager, A.; Duron, E.; Gabelle, A.; de la Sayette, V.; Volpe-Gillot, L.; et al. Primary Progressive Aphasia in the Network of French Alzheimer Plan Memory Centers. *J. Alzheimers Dis.* **2016**, *54*, 1459–1471. [CrossRef] [PubMed]
8. Bergeron, D.; Gorno-Tempini, M.L.; Rabinovici, G.D.; Santos-Santos, M.A.; Seeley, W.; Miller, B.L.; Pijnenburg, Y.; Keulen, M.A.; Groot, C.; van Berckel, B.N.M.; et al. Prevalence of Amyloid-β Pathology in Distinct Variants of Primary Progressive Aphasia. *Ann. Neurol.* **2018**, *84*, 729–740. [CrossRef] [PubMed]
9. Moyle, W.; Murfield, J.E. Health-Related Quality of Life in Older People with Severe Dementia: Challenges for Measurement and Management. *Expert Rev. Pharm. Outcomes Res.* **2013**, *13*, 109–122. [CrossRef] [PubMed]
10. Manuel, D.G.; Garner, R.; Finès, P.; Bancej, C.; Flanagan, W.; Tu, K.; Reimer, K.; Chambers, L.W.; Bernier, J. Alzheimer's and Other Dementias in Canada, 2011 to 2031: A Microsimulation Population Health Modeling (POHEM) Study of Projected Prevalence, Health Burden, Health Services, and Caregiving Use. *Popul. Health Metr.* **2016**, *14*, 344–362. [CrossRef]
11. Moore, K.J.; Goodison, H.; Sampson, E.L. The Role of the Memory Service in Helping Carers to Prepare for End of Life: A Mixed Methods Study. *Int. J. Geriatr. Psychiatry* **2019**, *34*, 360–368. [CrossRef]
12. Moore, K.J.; Davis, S.; Gola, A.; Harrington, J.; Kupeli, N.; Vickerstaff, V.; King, M.; Leavey, G.; Nazareth, I.; Jones, L.; et al. Experiences of End of Life amongst Family Carers of People with Advanced Dementia: Longitudinal Cohort Study with Mixed Methods. *BMC Geriatr.* **2017**, *17*, 135. [CrossRef]
13. Kansal, K.; Mareddy, M.; Sloane, K.L.; Minc, A.A.; Rabins, P.V.; McGready, J.B.; Onyike, C.U. Survival in Frontotemporal Dementia Phenotypes: A Meta-Analysis. *Dement. Geriatr. Cogn. Disord.* **2016**, *41*, 109–122. [CrossRef]
14. R Foundation for Statistical Computing; R Development Core Team R. *A Language and Environment for Statistical Computing*; R Development Core Team R: Vienna, Austria, 2008; ISBN 3900051070.
15. Hodges, J.R.; Davies, R.; Xuereb, J.; Kril, J.; Halliday, G. Survival in Frontotemporal Dementia. *Neurology* **2003**, *61*, 349–354. [CrossRef]

16. Kertesz, A.; Blair, M.; McMonagle, P.; Munoz, D.G. The Diagnosis and Course of Frontotemporal Dementia. *Alzheimer Dis. Assoc. Disord.* **2007**, *21*, 155–163. [CrossRef]

17. Nunnemann, S.; Last, D.; Schuster, T.; Förstl, H.; Kurz, A.; Diehl-Schmid, J. Survival in a German Population with Frontotemporal Lobar Degeneration. *Neuroepidemiology* **2011**, *37*, 160–165. [CrossRef] [PubMed]

18. Roberson, E.D.; Hesse, J.H.; Rose, K.D.; Slama, H.; Johnson, J.K.; Yaffe, K.; Forman, M.S.; Miller, C.A.; Trojanowski, J.Q.; Kramer, J.H.; et al. Frontotemporal Dementia Progresses to Death Faster than Alzheimer Disease. *Neurology* **2005**, *65*, 719–725. [CrossRef] [PubMed]

19. Ferrari, C.; Polito, C.; Vannucchi, S.; Piaceri, I.; Bagnoli, S.; Lombardi, G.; Lucidi, G.; Berti, V.; Nacmias, B.; Sorbi, S. Primary Progressive Aphasia: Natural History in an Italian Cohort. *Alzheimer Dis. Assoc. Disord.* **2019**, *33*, 42–46. [CrossRef] [PubMed]

20. Matias-Guiu, J.A.; Cabrera-Martín, M.N.; Moreno-Ramos, T.; García-Ramos, R.; Porta-Etessam, J.; Carreras, J.L.; Matías-Guiu, J. Clinical Course of Primary Progressive Aphasia: Clinical and FDG-PET Patterns. *J. Neurol.* **2015**, *262*, 570–577. [CrossRef] [PubMed]

21. Ulugut, H.; Stek, S.; Wagemans, L.E.E.; Jutten, R.J.; Keulen, M.A.; Bouwman, F.H.; Prins, N.D.; Lemstra, A.W.; Krudop, W.; Teunissen, C.E.; et al. The Natural History of Primary Progressive Aphasia: Beyond Aphasia. *J. Neurol.* **2021**, *831*, 442–468. [CrossRef]

22. Van Langenhove, T.; Leyton, C.E.; Piguet, O.; Hodges, J.R. Comparing Longitudinal Behavior Changes in the Primary Progressive Aphasias. *J. Alzheimers Dis.* **2016**, *53*, 1033–1042. [CrossRef] [PubMed]

23. Foxe, D.; Irish, M.; Hu, A.; Carrick, J.; Hodges, J.R.; Ahmed, R.M.; Burrell, J.R.; Piguet, O. Longitudinal Cognitive and Functional Changes in Primary Progressive Aphasia. *J. Neurol.* **2021**, *321*, 23–36. [CrossRef]

24. Rogalski, E.; Cobia, D.; Harrison, T.M.; Wieneke, C.; Weintraub, S.; Mesulam, M.-M. Progression of Language Decline and Cortical Atrophy in Subtypes of Primary Progressive Aphasia. *Neurology* **2011**, *76*, 1804–1810. [CrossRef] [PubMed]

25. Ash, S.; Nevler, N.; Phillips, J.; Irwin, D.J.; McMillan, C.T.; Rascovsky, K.; Grossman, M. A Longitudinal Study of Speech Production in Primary Progressive Aphasia and Behavioral Variant Frontotemporal Dementia. *Brain Lang.* **2019**, *194*, 46–57. [CrossRef] [PubMed]

26. Hsieh, S.; Hodges, J.R.; Leyton, C.E.; Mioshi, E. Longitudinal Changes in Primary Progressive Aphasias: Differences in Cognitive and Dementia Staging Measures. *Dement. Geriatr. Cogn. Disord.* **2012**, *34*, 135–141. [CrossRef] [PubMed]

27. Henry, M.L.; Grasso, S.M. Assessment of Individuals with Primary Progressive Aphasia. *Semin. Speech Lang.* **2018**, *39*, 231–241. [CrossRef] [PubMed]

Article

Verbal Short-Term Memory Disturbance in the Primary Progressive Aphasias: Challenges and Distinctions in a Clinical Setting

David Foxe [1,2], Sau Chi Cheung [1,2], Nicholas J. Cordato [2,3,4,5], James R. Burrell [2,6], Rebekah M. Ahmed [2,7], Cathleen Taylor-Rubin [8,9], Muireann Irish [1,2] and Olivier Piguet [1,2,*]

1 School of Psychology, The University of Sydney, Sydney, NSW 2006, Australia; david.foxe@sydney.edu.au (D.F.); sau.cheung@sydney.edu.au (S.C.C.); muireann.irish@sydney.edu.au (M.I.)
2 Brain and Mind Centre, The University of Sydney, Sydney, NSW 2050, Australia; Nicholas.Cordato@health.nsw.gov.au (N.J.C.); james.burrell@sydney.edu.au (J.R.B.); rebekah.ahmed@sydney.edu.au (R.M.A.)
3 St George Clinical School, University of New South Wales, Sydney, NSW 2217, Australia
4 The Department of Aged Care, St George Hospital, Sydney, NSW 2217, Australia
5 Calvary Health Care Kogarah, Sydney, NSW 2217, Australia
6 Concord Clinical School, Sydney Medical School, The University of Sydney, Sydney, NSW 2139, Australia
7 Central Medical School, The University of Sydney, Sydney, NSW 2006, Australia
8 Department of Cognitive Science, Faculty of Medicine, Health and Human Sciences, Macquarie University, Sydney, NSW 2109, Australia; Cathleen.Taylor@health.nsw.gov.au
9 Department of Speech Pathology, Uniting War Memorial Hospital, Sydney, NSW 2024, Australia
* Correspondence: olivier.piguet@sydney.edu.au; Tel.: +61-2-9114-4336

Citation: Foxe, D.; Cheung, S.C.; Cordato, N.J.; Burrell, J.R.; Ahmed, R.M.; Taylor-Rubin, C.; Irish, M.; Piguet, O. Verbal Short-Term Memory Disturbance in the Primary Progressive Aphasias: Challenges and Distinctions in a Clinical Setting. *Brain Sci.* **2021**, *11*, 1060. https://doi.org/10.3390/brainsci11081060

Academic Editors: Jordi A. Matias-Guiu, Robert Laforce Jr. and Rene L. Utianski

Received: 7 July 2021
Accepted: 9 August 2021
Published: 12 August 2021

Publisher's Note: MDPI stays neutral with regard to jurisdictional claims in published maps and institutional affiliations.

Abstract: Impaired verbal 'phonological' short-term memory is considered a cardinal feature of the logopenic variant of primary progressive aphasia (lv-PPA) and is assumed to underpin most of the language deficits in this syndrome. Clinically, examination of verbal short-term memory in individuals presenting with PPA is common practice and serves two objectives: (i) to help understand the possible mechanisms underlying the patient's language profile and (ii) to help differentiate lv-PPA from other PPA variants or from other dementia syndromes. Distinction between lv-PPA and the non-fluent variant of PPA (nfv-PPA), however, can be especially challenging due to overlapping language profiles and comparable psychometric performances on verbal short-term memory tests. Here, we present case vignettes of the three PPA variants (lv-PPA, nfv-PPA, and the semantic variant (sv-PPA)) and typical Alzheimer's disease (AD). These vignettes provide a detailed description of the short-term and working memory profiles typically found in these patients and highlight how speech output and language comprehension deficits across the PPA variants differentially interfere with verbal memory performance. We demonstrate that a combination of verbal short-term and working memory measures provides crucial information regarding the cognitive mechanisms underlying language disturbances in PPA. In addition, we propose that analogous visuospatial span tasks are essential for the assessment of PPA as they measure memory capacity without language contamination.

Keywords: primary progressive aphasia; frontotemporal dementia; Alzheimer's disease; neuropsychology; span; sentence repetition; working memory; phonological; visuospatial

1. Introduction

Impaired verbal 'phonological' short-term memory is considered a cardinal feature of the logopenic variant of primary progressive aphasia (lv-PPA) and is thought to underpin many of the language deficits in this syndrome [1]. Indeed, lv-PPA patients display impaired digit, letter, and word span on formal testing but perform normally on single-digit and -word repetition tasks [2–4]. Importantly, these deficits occur in the context of relatively preserved grammar and articulation, although phonological paraphasias may be present [1,3]. Poor verbal short-term memory performance also occurs in the non-fluent variant (nfv-PPA), although this impairment is typically due to motor speech or articulatory deficits [1,3,4]. In contrast, verbal short-term memory performance remains relatively spared in the early stages of the semantic variant of PPA (sv-PPA) [3,4]. These distinct verbal short-term memory profiles led the international consensus criteria for PPA to include 'impaired sentence repetition and phrases' as a core clinical feature of lv-PPA [1]—prompting clinicians to evaluate the verbal short-term memory system when assessing patients with a differential diagnosis of lv-PPA.

Despite these recommendations, multiple challenges exist for clinicians assessing these skills at the individual case level. For example, differentiating lv-PPA from nfv-PPA hinges on detecting motor speech and/or grammatical errors—a skill which requires considerable expertise in language assessment [5]. In addition, the presence and severity of these speech and language features are variable, especially in the early stages of the disease, making the distinction between lv-PPA and nfv-PPA challenging [6–10].

In clinical practice, the combination of language and short-term memory tests, however, can improve the clinician's ability to detect phonological impairment and delineate lv-PPA from the other PPA variants [3,11]. Evaluation of performance scores across tests, as well as awareness of the qualitative aspects of language (e.g., phonological disturbance, dysarthria, agrammatism), helps determine if impaired performance on verbal short-term memory measures is due to the breakdown of the verbal store and rehearsal system—indicating lv-PPA—or to a breakdown of other processes (e.g., nfv-PPA: motor speech programming deficits; sv-PPA: disrupted conceptual knowledge resulting in poor understanding/recollection of words or phrases) [3,11,12]. While these views are well-documented in PPA group comparison studies, attempts to implement this understanding through specific tests at an individual patient level have been limited. Investigations at the case level have several advantages over larger PPA group comparison studies, including the ability to: (i) interpret individual cases based on established norms tailored to the age and education of the individual; (ii) establish a differential diagnosis without referencing a demographic and disease severity matched PPA sample group, and; (iii) place emphasis on interpreting important qualitative aspects of language in conversational speech and on formal standardised testing.

In this study, we explored in detail the short-term memory profiles of individual patients with PPA (lv-PPA, nfv-PPA, sv-PPA) and, for comparison, Alzheimer's disease (AD). Using tests typically administered in secondary and tertiary clinics, we demonstrate how the language deficits of each PPA variant influence performance across various measures of verbal short-term memory and working memory. We also highlight how the breakdown of these performances can provide clinicians with qualitative insights into the core speech and memory mechanisms affected in an individual patient. Finally, we propose that the assessment of visuospatial short-term and working memory is relevant for the establishment of an accurate diagnosis of PPA.

2. Materials and Methods

The four patients presented here as case vignettes were seen at the FRONTIER Frontotemporal Dementia Research Group at the Brain and Mind Centre, The University of Sydney. They all underwent a comprehensive neurological (NJC, JRB, RMA), and systematic cognitive (DF, SCC) and speech assessment (CTR, DF, SCC), as well as structural brain magnetic resonance imaging (MRI). Sentence repetition was phonetically transcribed by CTR and qualitatively scored using the Hohlbaum, Dressel [12] scoring criteria. This study was approved by the Human Research Ethics Committee of the South-Eastern Sydney Local Area Health District (HREC 10/126). All participants provided written informed consent in accordance with the Declaration of Helsinki. Patient initials have been altered to protect the privacy of the individuals and their families.

3. Case Vignettes

3.1. lv-PPA Patient: NS

At presentation, NS was a 67-year-old, right-handed man (Table 1). He had 12 years of education and had been retired for 7 years, having previously worked in government services and in the tourism sector. He had also been heavily involved in managing the finances and building repairs at his local church but had ceased these duties approximately two years prior to his visit. His past medical history revealed a coronary stent 3 years prior to the assessment, and high cholesterol which was managed with medication. There was no known family history of dementia or other neurodegenerative conditions.

Table 1. Demographics and neuropsychological test scores.

Domain	Cognitive Test	Subtest (Max Score)	TN: AD Patient	NS: lv-PPA Patient	ML: nfv-PPA Patient	JC: sv-PPA Patient
Demographics						
Sex (m:f)			Male	Male	Female	Male
Age (y)			67	66	64	62
Handedness			Right	Right	Right	Right
Education (y)			9	12.25	12	16
Disease duration (y)			7.4	3.5	6.6	5.4
General cognition	ACE-III	Total (100)	68 **	66 **	81 *	67 **
Attention and executive functioning	Trails	A time (errors)	48 (0)	56 (0) *	65 (0) **	43 (0)
		B time (errors)	365 (3) **	460 (2) **	344 (0) **	99 (0)
		B-A time difference	317 **	404 **	279 **	56
	Letter fluency	F, A, S	36	14 **	12 **	36
Short-term and working memory	Digit Span	Raw Forward (longest)	9 (6)	4 (3) **	5 (4) *	12 (8)
		Raw Backward (longest)	5 (4)	2 (3) **	5 (4) *	6 (4)
		Raw Total (SS)	14 (9)	6 (4) **	10 (6) *	18 (11)
	Spatial Span	Raw Forward (longest)	6 (4)	4 (3) *	8 (6)	6 (6)
		Raw Backward (longest)	4 (4) *	5 (4)	6 (4)	8 (6)
		Raw Total (SS)	10 (6) *	9 (5) *	14 (10)	14 (10)
	Sentence Rep	Raw Total (14)	Nil	6 **	3 **	9 **
	Word Span	Raw Total (30)	Nil	9 **	11 **	24
Memory	RCFT	Copy (36)	12.5 **	29 *	30 *	36
		3-min recall (36)	1 **	7.5 *	18.5	22.5
Language	SYDBAT	Naming (30)	23 *	20 **	30	10 **
		Repetition (30)	30	24**	2 **	30
		Comprehension (30)	25 *	28	30	20 **
		Semantic Assoc. (30)	27	29	29	19 **
Visuospatial	Clock drawing	(5)	5	5	5	5
	ACE Visuospatial	(16)	15	15	14	16
	RCFT	Copy time (secs)	594 **	513 **	229	290
Mood	DASS-21	Depression	4 (Normal)	0 (Normal)	3 (Normal)	0 (Normal)
		Anxiety	2 (Normal)	1 (Normal)	3 (Normal)	0 (Normal)
		Stress	4 (Normal)	4 (Normal)	4 (Normal)	3 (Normal)
Functional capacity	FRS	Total Rasch	0.16 (Mod.)	2.86 (Mild)	5.39 (V. Mild)	2.19 (Mild)
	CDR-FTLD	Sums of boxes (SoB)	5 (Mild)	1.5 (Quest.)	4 (V. Mild)	2 (Quest.)

Notes: ACE-III: Addenbrooke's Cognitive Examination-Third Edition [13]; ACE Visuospatial: Visuospatial sub-score of the Addenbrooke's Cognitive Examination-III; CDR-FTLD: Frontotemporal Lobar Degeneration-Modified Clinical Dementia Rating Scale [14]; Clock drawing: Clock drawing subtest of the Addenbrooke's Cognitive Examination-III; DASS–21: Depression Anxiety Stress Scale–21 items [15,16]; Digit Span: Digit Span subtest of the Wechsler Adult Intelligence Scale–III WAIS-III; [17]; FRS: Frontotemporal Dementia Rating Scale [18]; Letter Fluency: Letters F, A and S [19]; Mod.: Moderate; Quest.: Questionable; RCFT: Rey Complex Figure Test [20]; Secs: seconds; Sentence Repetition: Sentence Repetition from the Multilingual Aphasia Examination [21]; Spatial Span: Spatial Span subtest of the Wechsler Memory Scale–III (WMS–III) [22]; SS: scaled score; SYDBAT: The Sydney Language Battery [23]; Trails: Trail Making Test [24]; V. Mild: Very Mild; Word Span: word span test from Leyton, Savage [3]. * indicates borderline performance: 1.3 < z-score < 2.0; 3 < percentile < 9; ** indicates extremely low performance: z-score < −2.0; percentile < 2.

NS was assessed following a 4-year history of speech and language difficulties. Initial symptoms included mispronouncing some words ("stunt" for 'stent'; "wiltered" for 'withered'), word substitutions, slowed reading rate, surface dyslexia, and spelling errors. Word-finding difficulties had reportedly become more apparent in the 2 years prior to the visit, particularly in instances that required rapid spontaneous speech. Cognitively, NS felt less confident about his memory and concentration. He also experienced some topographical disorientation in unfamiliar locations. No other cognitive or motor changes were reported, and he remained independent in all activities of daily living (ADLs). His wife had not noticed any behavioural or personality changes and there was no history of psychiatric features. NS did not report any symptoms of depression, anxiety, or stress, and he demonstrated appropriate emotional reactivity during the assessment.

3.1.1. Neuropsychological Assessment

Based on his educational and vocational history, NS's estimated premorbid level of functioning was average. On a measure of general cognitive ability, the Addenbrooke's Cognitive Examination-III (ACE-III), he scored 66/100 which was well-below normal limits (normal > 88), [13,25] (Table 1). His conversational speech was dysfluent with frequent word-finding pauses and phonological errors, though prosody was intact. Formal neuropsychological assessment revealed moderate to severe expressive language difficulties. Verbal fluency and confrontation naming were very impaired, and repetition of multisyllabic words was also reduced somewhat (Table 1, Figure 1). In contrast, comprehension (i.e., word-picture matching) and conceptual semantic knowledge were relatively preserved (Figure 1).

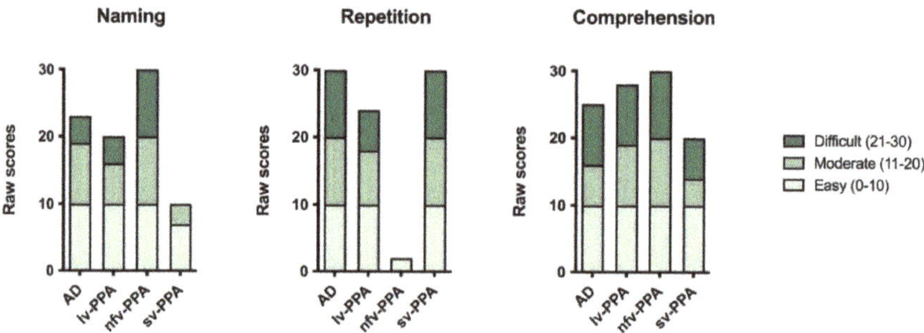

Figure 1. Raw scores on the Naming, Repetition, and Comprehension subtests of the Sydney Language Battery. Note, due to speech output difficulties, the nfv-PPA patient wrote their responses for the Naming subtest (Appendix A).

Verbal short-term memory (i.e., Digit Span Forward, Sentence Repetition, Word Span) and verbal working memory (i.e., Digit Span Backward) were extremely impaired (Tables 1 and 2, Figure 2). His performance on the visuospatial counterpart tasks was comparatively better although still fell in the borderline-impaired range (Table 1, Figure 3).

Executive functioning difficulties were also evident. Specifically, NS demonstrated impairments in proverb interpretation, and rapid set-shifting. Complex visuo-constructional planning was disorganised and extremely fragmented (Figure 4) in the context of intact basic visuo-perceptual skills and psychomotor speed. This was likely to have impacted on his visual memory performance which was borderline-impaired.

Table 2. PPA patients' responses on the Multilingual Aphasia Examination Sentence Repetition test.

		Analysis on Sentence-Level 0 = No; 1 = Yes						Frequency		Analysis on Word-Level					
Item	Produced Sentence	Correct 0/1	Erroneous Correct 0/1	Uncertain/Repaired Correct 0/1	Required Stimulus Repetition Correct 0/1	Misordered Words Correct 0/1	Violation of Syntactic Rules Correct 0/1	Omission	Semantic Substitution	Formal Errors	Phonological Errors	Phonological Errors Affecting Morphemes	Grammatical Errors	Sound Deviations	Not Classified Errors
NS: lv-PPA patient															
1	Take this home	1	0	0	0	0	0								
2	Where is the child?	1	0	0	0	0	0								
3	The car will not run	1	0	0	0	0	0								
4	Why are they not living here?	1	0	0	0	0	0								
5	The band (.) played and the/kraʊnd/(5) cheered	0	1	0	0	0	0				1				
6	Where are you going to work next summer?	1	0	0	0	0	0								
7	He sold his house/ɔn/moved to the farm	0	1	0	0	0	0	1							
8	Work in the garden until you've picked all the beans	1	0	0	0	0	0								
9	The artist/peɪtəd/painted (3) many pictures of the/fɑː/no sorry	0	1	0	0	0	0	4	1		2				
10	This doctor doesn't go to all of the towns	0	1	0	0	0	0	3	1 *						
11	She should be able to tell us when/ʃɜː/(2) when she (.) is (.) performing	0	1	0	0	0	0	4		1					
12	Why/dɒn/(.) that group (1) why doesn't that group apply (.) for (.) money	0	1	0	0	0	0	8	1 *		1				
13	Many/piːpəl/(5) they were not able to get work because of the (.) weather	0	1	0	0	0	0	6	2 *				1		
14	*Did not attempt*														
	Total	6	7	0	0	0	0	26	5	1	4	0	1	0	0

152

Table 2. *Cont.*

Item	Produced Sentence	Analysis on Sentence-Level 0 = No; 1 = Yes						Analysis on Word-Level Frequency							
		Correct 0/1	Erroneous Correct 0/1	Uncertain/Repaired Correct 0/1	Required Stimulus Repetition Correct 0/1	Misordered Words Correct 0/1	Violation of Syntactic Rules Correct 0/1	Omission	Semantic Substitution	Formal Errors	Phonological Errors	Phonological Errors Affecting Morphemes	Grammatical Errors	Sound Deviations	Not Classified Errors
ML: nfv-PPA patient															
1	/teɪki ðə/	0	1	0	0	0	0	1							
2	/wɜːʳ ɪz ðə/child	1	0	0	0	0	0								
3	/kɑ (2) ðəkɑ wɪl wʊz nɒʔ rʌn/	0	1	0	0	0	0		2 **						
4	why/ɑ neɪ nɒʔ/living here	1	0	0	0	0	0								
5	/ðɜːʳ beɪn peɪʔ əən tʃi (4) tʃeəʳz/	0	1	0	0	0	0	1							
6	/wɜ əju: goʊɪŋtu: wɜːʳʔ nəʔ zənənətʃə(3) zʌmə/	1	0	0	0	0	0								
7	/i: (.) sʊld ɪz haʊs əndəðəwi: (4) fɑm/	0	1	0	0	0	0	3							
8	/wɜːʳk ən ðəgɑːdən æn (4) bɪk tʃɒ gɒ ðəbiːnz/	0	1	0	0	0	0	4							
9	/ðɜːʳ ɑːdɜːʳzd peɪʔ (3)/um/peɪndəd/ … no	0	1	0	0	0	0	8							
10	/ɪz (3) ɒl ðɜːʳ kʊntri: (1) ɒl ðɜːʳ (1) dɒz ɪn ðəkʊntri:/	0	1	0	0	0	0	7							
11	/i: wɒz/no	0	1	0	0	0	0	13							
12	/waɪz du ðəgruf (3) waɪʔs (4) wəz/	0	1	0	0	0	0	10							
13	*Did not attempt*														
14	*Did not attempt*														
	Total	3	9	0	0	0	0	47	2	0	0	0	0	0	0

Table 2. *Cont.*

Item	Produced Sentence	Analysis on Sentence-Level 0 = No; 1 = Yes						Analysis on Word-Level Frequency							
		Correct 0/1	Erroneous Correct 0/1	Uncertain/Repaired Correct 0/1	Required Stimulus Repetition Correct 0/1	Misordered Words Correct 0/1	Violation of Syntactic Rules Correct 0/1	Omission	Semantic Substitution	Formal Errors	Phonological Errors	Phonological Errors Affecting Morphemes	Grammatical Errors	Sound Deviations	Not Classified Errors
JC: sv-PPA patient															
1	Take this home	1	0	0	0	0	0								
2	Where is the child?	1	0	0	0	0	0								
3	The car will not run	1	0	0	0	0	0								
4	Why are they not living here?	1	0	0	0	0	0								
5	The band played and the crowd cheered	1	0	0	0	0	0								
6	Where are you going to work next summer?	1	0	0	0	0	0								
7	He sold his house and they moved to the farm.	1	0	0	0	0	0								
8	Work in the garden until you have picked all the beans	1	0	0	0	0	0								
9	The artist painted many of the beautiful scenes in this valley	1	0	0	0	0	0								
10	This doctor does not travel to all the towns in this country	0	1	0	0	0	0		1						
11	He should actually be able to tell us when she will actually be performing here	0	1	0	0	0	0		2 **						
12	Why do members of that group never write to their representatives of their group for aid?	0	1	0	0	0	0		3 **						
13	Many men and women were not able to get to their work because of the severe snowstorm	0	1	0	0	0	0		1 **						
14	The members of the committee have agreed to hold their meeting on the first Tuesday of every month	0	1	0	0	0	0		1						
	Total	9	5	0	0	0	0	0	8	0	0	0	0	0	0

Notes, Sentences were transcribed using the international phonetic transcription (IPA). Sentences were scored using the Hohlbaum, Dressel [12] scoring criteria. Notes, ?: glottal stop; (2) represents pause in seconds; (.) represents pause of <1 s; *: These errors are coded as semantic substitutions but in fact they do not appear to arise as a result of impaired lexical retrieval. Rather the lv-PPA patient decodes the meaning but cannot repeat the content word by word. So, he paraphrases, e.g., people for men and women; **: These errors are additions. The Hohlbaum system codes them as semantic substitutions, although there is no specific code for such errors.

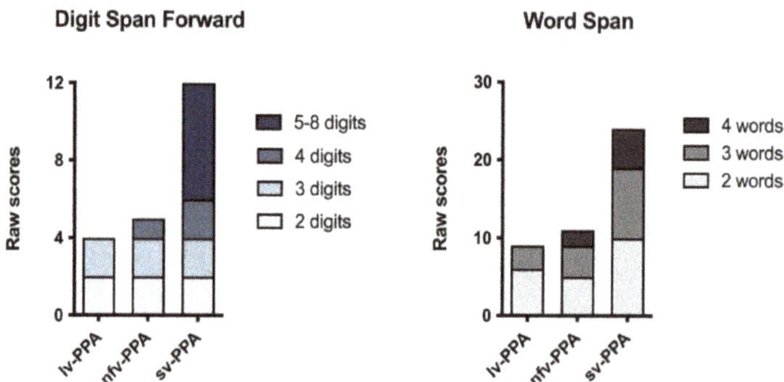

Figure 2. Raw scores on the Digit Span Forward and Word Span tests. The gradient colours/shading within the bar charts represent the gradient levels of difficulty (i.e., Digit Span: raw scores at the 2–, 3–, 4–, and 5–8-digit item levels; Word Span: raw scores at the 2–, 3–, and 4–word item levels).

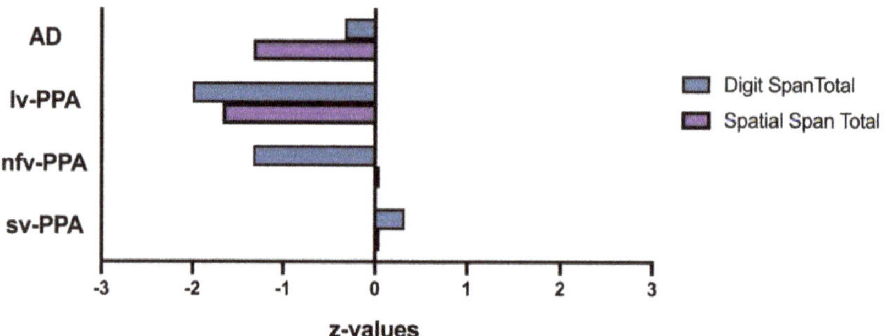

Figure 3. Digit and Spatial Span total raw scores graphically represented as age-adjusted z-values. Overall performance on Digit Span was very impaired for the lv-PPA patient, borderline-impaired for the nfv-PPA patient, and average for the sv-PPA and AD patients. Overall performance on Spatial Span was borderline-impaired for the lv-PPA and AD patients, and average for the nfv-PPA and sv-PPA patients. Overall Digit Span was significantly (0.05) worse than Spatial Span for the nfv-PPA patient; with the reverse pattern (i.e., Spatial < Digits) found for the AD patient. There was no statistical difference between test modality for lv-PPA and sv-PPA patients (statistical thresholds were taken from Table F4 [Appendix F] of the WMS III Scoring Manual).

3.1.2. Clinical Opinion about the Patient's Verbal and Visuospatial Short-Term and Working Memory Profile

NS demonstrates the hallmark verbal short-term memory disorder characteristic of lv-PPA, evidenced by impaired digit span, word span, and sentence repetition, in the context of relatively spared repetition of high frequency (i.e., lower cognitive load) multisyllabic words. Qualitatively, NS's intact repetition of multisyllabic words but impaired sentence repetition suggests that the latter arises from difficulties accessing and rehearsing verbal information in mind—that is, a dissolution of the verbal short-term memory system, rather than from deficits in motor speech production (Table 2) [11,12]. His mild articulatory errors and adequate prosody during speech, with well-articulated sound substitutions and lack of distortions further supports this position. Notably, NS's short-term and working memory impairments appear to extend beyond the verbal domain, evidenced by his impaired visuospatial span.

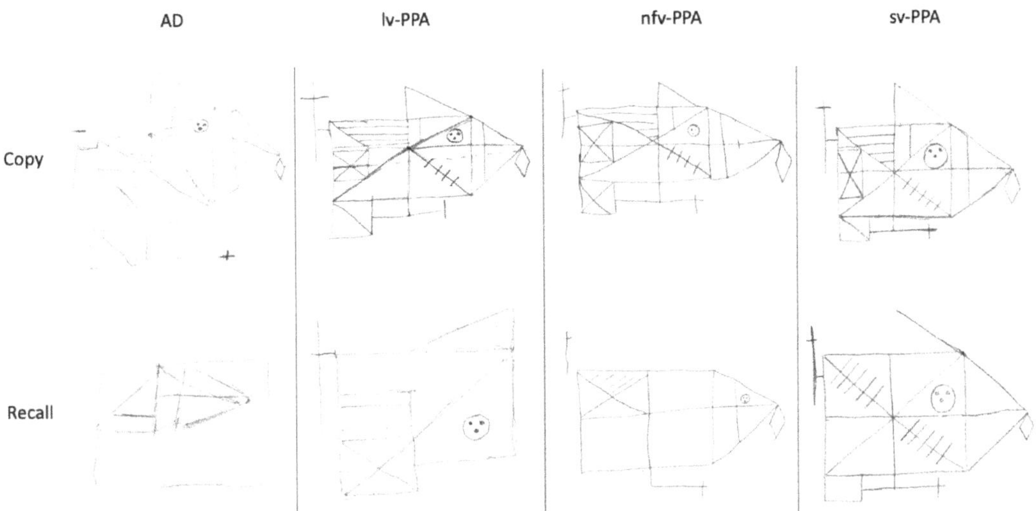

Figure 4. Rey Complex Figure copy and 3-minute recall. Copy performance was very impaired for the AD patient, borderline-impaired for the lv-PPA and nfv-PPA patients, and within normal limits for the sv-PPA patient. Three-minute recall performance was very impaired for the AD patient, borderline-impaired for the lv-PPA patient, and within normal limits for the nfv-PPA and sv-PPA patients.

3.1.3. Brain MRI and Clinical Diagnosis

T1 coronal brain MR images revealed mild generalised cortical atrophy, slightly more prominent on the left than the right, extending posteriorly to involve the parietal lobes (Figure 5). There was marginally greater atrophy in the peri-insular region on the left than the right. T2 weighted MRI images showed occasional hyperintensities in the cerebral hemispheres which were within normal limits for NS's age. The pattern of brain atrophy, clinical history, language, and neuropsychological profile were consistent with a diagnosis of lv-PPA (Table 3a). Pathological confirmation was unavailable as NS is still alive.

3.2. nfv-PPA Patient: ML

At presentation, ML was a 64-year-old, right-handed woman with 12 years of education (Table 1). She had been retired for five years, having previously worked as a shop owner and a public servant. She had a past history of liver disease due to hepatitis C, hepatic cirrhosis, and long-term alcohol consumption. At the time of the assessment, ML had been abstinent from alcohol for 7 years. Her father had been diagnosed with dementia (type unknown) in his 80s and died at the age of 87.

ML was seen following a 5-year history of progressively deteriorating speech which had worsened noticeably over the last 12 months. Initial symptoms also included frequent spelling errors and incorrect sentence construction. Her husband reported that her text messages often had errors but remained largely understandable. She reported occasional Yes/No and Hi/Bye confusion but had no trouble using corresponding non-verbal gestures. No other cognitive changes were reported. She described no swallowing difficulties and no Parkinsonian symptoms. She did not report any symptoms of depression, anxiety, or stress on a self-report measure of recent mood, and she demonstrated appropriate emotional reactivity during the assessment. According to her husband, her ADLs were mildly impaired.

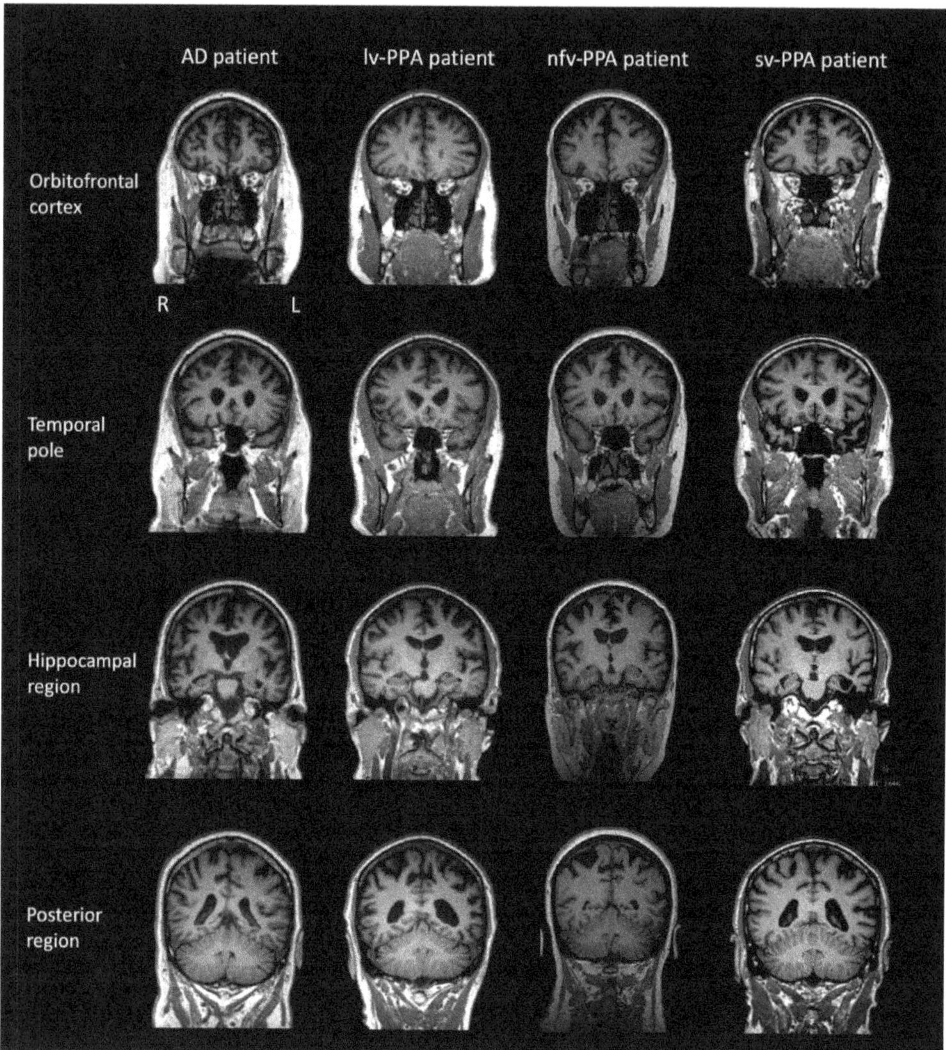

Figure 5. Brain T1 magnetic resonance images (MRI) of patients at the time of neuropsychological testing. Note: L = left; R = right. Brain images are presented in the coronal plane. lv-PPA patient: Mild generalised cortical atrophy was evident, slightly more prominent on the left than the right, extending posteriorly to involve the parietal lobes. There was marginally greater atrophy in the peri-insular region on the left than the right. nfv-PPA patient: Mild generalised cortical atrophy with particular involvement of the left peri-insular region anteriorly. sv-PPA patient: Severe atrophy of the anterior temporal pole bilaterally, but much worse on the left than the right. AD patient: Mild-moderate generalised cortical atrophy with involvement of the mesial temporal lobes.

Table 3. (a–c) PPA patients according to the Gorno-Tempini, Hillis [1] international consensus criteria of primary progressive aphasia.

(a) Diagnostic Criteria for Logopenic Variant of PPA	Patient NS	(b) Diagnostic Criteria for Non-Fluent Variant of PPA	Patient ML	(c) Diagnostic Criteria for Semantic Variant of PPA	Patient JC
I. Clinical diagnosis of logopenic variant PPA **Both of the following core features must be present:**		**I. Clinical diagnosis of non-fluent/agrammatic variant PPA** **At least one of the following core features must be present:**		**I. Clinical diagnosis of semantic variant PPA** **Both of the following core features must be present:**	
1. Impaired single-word retrieval in spontaneous speech and naming	✓	1. Agrammatism in language production	✓	1. Impaired confrontation naming	✓
2. Impaired repetition of sentences and phrases	✓	2. Effortful, halting speech with inconsistent speech sound errors and distortions (apraxia of speech)	✓	2. Impaired single-word comprehension	✓
At least 3 of the following other features must be present:		**At least 2 of 3 of the following other features must be present:**		**At least 3 of the following other features must be present:**	
1. Speech (phonologic) errors in spontaneous speech ,	✓	1. Impaired comprehension of syntactically complex sentences	Unknown	1. Impaired object knowledge, particularly for low-frequency or low-familiarity items	✓
2. Spared single-word comprehension and object knowledge	✓	2. Spared single-word comprehension	✓	2. Surface dyslexia or dysgraphia	✓
3. Spared motor speech	✓	3. Spared object knowledge	✓	3. Spared repetition	✓
4. Absence of frank agrammatism	✓	**II. Imaging-supported non-fluent/agrammatic variant diagnosis** **Both of the following criteria must be present:**		4. Spared speech production (grammar and motor speech)	✓
II. Imaging-supported logopenic variant diagnosis **Both of the following criteria must be present:**		1. Clinical diagnosis of non-fluent/agrammatic variant PPA	✓	**II. Imaging-supported semantic variant diagnosis** **Both of the following criteria must be present:**	
1. Clinical diagnosis of logopenic variant PPA	✓	2. Imaging must show one or more of the following results:		1. Clinical diagnosis of semantic variant PPA	✓
2. Imaging must show at least one of the following results:		a. Predominant left posterior fronto-insular atrophy on MRI or	✓	2. Imaging must show one or more of the following results:	
a. Predominant left posterior perisylvian or parietal atrophy on MRI	✓	b. Predominant left posterior fronto-insular hypoperfusion or hypometabolism on SPECT or PET	Not available	a. Predominant anterior temporal lobe atrophy	✓
b. Predominant left posterior perisylvian or parietal hypoperfusion or hypometabolism on SPECT or PET	Not available	**III. Non-fluent/agrammatic variant PPA with definite pathology**		b. Predominant anterior temporal hypoperfusion or hypometabolism on SPECT or PET	Not available
III. Logopenic variant PPA with definite pathology **Clinical diagnosis (criterion 1 below) and either criterion 2 or 3 must be present:**		**Clinical diagnosis (criterion 1 below) and either criterion 2 or 3 must be present:** 1. Clinical diagnosis of non-fluent/agrammatic variant PPA	✓	**III. Semantic variant PPA with definite pathology** **Clinical diagnosis (criterion 1 below) and either criterion 2 or 3 must be present:**	
1. Clinical diagnosis of logopenic variant PPA	✓	2. Histopathologic evidence of a specific neurodegenerative pathology (e.g., FTLD-tau, FTLDTDP, AD, other)	Not available	1. Clinical diagnosis of semantic variant PPA	✓
2. Histopathologic evidence of a specific neurodegenerative pathology (e.g. AD, FTLD-tau, FTLD-TDP, other)	Not available	3. Presence of a known pathogenic mutation	Not available	2. Histopathologic evidence of a specific neurodegenerative pathology (e.g., FTLD-tau, FTLDTDP, AD, other)	Not available
3. Presence of a known pathogenic mutation	Not available			3. Presence of a known pathogenic mutation	Not available

Notes, Abbreviations: AD = Alzheimer's disease; FTLD = frontotemporal lobar degeneration; PPA = primary progressive aphasia; TDP = TAR DNA-binding protein.

3.2.1. Neuropsychological Assessment

Based on her educational and vocational history, ML's premorbid intellectual ability was estimated to be average. She scored 81/100 on the ACE-III, which was below established cut-off scores for normal performance (normal > 88; Table 1). Formal neuropsychological testing revealed a primary impairment in expressive language. Qualitatively, her speech was markedly dysfluent, and contained articulatory and occasional grammatical errors. Single-word repetition and verbal fluency were extremely impaired on testing. Her other language abilities (comprehension, semantic knowledge) and writing, however, remained preserved (Table 1, Figure 1). Indeed, she could provide a reliable history of her difficulties by writing her responses. Performance on verbal short-term and working memory measures were also reduced, but likely due to her dysfluent speech (Table 2, Figure 2). Visuospatial short-term and working memory, on the other hand, was sound (Figure 3). Encoding and retention of verbal and visual information was preserved. While basic visuo-perceptual abilities were intact, ML had subtle visuo-constructional difficulties evidenced by an imprecise and slightly disorganised copy of the Rey Complex Figure (Figure 4). Other aspects of executive functioning (rapid set-shifting, inhibitory control) as well as psychomotor speed were impaired.

3.2.2. Clinical Opinion about the Patient's Verbal and Visuospatial Short-Term and Working Memory Profile

ML's profile is characteristic of nfv-PPA. Her verbal short-term and working memory span and sentence repetition were markedly impaired; however, frank motor speech deficits largely contributed to her impaired performance (Tables 1 and 2). Notably, ML had greater difficulty with repeating sentences and multisyllabic words or phrases than (predominantly monosyllabic) digits. In contrast, her visuospatial short-term and working memory performance appeared relatively intact (Figure 3).

3.2.3. Brain MRI and Clinical Diagnosis

The T1 coronal brain MR images revealed mild generalised cortical atrophy with particular involvement of the left peri-insular region anteriorly (Figure 5). Cerebral atrophy over the convexity was also present with widening of the interhemispheric fissure. The pattern of brain atrophy, clinical history, language, and neuropsychological profile were consistent with a diagnosis of nfv-PPA (Table 3b). Pathological confirmation was unavailable as ML remains alive.

3.3. sv-PPA Patient: JC

At presentation, JC was a 62-year-old, right-handed man (Table 1). He completed 16 years of education and worked as a principal of a primary school before retiring 2 years prior to the visit. His medical history was significant for a parathyroid cancer 13 years prior which was treated with a thyroidectomy and subsequently managed with levothyroxine, and ischaemic heart disease, with a myocardial infarction and 4-vessel coronary artery bypass graft surgery 4 years prior to the assessment. At the time of the assessment, JC was on antiplatelet and cholesterol-lowering medications. There was no report of symptoms of depression, anxiety, or stress on a self-report measure of recent mood and there were no significant periods of mood disorder noted. He demonstrated appropriate emotional reactivity during the assessment. There was no known family history of dementia.

JC was assessed following a 6-year history of speech and language difficulties with initial symptoms including forgetfulness and difficulties learning students' names at work. He reported progressive difficulties recalling names of people and objects (i.e., plants, animals) in the past 3 years, as well as a decline in his language comprehension and semantic knowledge. He was reportedly an avid reader previously though this had declined due to difficulties understanding the meaning of words. No issues with reading the actual words or recognising letters were reported. According to JC's wife, there was no change in behaviour or personality, eating habits, appetite, or weight. JC remained physically well and had no weakness or motor dysfunction. According to his wife, his ADLs were mildly impaired.

3.3.1. Neuropsychological Assessment

Based on his education and vocational history, JC's premorbid intellectual ability was estimated to lie within the average to high average range. He scored 67/100 on the ACE-III which was well below normal limits (normal > 88, Table 1). Qualitatively, his speech output was fluent with no phonological errors or substitutions, though occasional word-finding problems were noted, and he was slightly circumlocutory. Formal neuropsychological testing revealed intact visuospatial short-term and working memory (Table 1, Figure 3). Whilst his verbal working memory for numerical information and single words was intact, sentence repetition was compromised (Figure 2, Table 2). In terms of his language, JC demonstrated impaired confrontation naming, single-word comprehension and semantic knowledge; single-word repetition and verbal fluency, however, remained intact (Figure 1). Aside from suboptimal set-shifting, no significant executive functioning impairments were evident. JC's visuo-constructional planning and organisation remained intact. No visual memory deficits were apparent (Figure 4); detailed assessment of JC's verbal learning and memory, however, was not conducted on this occasion.

3.3.2. Clinical Opinion about the Patient's Verbal and Visuospatial Short-Term and Working Memory Profile

Overall, JC's cognitive profile is consistent with the characteristic sv-PPA profile. Whilst his short-term and working memory for digits, words and visuospatial information remained relatively spared, his repetition of complex sentences was more problematic. Notably and consistent with his intact basic verbal short-term memory span, qualitative appraisal of his sentence repetition (Table 2) suggests that his poor performance results from inclusions of grammatically correct but superfluous words (see Discussion).

3.3.3. Brain MRI and Clinical Diagnosis

T1 coronal brain MR images showed bilateral atrophy in the temporal lobes (left more markedly than right) particularly affecting the left hippocampal region and peri-insular region. Mild atrophy of the frontal lobes bilaterally was also evident (Figure 5). T2 weighted MRI images revealed evidence of scattered white matter hyperintensities in both hemispheres in keeping with small vessel ischaemic change. No established territorial infarcts were evident. The pattern of brain atrophy, clinical history, language, and neuropsychological profile were in keeping with a diagnosis of sv-PPA (Table 3c). JC underwent a Pittsburgh compound B (PiB) positron emission tomography scan (PiB-PET), which uses a radio-ligand of amyloid protein as a biomarker for AD [26]. The patient showed a low uptake of the PiB tracer, suggesting the absence of underlying Alzheimer pathology [26].

3.4. Typical AD Patient: TN

TN presented as a right-handed 67-year-old male. He completed nine years of formal education and worked in the government services for three years before working in hospitality and owning a business. At the time of assessment, he had been retired for 4 years. His medical history included ischaemic heart disease, with coronary artery bypass graft surgery 19 years prior to the assessment. TN also had a history of post-traumatic stress disorder arising from his previous employment, though this was well-managed at the time of assessment. He did not report significant symptoms of depression, anxiety, or stress on a self-report measure of recent mood. Regular medications included telmisartan for hypertension, antiplatelet medication, and selective serotonin reuptake inhibitors (SSRIs) for depressive symptoms. There was no known family history of neurodegenerative disease.

TN presented for assessment following a 7-year history of insidious memory decline, which had worsened considerably in the last 3 years. Both TN and his wife reported difficulties with his memory, specifically with names and topographical memory. Some organisational and planning difficulties were also noted which impacted his instrumental ADLs. TN's wife had noticed mild apathy but there were otherwise no other personality or behavioural changes.

3.4.1. Neuropsychological Assessment

Based on his education and vocational history, TN's estimated premorbid intellectual ability fell in the average range. He scored 68/100 on the ACE-III, which was well below normal limits (normal > 88; Table 1). Consistent with his diagnosis, TN demonstrated prominent verbal and visual memory impairment on testing. He was unable to learn a word list over repeated trials, and his recall of a previously copied two-dimensional complex geometric figure after a 3-minute delay was extremely poor (Figure 4). Whilst verbal short-term and working memory was intact, visuospatial short-term memory was reduced, and his visuospatial working memory (i.e., Spatial Span Backward) was borderline-impaired (Table 1). Other executive functioning abilities were variable—he was extremely slow and made several errors on a set-shifting task. Basic visuo-perceptual skills and visuo-constructional abilities (e.g., drawing simple objects) were preserved, although some higher-level visuo-constructional difficulties were present (Figure 4). Psychomotor speed was intact. Finally, no overt expressive language issues were noted in conversation although confrontation naming and comprehension (i.e., word-picture matching) were below expectations on testing (Figure 1). Other aspects of language (repetition, verbal fluency, and higher-level semantic knowledge) were relatively intact.

3.4.2. Clinical Opinion about the Patient's Verbal and Visuospatial Short-Term and Working Memory Profile

TN's verbal and visuospatial span profile was consistent with typical AD (Figure 3). As expected from a typical AD diagnosis, TN's visuospatial working memory ability (as demonstrated on Spatial Span Backward) was poor, whilst basic verbal and visuospatial short-term memory span remained relatively preserved.

3.4.3. Brain MRI and Clinical Diagnosis

T1 coronal brain MR images revealed atrophy of the medial temporal region bilaterally including the hippocampus, as well as diffuse frontal and parietal atrophy (Figure 5). Ventricular enlargement was also evident. T2 weighted MRI images revealed scattered white matter hyperintensities in both hemispheres in keeping with small vessel ischaemic change. No established territorial infarcts were evident.

Two years after his assessment, TN underwent in vivo amyloid-PET imaging, which showed uptake of the amyloid ligand above the cut-off for an amyloid based pathology, indicating the presence of underlying Alzheimer disease [26]. The clinical history, language, neuropsychological profile, and confirmation of underlying Alzheimer pathology were consistent with a clinical diagnosis of typical AD [27].

3.4.4. Summary of the Short-Term and Working Memory Profiles across Patients and Relative to a Matched Control Group

These short-term and working memory profiles were established based on standardised norms from various population groups and sizes: WAIS-III and WMS-III Digit and Spatial Span [17,22], and Word Span and Sentence Repetition norms [4]. To ensure that our findings were not due to differences across normative populations, performance profiles of the case vignettes were compared to one sample of Australian matched controls [4] and are displayed as percentage scores (Figure 6).

Inspection of these scores confirm that the lv-PPA and nfv-PPA patients were disproportionately impaired on Digit and Word Span relative to Spatial Span and overall cognitive ability (i.e., ACE total). By contrast, performance differences across verbal and visuospatial modalities were less evident for the sv-PPA and AD patients. Sentence Repetition performance across patients and within individual performance profiles was variable and uninterpretable across patients based on raw scores alone.

Regarding the specific short-term and working memory profiles, the lv-PPA, sv-PPA and AD patients demonstrated greater impairment on the Digit Span Backward than Forward tasks compared to controls. The reverse pattern was found for the nfv-PPA patient. These findings suggest that, over and above the inherent general difficulty associated with verbal working memory, the lv-PPA, sv-PPA and AD patients displayed more difficulty on this task relative to their verbal short-term memory capacity. As previously discussed, the nfv-PPA patient's impoverished speech is likely to have contributed to their performance on this task.

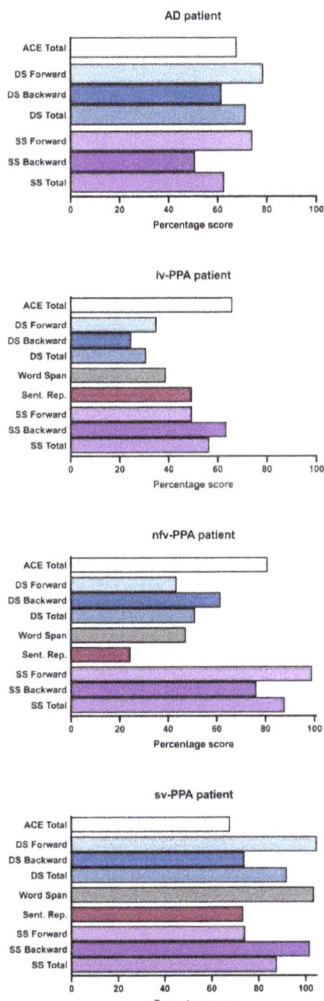

Figure 6. Short-term and working memory performances represented as percentage scores of normal performance (i.e., the patient's score divided by a control mean, times by 100). The ACE total bar charts are the patient's raw score on this measure. Control means taken from Foxe, Irish [4]. Notes, ACE: Addenbrooke's Cognitive Examination-III; DS: Digit Span; Sent. Rep: Sentence Repetition; SS: Spatial Span.

Overall Spatial Span profiles were distinct across patients. Relative to overall cognitive ability (i.e., ACE Total), the sv-PPA and nfv-PPA patients' Spatial Span performance was relatively spared. By contrast, the lv-PPA and AD patient's Spatial Span performance was reduced somewhat. These findings were in keeping with other measures of visuospatial episodic memory (i.e., RCFT 3-minute recall) and visuo-construction (RCFT Copy).

4. Discussion

While significant advancements have been made in the classification of PPA and its variants, challenges remain in clinical practice in differentiating PPA profiles in individual patients. Using case vignettes, we demonstrate that the canonical phonological disturbance displayed in lv-PPA is distinguishable from other PPA language profiles when using a selection of basic language and short-term memory measures. Importantly, we highlight how speech output and language deficits differentially interfere with verbal memory performance across the PPA variants, and how these differences provide insights into the underlying cognitive processes affected in these syndromes. Further, we demonstrate that visuospatial span tasks are essential for the assessment of PPA as they measure memory capacity without language contamination.

The lv-PPA patient demonstrates the typical verbal short-term memory deficit observed in this syndrome: very impaired digit and word span, and markedly reduced sentence repetition, in the context of relatively spared repetition of single two- and three-syllabic words [1,2]. Importantly, this occurred in the absence of frank motor speech deficits. Phonological paraphasias were, however, present. In contrast the nfv-PPA patient also performed poorly on verbal repetition tasks, but this performance was contaminated by frank motor speech deficits. Notably, and consistent with previous studies, we found that the nfv-PPA patient's verbal short-term memory performance declined as the motor speech sequencing requirements increased (i.e., repeating monosyllabic digits compared to a span of multisyllabic words and/or phrases [3]). Considering the sv-PPA patient's profile, we found digit and word span were spared whereas sentence repetition was compromised, likely to be due to his degraded semantic store [3]. Specifically, it is thought that the dissolving semantic knowledge and ability to form conceptual representations in sv-PPA may impact on the capacity to 'chunk' verbal material into meaningful components—a skill necessary for holding larger quantities of verbal information [3,11,28]. While not systematically verified in the current study, the sv-PPA patient's occasional circumlocutory responses and/or word inclusions on the Sentence Repetition task would support this interpretation.

Taken together, we propose that, when assessing for lv-PPA, Digit Span Forward and Word Span tasks are more robust measures of verbal short-term memory than Sentence Repetition or Digit Span Backward tasks, as the former tasks are less susceptible to language and/or dysexecutive contamination [4,11,12]. This observation notwithstanding, important qualitative information can be gained from Sentence Repetition that will help with the distinction between the two non-fluent PPA syndromes; specifically, its ability to elicit phonological paraphasias (in lv-PPA) or motor speech deficits (in nfv-PPA) [11,12]. For the assessment of sv-PPA, we caution that some patients may perform poorer on Sentence Repetition than other span tests as their degraded semantic knowledge may preclude their ability to 'chunk' and/or form meaningful representations in mind [3,11,28].

In contrast to their similar verbal span and sentence repetition performance, the lv-PPA and nfv-PPA patients showed distinct visuospatial span profiles. The disproportionately compromised visuospatial short-term and working memory in lv-PPA relative to the other PPA variants is consistent with a growing body of research and alludes to their distinct neuroanatomical profile [4,29–32]. Briefly, brain regions involved in visuospatial short-term and working memory, including bilateral posterior temporal and parietal brain structures, are more compromised in lv-PPA and AD than in nfv-PPA and sv-PPA [4,33–35]. Undoubtedly, and unlike in nfv-PPA where the deficit remains primarily verbal, the impaired short-term memory of lv-PPA extends to the non-verbal domain, even at low levels of difficulty [30,31,36,37]. Based on these findings, we propose that when assessing

PPA patients without motor features, difficulty on basic/lower-load visuospatial short-term memory tests strongly indicates lv-PPA [31,38,39]. This distinction will assist the diagnostic process, particularly in the presence of a mixed language profile.

It is notable that short-term and working memory deficits are not limited to PPA. Indeed, we found that the AD patient showed impairments across these domains. Unlike the PPA patients, however, the AD patient's overall visuospatial span was significantly worse than their overall verbal span profile. These findings are consistent with the commonly held view that multiple components of visuospatial memory, including processing, integration, storing, and retaining visual material, break down in the earlier stages of typical AD [4,31,33,40]. Consistent with this view, the visuospatial difficulties of the AD patient extended beyond Spatial Span—with deficits also noted on complex visuo-constructional, visuospatial episodic memory, and attentional tasks (i.e., Rey Complex Figure, Trails A and B). Importantly, the widespread memory and visuospatial deficits of the AD patient supports the opinion that AD is distinct from lv-PPA [31,41]. That is, lv-PPA is not simply typical AD with additional language deficits [42]. While this nuanced distinction may seem unnecessary, it has clinical implications when addressing the care needs, treatment options, and estimated survival of either AD or lv-PPA [43]. To that end, we propose requisite assessment of both verbal and visuospatial cognitive domains for the differential diagnosis of AD and lv-PPA.

In clinical practice, awareness that the cognitive profiles of PPA and AD vary across individuals is vital. Most studies compare matched PPA subgroups at a single point in time (typically at the mild to moderate disease severity stage) and provide findings which typically overemphasise the differences across variants but underemphasise the differences within each syndrome. Studies that have investigated within syndromes, however, suggest that the language and cognitive profile of lv-PPA varies considerably [6–10,44]. For example, it is reported that while most lv-PPA patients present with multi-domain cognitive impairment at baseline assessment, a subset of lv-PPA patients present with relatively circumscribed language deficits with relatively mild cognitive deficits in other domains [6]. Important to this topic is the awareness that, with disease progression, language and cognitive abilities of PPA and AD inevitably decline, eventuating in a manifold dissolution of functional abilities [45–47]. As such, the distinct cognitive profiles observed in the earlier stages of these diseases may become less apparent in later disease stages. To illustrate this point, Table 4 demonstrates the decline of Spatial and Digit Span performance across PPA and AD patients, stratified by overall cognitive ability. Put together, clinicians should take a gestalt approach to assessing PPA and AD in clinical practice and consider the 'moving parts' of language and cognitive deficits, as well as overall cognitive ability, before forming a formal clinical diagnosis.

Several caveats warrant attention. Pathological confirmation was not available in any of the PPA cases. Nonetheless, the clinical, cognitive, and imaging information provided in the current study is typical of what is commonly available at baseline assessment in routine clinical practice (i.e., non-tertiary/specialised centres) and was sufficient to establish a clinical diagnosis of each PPA variant [1]. It is reported that ~70% of lv-PPA patients have underlying Alzheimer pathology, ~70% of nfv-PPA have tauopathy, and ~85% of sv-PPA have TDP−43, and that detailed and careful clinical, cognitive, and imaging examination improves this pathological correspondence [48–50]. More research, however, is warranted to determine if specific cognitive profiles within PPA syndromes can further discern the pathological course.

References

1. Gorno-Tempini, M.L.; Hillis, A.E.; Weintraub, S.; Kertesz, A.; Mendez, M.; Cappa, S.F.; Ogar, J.M.; Rohrer, J.D.; Black, S.; Boeve, B.F.; et al. Classification of primary progressive aphasia and its variants. *Neurology* **2011**, *76*, 1006–1014. [CrossRef]
2. Gorno-Tempini, M.L.; Brambati, S.M.; Ginex, V.; Ogar, J.; Dronkers, N.F.; Marcone, A.; Perani, D.; Garibotto, V.; Cappa, S.F.; Miller, B.L. The logopenic/phonological variant of primary progressive aphasia. *Neurology* **2008**, *71*, 1227–1234. [CrossRef]
3. Leyton, C.E.; Savage, S.; Irish, M.; Schubert, S.; Piguet, O.; Ballard, K.J.; Hodges, J.R. Verbal repetition in primary progressive aphasia and Alzheimer's disease. *J. Alzheimer's Dis.* **2014**, *41*, 575–585. [CrossRef] [PubMed]
4. Foxe, D.; Irish, M.; Roquet, D.; Scharfenberg, A.; Bradshaw, N.; Hodges, J.R.; Burrell, J.R.; Piguet, O. Visuospatial short-term and working memory disturbance in the primary progressive aphasias: Neuroanatomical and clinical implications. *Cortex* **2020**, *132*, 223–237. [CrossRef]
5. Croot, K.; Ballard, K.; Leyton, C.E.; Hodges, J.R. Apraxia of speech and phonological errors in the diagnosis of nonfluent/agrammatic and logopenic variants of primary progressive aphasia. *J. Speech Lang. Hear. Res.* **2012**, *55*, s1562–s1572. [CrossRef]
6. Owens, T.E.; Machulda, M.M.; Duffy, J.R.; Strand, E.A.; Clark, H.M.; Boland, S.; Martin, P.R.; Lowe, V.J.; Jack, C.R.; Whitwell, J.L.; et al. Patterns of neuropsychological dysfunction and cortical volume changes in logopenic aphasia. *J. Alzheimer's Dis.* **2018**, *66*, 1015–1025. [CrossRef]
7. Sajjadi, S.A.; Patterson, K.; Nestor, P.J. Logopenic, mixed, or Alzheimer-related aphasia? *Neurology* **2014**, *82*, 1127–1131. [CrossRef]
8. Murley, A.G.; Coyle-Gilchrist, I.; Rouse, M.A.; Jones, P.S.; Li, W.; Wiggins, J.; Lansdall, C.; Rodriguez, P.V.; Wilcox, A.; Tsvetanov, K.A.; et al. Redefining the multidimensional clinical phenotypes of frontotemporal lobar degeneration syndromes. *Brain* **2020**, *143*, 1555–1571. [CrossRef]
9. Sajjadi, S.A.; Patterson, K.; Arnold, R.J.; Watson, P.C.; Nestor, P.J. Primary progressive aphasia: A tale of two syndromes and the rest. *Neurology* **2012**, *78*, 1670–1677. [CrossRef]
10. Vandenberghe, R. Classification of the primary progressive aphasias: Principles and review of progress since 2011. *Alzheimer's Res. Ther.* **2016**, *8*, 1–9. [CrossRef]
11. Beales, A.; Whitworth, A.; Cartwright, J.; Panegyres, P.K.; Kane, R.T. Profiling sentence repetition deficits in primary progressive aphasia and Alzheimer's disease: Error patterns and association with digit span. *Brain Lang.* **2019**, *194*, 1–11. [CrossRef]
12. Hohlbaum, K.; Dressel, K.; Lange, I.; Wellner, B.; Saez, L.E.; Huber, W.; Grande, M.; Amunts, K.; Grodzinsky, Y.; Heim, S. Sentence repetition deficits in the logopenic variant of PPA: Linguistic analysis of longitudinal and cross-sectional data. *Aphasiology* **2018**, *32*, 1445–1467. [CrossRef]
13. Hsieh, S.; Schubert, S.; Hoon, C.; Mioshi, E.; Hodges, J.R. Validation of the Addenbrooke's Cognitive Examination III in frontotemporal dementia and Alzheimer's disease. *Dement. Geriatr. Cogn. Disord.* **2013**, *36*, 242–250. [CrossRef]
14. Knopman, D.S.; Kramer, J.H.; Boeve, B.F.; Caselli, R.J.; Graff-Radford, N.R.; Mendez, M.F.; Miller, B.L.; Mercaldo, N. Development of methodology for conducting clinical trials in frontotemporal lobar degeneration. *Brain* **2008**, *131*, 2957–2968. [CrossRef]
15. Ng, F.; Trauer, T.; Dodd, S.; Callaly, T.; Campbell, S.; Berk, M. The validity of the 21-item version of the depression anxiety stress scales as a routine clinical outcome measure. *Acta Neuropsychiatr.* **2007**, *19*, 304–310. [CrossRef] [PubMed]
16. Lovibond, P.F.; Lovibond, S.H. The structure of negative emotional states—comparison of the depression anxiety stress scales (dass) with the beck depression and anxiety inventories. *Behav. Res. Ther.* **1995**, *33*, 335–343. [CrossRef]
17. Wechsler, D. *Weschler Adult Intelligence Scale—Third Edition: Administration and Scoring Manual*; Psychological Corporation: San Antonio, TX, USA, 1997.
18. Mioshi, E.; Hsieh, S.; Savage, S.; Hornberger, M.; Hodges, J.R. Clinical staging and disease progression in frontotemporal dementia. *Neurology* **2010**, *74*, 1591–1597. [CrossRef] [PubMed]
19. Strauss, E.; Sherman, E.M.S.; Spreen, O. *A Compendium of Neuropsychological Tests: Administration, Norms, And Commentary*, 3rd ed.; Oxford University Press: New York, NY, USA, 2006.
20. Rey, A. L'examen psychologique dans les cas d'encéphalopathie traumatique. *Arch. Psychol.* **1941**, *28*, 286–340.
21. Benton, A.L.; Hamsher, K.d.; Sivian, A.B. *Multilingual Aphasia Examination*, 3rd ed.; AJA Associates, Inc.: Iowa City, IA, USA, 1994.
22. Wechsler, D. *Weschler Memory Scale—Third Edition: Administration and Scoring Manual*; Psychological Corporation: San Antonio, TX, USA, 1997.
23. Savage, S.; Hsieh, S.; Leslie, F.; Foxe, D.; Piguet, O.; Hodges, J.R. Distinguishing subtypes in primary progressive aphasia: Application of the Sydney language battery. *Dement. Geriatr. Cogn. Disord.* **2013**, *35*, 208–218. [CrossRef] [PubMed]
24. Reitan, R.M. The relation of the trail making test to organic brain damage. *J. Consult. Psychol.* **1955**, *19*, 393–394. [CrossRef]
25. So, M.; Foxe, D.; Kumfor, F.; Murray, C.; Hsieh, S.; Savage, G.; Ahmed, R.M.; Burrell, J.R.; Hodges, J.R.; Irish, M.; et al. Addenbrooke's cognitive examination III: Psychometric characteristics and relations to functional ability in dementia. *J. Int. Neuropsychol. Soc.* **2018**, *24*, 854–863. [CrossRef] [PubMed]
26. Klunk, W.E.; Engler, H.; Nordberg, A.; Wang, Y.; Blomqvist, G.; Holt, D.P.; Bergstrom, M.; Savitcheva, I.; Huang, G.F.; Estrada, S.; et al. Imaging brain amyloid in Alzheimer's disease with Pittsburgh compound-B. *Ann. Neurol.* **2004**, *55*, 306–319. [CrossRef] [PubMed]
27. McKhann, G.M.; Knopman, D.S.; Chertkow, H.; Hyman, B.T.; Jack, C.R., Jr.; Kawas, C.H.; Klunk, W.E.; Koroshetz, W.J.; Manly, J.J.; Mayeux, R.; et al. The diagnosis of dementia due to Alzheimer's disease: Recommendations from the national institute

on aging-Alzheimer's association workgroups on diagnostic guidelines for Alzheimer's disease. *Alzheimer's Dement.* **2011**, *7*, 263–269. [CrossRef]

28. Jefferies, E.; Patterson, K.; Jones, R.W.; Bateman, D.; Lambon Ralph, M.A. A category-specific advantage for numbers in verbal short-term memory: Evidence from semantic dementia. *Neuropsychologia* **2004**, *42*, 639–660. [CrossRef] [PubMed]

29. Eikelboom, W.S.; Janssen, N.; Jiskoot, L.C.; van den Berg, E.; Roelofs, A.; Kessels, R.P.C. Episodic and working memory function in primary progressive aphasia: A meta-analysis. *Neurosci. Biobehav. Rev.* **2018**, *92*, 243–254. [CrossRef] [PubMed]

30. Kamath, V.; Sutherland, E.R.; Chaney, G.A. A meta-analysis of neuropsychological functioning in the logopenic variant of primary progressive aphasia: Comparison with the semantic and non-fluent variants. *J. Int. Neuropsychol. Soc.* **2020**, *26*, 322–330. [CrossRef]

31. Foxe, D.; Irish, M.; D'Mello, M.; Barhon, L.; Burrell, J.R.; Kessels, R.P.C.; Piguet, O. The box task: A novel tool to differentiate the primary progressive aphasias. *Eur. J. Neurol.* **2021**. [CrossRef]

32. Foxe, D.G.; Irish, M.; Hodges, J.R.; Piguet, O. Verbal and visuospatial span in logopenic progressive aphasia and Alzheimer's disease. *J. Int. Neuropsychol. Soc.* **2013**, *19*, 247–253. [CrossRef]

33. Foxe, D.; Leyton, C.E.; Hodges, J.R.; Burrell, J.R.; Irish, M.; Piguet, O. The neural correlates of auditory and visuospatial span in logopenic progressive aphasia and Alzheimer's disease. *Cortex* **2016**, *83*, 39–50. [CrossRef]

34. Wager, T.D.; Smith, E.E. Neuroimaging studies of working memory: A meta-analysis. *Cogn. Affect. Behav. Neurosci.* **2003**, *3*, 255–274. [CrossRef]

35. Smith, E.E.; Jonides, J. Working memory: A view from neuroimaging. *Cogn. Psychol.* **1997**, *33*, 5–42. [CrossRef]

36. Tippett, D.C.; Breining, B.; Goldberg, E.; Meier, E.; Sheppard, S.M.; Sherry, E.; Stockbridge, M.; Suarez, A.; Wright, A.E.; Hillis, A.E. Visuomotor figure construction and visual figure delayed recall and recognition in primary progressive aphasia. *Aphasiology* **2019**, *34*, 1456–1470. [CrossRef]

37. Watson, C.L.; Possin, K.; Allen, I.E.; Hubbard, H.I.; Meyer, M.; Welch, A.E.; Rabinovici, G.D.; Rosen, H.; Rankin, K.P.; Miller, Z.; et al. Visuospatial functioning in the primary progressive aphasias. *J. Int. Neuropsychol. Soc.* **2018**, *24*, 259–268. [CrossRef] [PubMed]

38. Possin, K.L. Visual spatial cognition in neurodegenerative disease. *Neurocase* **2010**, *16*, 466–487. [CrossRef] [PubMed]

39. Bak, T.H.; Caine, D.; Hearn, V.C.; Hodges, J.R. Visuospatial functions in atypical parkinsonian syndromes. *J. Neurol. Neurosurg. Psychiatry* **2006**, *77*, 454–456. [CrossRef]

40. Huntley, J.D.; Howard, R.J. Working memory in early Alzheimer's disease: A neuropsychological review. *Int. J. Geriatr. Psychiatry* **2010**, *25*, 121–132. [CrossRef]

41. Magnin, E.; Chopard, G.; Ferreira, S.; Sylvestre, G.; Dariel, E.; Ryff, I.; Mertz, C.; Lamidieu, C.; Hidalgo, J.; Tio, G.; et al. Initial neuropsychological profile of a series of 20 patients with logopenic variant of primary progressive aphasia. *J. Alzheimer's Dis.* **2013**, *36*, 799–808. [CrossRef] [PubMed]

42. Ahmed, S.; de Jager, C.A.; Haigh, A.M.; Garrard, P. Logopenic aphasia in Alzheimer's disease: Clinical variant or clinical feature? *J. Neurol. Neurosurg. Psychiatry* **2012**, *83*, 1056–1062. [CrossRef] [PubMed]

43. Taylor-Rubin, C.; Croot, K.; Nickels, L. Speech and language therapy in primary progressive aphasia: A critical review of current practice. *Expert Rev. Neurother.* **2021**, *21*, 419–430. [CrossRef] [PubMed]

44. Ramanan, S.; Roquet, D.; Goldberg, Z.L.; Hodges, J.R.; Piguet, O.; Irish, M.; Lambon Ralph, M.A. Establishing two principal dimensions of cognitive variation in logopenic progressive aphasia. *Brain Commun.* **2020**, *2*, fcaa125. [CrossRef]

45. Foxe, D.; Irish, M.; Hu, A.N.; Carrick, J.; Hodges, J.R.; Ahmed, R.M.; Burrell, J.R.; Piguet, O. Longitudinal cognitive and functional changes in primary progressive aphasia. *J. Neurol.* **2021**, *268*, 1951–1961. [CrossRef] [PubMed]

46. O'Connor, C.M.; Clemson, L.; Flanagan, E.; Kaizik, C.; Brodaty, H.; Hodges, J.R.; Piguet, O.; Mioshi, E. The relationship between behavioural changes, cognitive symptoms, and functional disability in primary progressive aphasia: A longitudinal study. *Dement. Geriatr. Cogn. Disord.* **2016**, *42*, 215–226. [CrossRef] [PubMed]

47. Jang, J.; Cushing, N.; Clemson, L.; Hodges, J.R.; Mioshi, E. Activities of daily living in progressive non-fluent aphasia, logopenic progressive aphasia and Alzheimer's disease. *Dement. Geriatr. Cogn. Disord.* **2012**, *33*, 354–360. [CrossRef]

48. Chare, L.; Hodges, J.R.; Leyton, C.E.; McGinley, C.; Tan, R.H.; Kril, J.J.; Halliday, G.M. New criteria for frontotemporal dementia syndromes: Clinical and pathological diagnostic implications. *J. Neurol. Neurosurg. Psychiatry* **2014**, *85*, 865–870. [CrossRef]

49. Harris, J.M.; Gall, C.; Thompson, J.C.; Richardson, A.M.; Neary, D.; du Plessis, D.; Pal, P.; Mann, D.M.; Snowden, J.S.; Jones, M. Classification and pathology of primary progressive aphasia. *Neurology* **2013**, *81*, 1832–1839. [CrossRef]

50. Mesulam, M.; Wicklund, A.; Johnson, N.; Rogalski, E.; Leger, G.C.; Rademaker, A.; Weintraub, S.; Bigio, E.H. Alzheimer and frontotemporal pathology in subsets of primary progressive aphasia. *Ann. Neurol.* **2008**, *63*, 709–719. [CrossRef]

51. Rowe, J.B. Parkinsonism in frontotemporal dementias. *Int. Rev. Neurobiol.* **2019**, *149*, 249–275. [CrossRef] [PubMed]

Review

Contribution of the Cognitive Approach to Language Assessment to the Differential Diagnosis of Primary Progressive Aphasia

Joël Macoir [1,2,*], **Annie Légaré** [1] and **Monica Lavoie** [3]

[1] Département de Réadaptation, Faculté de Médecine, Université Laval, Québec, QC G1V 0A6, Canada; annie.legare@fmed.ulaval.ca

[2] Centre de Recherche CERVO (CERVO Brain Research Centre), Québec, QC G1J 2G3, Canada

[3] Chaire de Recherche sur les Aphasies Primaires Progressives—Fondation de la Famille Lemaire, Québec, QC G1J 1Z4, Canada; monica.lavoie.1@ulaval.ca

* Correspondence: joel.macoir@fmed.ulaval.ca; Tel.: +1-418-656-2131 (ext. 412190)

Abstract: Diagnosis of primary progressive aphasia (PPA) is essentially based on the identification of progressive impairment of language abilities while other cognitive functions are preserved. The three variants of PPA are characterized by core and supportive clinical features related to the presence or absence of language impairment in different linguistic domains. In this article, we review the cognitive neuropsychological approach to the assessment of PPA and its contribution to the differential diagnosis of the three variants. The main advantage of this assessment approach is that it goes beyond the mere description and classification of clinical syndromes and identifies impaired and preserved cognitive and linguistic components and processes. The article is structured according to the main language domains: spoken production, language comprehension, and written language. Each section includes a brief description of the cognitive processes involved in the assessment tasks, followed by a discussion of typical characteristics for each PPA variant and common pitfalls in the interpretation of the results. In addition, the clinical benefit of the cognitive neuropsychological approach for the behavioral management of PPA is briefly sketched out in the conclusion.

Keywords: primary progressive aphasia; assessment; diagnosis; cognitive approach; dementia

Citation: Macoir, J.; Légaré, A.; Lavoie, M. Contribution of the Cognitive Approach to Language Assessment to the Differential Diagnosis of Primary Progressive Aphasia. *Brain Sci.* **2021**, *11*, 815. https://doi.org/10.3390/brainsci11060815

Academic Editors: Rene L. Utianski, Robert Laforce Jr. and Jordi A. Matias-Guiu

Received: 25 May 2021
Accepted: 16 June 2021
Published: 19 June 2021

Publisher's Note: MDPI stays neutral with regard to jurisdictional claims in published maps and institutional affiliations.

1. Introduction

Dementia is a common condition that mainly occurs in older people. It is characterized by a decline in cognitive functioning that is severe enough to impact activities of daily living and social functioning [1]. The loss of cognitive functioning in dementia may affect long- and short-term memory, attention, visual perception, executive functions, motor planning and execution, problem solving, and language [2]. Dementia can be caused by a wide variety of pathological entities, including Alzheimer's disease, which is the most common one. Other types of dementia include vascular dementia, dementia in atypical parkinsonian syndromes, such as Lewy body dementia and corticobasal degeneration, and frontotemporal dementia [3]. They are not only commonly associated with episodic memory impairment but also usually characterized by language deficits that may affect word and sentence comprehension and production abilities [4]. Clinical language profiles that are generally associated with common forms of dementia have been described, some in more detail than others. Neurolinguistic studies go beyond the mere description of symptoms to identify the functional localization of impaired and preserved linguistic processes in dementia.

Primary progressive aphasia (PPA) is a neurodegenerative syndrome associated with atrophy of the frontal, temporal, and parietal regions of the left hemisphere of the brain. PPA is a heterogeneous condition; the most prominent clinical feature is difficulty with language, while other cognitive domains are not affected at onset or in the early stages of the disease [5]. In 2011, an international group of experts proposed recommendations for PPA diagnosis and classification [6]. According to those recommendations, there are three main PPA variants: the nonfluent/agrammatic variant (nfvPPA), the semantic variant (svPPA), and the logopenic variant (lvPPA).

At least one of the following core features must be present to detect nfvPPA: (1) effortful, halting speech with inconsistent speech sound errors and distortions (apraxia of speech) and/or (2) agrammatism in language production. Moreover, at least two of the following features must also be present: (1) impaired comprehension of syntactically complex sentences, (2) spared single-word comprehension, and/or (3) spared object knowledge [6]. Imaging abnormalities in the left posterior frontoinsular region support the diagnosis of nfvPPA. Meanwhile, svPPA is a clinical syndrome caused by atrophy of the temporal lobes, leading to the selective impairment of semantic memory. The following core features must be present to establish a diagnosis of svPPA: (1) impaired confrontation naming and (2) impaired single-word comprehension. Moreover, at least three of the following features must also be present: (1) impaired object knowledge, (2) surface dyslexia or dysgraphia, (3) spared repetition, and/or (4) spared speech production (grammar and motor speech) [6]. Finally, lvPPA, the most recently identified PPA variant, is caused by predominant left posterior perisylvian or parietal atrophy. According to clinical criteria established in 2011, following core features are essential to the diagnosis of lvPPA: (1) the presence of anomia in spontaneous speech and (2) confrontation naming and impaired repetition of sentences and phrases [6]. At least three of the following features must also be present: (1) production of phonological errors, (2) preservation of semantic memory, (3) preservation of articulation and prosody, and/or (4) absence of frank agrammatism.

The initial evaluation is the first significant step toward the clinical management of dementia, and it is based on consensual diagnostic criteria. In some dementia syndromes, such as PPA, language deficit characterization is of major importance for the differential diagnosis. A language function assessment is part of the general diagnosis process for neurodegenerative diseases affecting language, and it generally includes medical history, mental status tests, physical and neurological exams, diagnostic tests, and brain imaging. The present article focuses on the contribution of a specific assessment of language abilities to the differential diagnosis of PPA. We first briefly present the cognitive neuropsychological approach to language assessment. Then, in sections addressing the main domains of language (i.e., spoken production, comprehension, written language), we briefly present the cognitive processes involved in the assessment tasks, the typical characteristics of each PPA variant, and common pitfalls in the interpretation of the results.

2. The Cognitive Neuropsychological Approach to Language Assessment

Compared to the clinicopathological approach to assessment, which aims to identify the diagnostic label that best corresponds to the observed language deficits (e.g., anomia; agrammatism), the cognitive neuropsychological approach aims to identify the impaired and preserved language abilities and localize their functional underlying origin [7]. This approach is derived from information processing theories in which cognitive functions, including language, are sustained by specialized, interconnected processing components. The assessment is conceived as an investigation based on the administration of specific tests in which stimuli are controlled or manipulated for psycholinguistic variables (e.g., length, frequency, familiarity) that are known to influence language processes. Error analysis in these tests is another source of information. For example, anomia may arise from distinct underlying deficits (e.g., in the activation of conceptual semantic representations or the retrieval of phonological forms of words in the lexicon), leading to distinct types of errors (e.g., semantic substitutions, phonemic errors). The main advantage of this assessment approach is that it goes beyond the mere description and classification of clinical syndromes and identifies impaired vs. preserved cognitive and linguistic components and processes. Furthermore, with a comprehensive portrait of the patient's communication abilities, the clinician can better tailor the behavioral treatment to address impaired language processes with a restorative or compensatory objective.

3. The Assessment of Spoken Production

According to cognitive models of spoken word production [8], words are retrieved and produced through the activation of specialized and interconnected components. In these models, word production is conceived as a staged process in which the activation flow is initiated in a conceptual-semantic component, continues through the activation of phonological lexical representations, and ends with the execution of articulation mechanisms. Spoken production processes are usually assessed with tests exploring the ability to retrieve words in long-term memory (e.g., picture naming); repeat words, nonwords, and sentences [9]; and provide information in a discourse, conversational exchange, or interview [10]. A summary of the underlying cognitive deficits of language impairment and the salient characteristics of spoken production disorders in the three PPA variants are presented in Table 1.

Table 1. Underlying cognitive deficits and salient characteristics of spoken production disorders in the three PPA variants.

Spoken Production	svPPA	nfvPPA	lvPPA
Underlying deficit	Semantic memory	- Lexicon: Activation of phonological forms - Phonetic encoding: Activation of motor representations for articulation	- Lexicon: Activation of phonological forms - Phonological short-term memory
Influence of psycholinguistic variables	- -Concept familiarity - -Semantic category * - -Visual complexity of pictures *	- -Syllable complexity - -Stimulus length - -Syntactic complexity	- Stimulus length
Word production: Picture naming	Impaired: No responses, semantic paraphasias, and vague circumlocutions	Impaired: Apraxia of speech, phonological errors, no responses, and specific circumlocutions	Impaired: Phonological errors, no responses, and specific circumlocutions
Repetition - Words - Nonwords - Sentences	- -Preserved - -Preserved - -Mild impairment	Impaired for all types of stimuli: Apraxia of speech, phonological errors	Impaired for all types of stimuli: Phonological errors
Spontaneous speech and narrative discourse	Word-finding difficulties: Aborted sentences, latencies, circumlocutions, and occasional semantic paraphasias	Slow, hesitant, and effortful; phonetic and phonological errors; and agrammatism	Impaired: Word-finding difficulties and phonological errors

* Potential but nonessential psycholinguistic variable.

3.1. The Assessment of Word Production

The easiest way to assess the ability to retrieve and produce spoken words is through picture naming tests, such as the Boston Naming Test in English [11] or the TDQ30 in French [12]. Theoretical models of spoken word production can facilitate the identification of the functional origins of deficits. These deficits may result from a loss of semantic representations or difficulty retrieving them. A breakdown at this level leads to semantic-based anomia. This is the case in svPPA; difficulty retrieving words manifests as no responses, semantic paraphasias, or vague circumlocutions [13]. In this variant, performance in naming tasks may be influenced by the familiarity and semantic category of the concepts (e.g., better or worse performance for natural vs. man-made concepts) as well as the visual complexity of the pictures due to possible concomitant associative visual agnosia [14].

Disruption in the activation of the phonological forms of words is responsible for the production of phonological errors in lvPPA [15] and nfvPPA [16]. Furthermore, in these variants, anomia manifests as no responses and specific circumlocutions. Studies have also shown that phonological short-term memory impairment contributes to spoken word production impairment in lvPPA, e.g., [15]. This process is responsible for the temporary storage of activated phonological representations until the actual execution of articulation mechanisms. In this case, performance in picture naming might be influenced by word length. Finally, impairment of the phonetic encoding stage, in which a motor representation for articulation is generated, leads to apraxia of speech in nfvPPA [17,18]. The confounding variables at this processing stage are syllable complexity [19].

In picture naming, the error type makes it easy to distinguish svPPA from the other variants. However, for an accurate differential diagnosis, differentiation between the phonological errors (i.e., substitution, deletion, displacement, and addition of phonemes) produced in both nfvPPA and lvPPA, and the phonetic errors (i.e., production of distorted phonemes and alteration of transitions between phonemes) associated exclusively to apraxia of speech in nfvPPA, is essential. To this respect, the advantage of the cognitive approach lies in the ability to differentiate the underlying origin of phonological errors, being the phonological short-term memory in the vlAPP and the activation of motor representations for articulation in the nfvAPP.

3.2. The Assessment of Repetition Abilities

The assessment of repetition abilities plays an important role in PPA differential diagnosis. For single words, these seemingly simple abilities involve linguistic processes devoted to the auditory analysis of stimuli and the activation of their lexical and semantic representations, followed by the activation of spoken production processes, including the maintenance of phonological forms of words in short-term memory. Episodic memory and semantic memory are added to these processes for the repetition of sentences [20]. For nonsense stimuli, such as nonwords or pseudowords, theoretical models of spoken production include a nonlexical/semantic route, which links auditory analysis to phonological short-term memory through an auditory-to-phonological conversion route.

Word and nonword repetition is usually preserved in svPPA, while the performance of individuals with nfvPPA is affected by apraxia of speech and marked by the production of phonological and phonetic errors [21]. The performance of these tasks by individuals with lvPPA is also affected; specifically, the production of phonological errors increases as word and/or nonword length increases [22]. Finally, sentence repetition is particularly important for PPA assessment, especially with respect to the core criteria proposed by Gorno-Tempini et al. for lvPPA diagnosis [6]. Sentence repetition is impaired in the three PPA variants, although there are distinct manifestations and severity levels. Individuals with lvPPA show significant impairment in sentence repetition, with performance negatively influenced by stimulus length but not by syntactic complexity [23]. In this task, their performance is notably marked by word omissions, semantic substitutions (replacement of one or more sentence words with words having similar or closely similar meanings), and phonological errors [24]. This profile is attributed to phonological short-term memory impairment. Meanwhile, sentence repetition is mildly impaired in svPPA [21]. In this variant, performance is not dependent on sentence length or syntactic complexity but on comprehension of the words of the sentence [25]. Finally, in nfvPPA, sentence repetition is disrupted due to impairment of the phonetic encoding stage of spoken production [26]. A deficit in the rehearsal mechanism of encoded verbal information has also been suggested to explain the production of phonological errors in sentence repetition in nfvPPA [21]. In this variant, performance might be influenced by the syntactic complexity of sentences [27] due to associated agrammatism [28].

To conclude, although repetition is disrupted in the three PPA variants (except for single words and nonwords in svPPA), the functional origins of the impairments provide essential clues for the differential diagnosis. However, a prerequisite for this is the use of adequate tests in which confounding variables, such as stimulus length and syntactic complexity, are controlled and manipulated.

3.3. The Assessment of Language Production in Spontaneous Speech and Narrative Discourse

Traditional tests provide useful information on linguistic abilities and language impairments in PPA. However, performance on these tests does not necessarily predict how a person will communicate in more naturalistic settings and everyday life. Functionally, spontaneous speech is the best way to appreciate the verbal and nonverbal communication of individuals with PPA. This simple everyday life ability involves the execution and interaction of various cognitive (episodic memory, semantic memory, short-term memory, working memory, executive functions, attentional ability) and linguistic (speech production, speech comprehension, pragmatics) processes [29], making it particularly vulnerable in PPA. The functional origin of language deficits in the three PPA variants manifests in different ways in spontaneous speech. Spontaneous speech in patients with svPPA is fluent, well-articulated, and grammatically correct; semantic impairment primarily causes word-finding difficulties in the form of aborted sentences, latencies, circumlocutions, and occasional semantic paraphasias [30]. In nfvPPA, phonetic encoding impairment makes spontaneous speech slow, hesitant, and effortful [31]. As the disease progresses, speech fluency decreases, and articulation and prosody become more affected. Moreover, disruption in the activation of the phonological forms of words is responsible for hesitations and the production of phonological errors and contributes to the slow rate and abnormal pauses in connected speech [31]. Agrammatism, the second core feature of nfvPPA, can be subtle and may go unnoticed in connected speech. When apparent, agrammatism in spontaneous speech is marked by difficulty with inflecting verbs, the omission or substitution of closed-class words, and difficulty in sentence construction [32]. Finally, in lvPPA, impairment is localized in the activation of phonological forms and phonological short-term memory and causes anomia and the production of phonological errors in spontaneous speech.

However, it is important to be aware that spontaneous speech may not be particularly useful for examining linguistic variables such as word retrieval and morphosyntax because deficits can be masked when individuals manipulate the complexity of their utterances and the specific lexical items they select. By contrast, narrative tasks such as storytelling or scene description which constrain an individual to certain vocabulary items and discourse structures can be highly informative over and above unconstrained conversation. In these tasks, similar manifestations to those mentioned previously for spontaneous speech might be observed in the three variants of PPA.

4. The Assessment of Comprehension

In PPA, the assessment of comprehension usually includes tests that explore word and sentence comprehension as well as object knowledge, which is stored in semantic memory. For differential diagnosis, the assessment of oral comprehension is usually sufficient. However, in the presence of noncompensated hearing loss, a written assessment can be useful. A summary of the salient comprehension deficits and their underlying cognitive impairment in the three PPA variants are presented in Table 2.

Table 2. Underlying cognitive deficits and salient characteristics of comprehension disorders in the three PPA variants.

Comprehension	svPPA	nfvPPA	lvPPA
Underlying deficit	Semantic memory	Grammar and working memory	Phonological short-term memory
Influence of psycholinguistic variables	- Concept familiarity - Concept typicality - Semantic category * - Visual complexity of pictures *	- Syntactic complexity	- Sentence length
Word comprehension	Impaired: Errors on semantic distractors	Preserved	Preserved
Object knowledge	Impaired in verbal and nonverbal modalities	Preserved	Preserved
Sentence comprehension	Preserved	Impaired in syntactically complex sentences (e.g., passive and relative sentences)	Preserved but can be impaired in long sentences

* Potential but nonessential psycholinguistic variable.

4.1. The Assessment of Word Comprehension and Object Knowledge

The assessment of word comprehension is crucial to the differential diagnosis of PPA. Word–picture matching tests, such as the Peabody Picture Vocabulary Test [33] in English or the spoken word-to-picture matching subtest of the BECLA battery [34] in French, are usually used to assess single-word comprehension. From the cognitive neuropsychology point of view, the processes involved in these tests include auditory analysis of the stimulus, activation of its lexical representation, and activation of the corresponding semantic representation within the semantic memory. Single-word comprehension is usually well-preserved in lvPPA and nfvPPA, whereas its impairment is one of the core features in svPPA [6]. In this variant, the deficit arises directly from semantic memory impairment, and errors are mostly made on semantic distractors.

The assessment of object knowledge directly recruits the activation of semantic information in semantic memory as well as links between semantic concepts. Picture association tasks, such as the Pyramids and Palm Trees Test [35], are usually used to assess nonverbal semantic processing. However, when visual impairment or visual agnosia is present, tests that use written words, such as the written word-to-written word semantic matching subtest of the BECLA battery, are preferable [34]. In the presence of a semantic memory deficit, semantic access to both pictures and words should be impaired. It is important to consider that this type of task can be especially challenging in the presence of executive deficits, which are often found in PPA, and could lead to misleading results. In this case, a simpler task, such as a semantic questionnaire (e.g., QueSQ in French [36]) can be used. Given the core impairment of semantic memory, object knowledge in nonverbal and verbal modalities is impaired in svPPA. However, it is usually largely spared in lvPPA and nfvPPA.

Psycholinguistic factors, such as familiarity and typicality, are particularly important when it comes to semantic memory. In svPPA, word comprehension and object knowledge performance are usually better preserved for concepts that are familiar to the person (e.g., objects used daily) [37]. Typicality is also important, as more typical items of a semantic category (e.g., apple) are processed faster than less typical items of the same category (e.g., mango) [38].

4.2. The Assessment of Sentence Comprehension

Sentence comprehension is usually assessed using a sentence-picture matching task, such as the Northwestern Assessment of Verbs and Sentences in English [39] or the Batterie d'évaluation de la compréhension syntaxique [40] in French. Sentence comprehension is typically well preserved in svPPA (if all words in the sentence are understood) and lvPPA [41]. However, in lvPPA, phonological short-term memory impairment could lead to difficulty understanding long sentences [41]. In this context, assessing sentence comprehension using written material could be particularly relevant for differential diagnosis, as visual support is likely to reduce the load on phonological short-term memory and lead to better performance. In nfvPPA, sentence comprehension is impaired, particularly for syntactically complex sentences [42]. In addition to the core impairment of grammar, working memory deficits have been documented in this clinical population [41] and could contribute to difficulties with syntactically complex sentences. Syntax complexity (e.g., active, passive, relative sentences) and sentence length are key parameters and should be controlled or manipulated in sentence comprehension tests.

5. The Assessment of Written Language

The assessment of written language involves the administration of reading and written spelling tests using different types of material: words, nonwords, sentences, texts, and narrative discourse. A summary of the underlying cognitive deficits of written language impairment and the salient characteristics of reading and written spelling disorders in the three PPA variants are presented in Table 3.

Table 3. Underlying cognitive deficits and salient characteristics of written language disorders in the three PPA variants.

Reading	svPPA	nfvPPA	lvPPA
Underlying deficit	Lexical-semantic route: Semantic memory	- Lexical-semantic route: Activation of phonological forms - Sublexical route: Activation of grapheme-to-phoneme conversion rules	- Lexical-semantic route: Partial impairment in the activation of phonological forms - Sublexical route: Activation of grapheme-to-phoneme conversion rules
Influence of psycholinguistic variables	- Orthographic consistency	- Orthographic consistency - Lexicality: Words and nonwords - Lexical frequency	- Lexicality: Words and nonwords - Lexical frequency
Reading	Surface dyslexia: Regularization errors	Phonological dyslexia: Phonological errors and impact of apraxia of speech	Mixed (deep/phonological) dyslexia: Phonological, semantic, and visual paralexias
Writing			
Underlying deficit	Lexical-semantic route: Semantic memory	- Lexical-semantic route: Activation of orthographic forms - Sublexical route: Activation of phonological-to-orthographic conversion rules	- Lexical-semantic route: Partial impairment in the activation of orthographic forms - Sublexical route: Activation of phonological-to-orthographic conversion rules
Influence of psycholinguistic variables	- Orthographic consistency	- Orthographic consistency - Lexicality: Words and nonwords - Lexical frequency	- Lexicality: words and nonwords - Lexical frequency
Word production: Picture naming and writing-to-dictation	Surface agraphia: No responses, semantic paragraphias, and phonologically plausible errors	Mixed agraphia: Phonologically and nonphonologically plausible errors	Phonological agraphia: Nonphonologically plausible errors and possible phonologically plausible errors
Spontaneous writing	Word-finding difficulties and surface agraphia: Aborted sentences, phonologically plausible errors, and occasional semantic paragraphias	Mixed agraphia and agrammatism: Phonologically and nonphonologically plausible errors and syntactic errors	Phonological agraphia: Nonphonologically plausible errors and possible phonologically plausible errors

5.1. The Assessment of Reading Abilities

In cognitive models, such as the dual-route cascaded model [43], reading is mediated by the computation of orthographic, phonological, and semantic information via two distinct routes: the lexical-semantic route and the sublexical route. The lexical-semantic route involves reading words with inconsistent orthography-to-phonology mappings (e.g., *yacht*), while the sublexical route, mediated by grapheme-to-phoneme conversion rules, mainly involves reading words with consistent orthography-to-phonology mappings (e.g., *banana*). Therefore, the control and manipulation of psycholinguistic variables of stimuli are essential in reading tests. According to the dual-route cascaded model, impairment of the lexical-semantic route alone causes surface dyslexia, which is characterized by difficulty reading inconsistent words [44]. Meanwhile, impairment of the sublexical route alone results in phonological dyslexia [45]. However, the disruption of both reading pathways results in deep dyslexia, which is characterized by difficulty reading words and nonwords and the production of semantic and visual paralexias [46]. Differentiation between the two reading pathways is particularly important for the differential diagnosis of PPA.

Semantic memory impairment affects the lexical-semantic route of reading in individuals with svPPA. This impairment causes surface dyslexia, one of the clinical features of this variant [6]. These individuals are better at reading orthographically consistent words than inconsistent words, and they show a preserved ability to read nonwords. Most of their reading errors consist of regularizations (e.g., bread→/brid/). Their performance is also influenced by the lexical frequency of words [47] and is directly linked to the extent of semantic loss [48]. Reading ability impairment is not part of the clinical criteria for nfvPPA or lvPPA. Although reading abilities are considered to be preserved in nfvPPA, impairment may emerge with disease progression [49]. Reading is characterized by phonological dyslexia, which is specifically affected by unfamiliar words and nonwords and marked by the production of phonological errors [50]. This profile suggests impairment of the sublexical route of reading. Reading is also characterized by manifestations of apraxia of speech in nfvPPA. In lvPPA, the underlying impairment of the activation of phonological lexical representation and phonological short-term memory causes difficulty in reading words and nonwords. Errors consist of a mix of phonological, semantic, and visual paralexias, which is suggestive of impairment of both routes of reading (deep/phonological dyslexia) [51]. The overlap of the manifestations makes it difficult to make a differential diagnostic between nfvPPA and lvPPA based on reading impairment alone. The manifestations of reading impairment in the three PPA variants are qualitatively similar but can be quantitatively exacerbated when abilities are tested with sentences or texts.

5.2. The Assessment of Writing Abilities

Cognitive models of writing, such as the dual-route model, also involve two distinct routes: the lexical-semantic route and the sublexical route [52]. The lexical-semantic route is used to write familiar words and includes processes that are initiated in the conceptual-semantic component, continue with the activation of orthographic lexical representations, and end with the execution of writing mechanisms. In a writing-to-dictation task, this sequential process is preceded by recognition of the spoken word in the phonological lexicon. The sublexical route is used to write unfamiliar words and nonwords through the activation of phonological-to-orthographic conversion rules. Impairment of the lexical-semantic route alone causes surface agraphia in which the use of the sublexical route leads to the production of phonologically plausible errors (e.g., phone → FONE). When only the sublexical route is disrupted, the resulting deficit, which is called phonological agraphia, affects nonword spelling. Finally, impairment of both routes causes deep agraphia, which is characterized by difficulty writing words and nonwords and the production of semantic and visual errors. Phonologically implausible errors (i.e., insertions, deletions, or transpositions of letters) are also possible due to partially impaired access to orthographic word forms.

Surface agraphia is one of the clinical features of svPPA [6]. The deficit is caused by impairment of the lexical-semantic route and is directly linked to semantic loss and difficulty activating orthographic forms in the lexicon [53]. The impairment is apparent regardless of the nature of the written task (e.g., spontaneous writing, writing-to-dictation, picture naming). Written production is usually more impaired than reading in nfvPPA. In this variant, writing impairment is suggestive of deep agraphia [54] due to difficulty retrieving orthographic forms of words in the lexicon, combined with disruption of the sublexical route [55]. Writing performance is also negatively influenced by orthographic inconsistency in nfvPPA but to a lesser degree than in svPPA [56]. The production of phonologically plausible errors is not exceptional in this variant [57]. Agrammatism is also generally apparent in nfvPPA in narrative discourse and spontaneous writing [58]. Finally, patients with lvPPA usually present with phonological agraphia, which is characterized by an impaired sublexical route and partially impaired access to orthographic word forms in the lexical-semantic route [55]. Other forms of agraphia, such as surface agraphia, are also possible in lvPPA [54]. As with reading, there is a partial overlap in the manifestations of agraphia in nfvPPA and lvPPA.

6. Conclusions

As shown in this article, PPA is a heterogeneous syndrome in terms of its clinical manifestations. The aforementioned diagnostic criteria aid in the differentiation of the three PPA variants [6]; however, they are very broad and are the subject of controversy. Their limits are important and relate to various aspects of linguistic semiology [59]. For example, Sajjadi et al. [60] performed a factor analysis of the language tasks results of 46 patients with PPA. The results were consistent with the existence of two variants, one characterized by semantic deficits (23% of cases) and the other by agrammatism and apraxia of speech (26% of cases). However, the analysis did not identify a cluster of measures that were compatible with the clinical profile of lvPPA. A few years later, Hoffman et al. [61] reanalyzed the data from those patients without taking their initial clinical diagnoses into account; they identified a distinct cluster for svPPA but not for the other variants.

We have also shown that an assessment process based on cognitive neuropsychological models allows clinicians to understand patients' deficits (i.e., surface manifestations, underlying origins, affected components) and identify strengths and weaknesses in their communication abilities. Although some surface manifestations of language impairments might overlap (e.g., anomia, repetition deficits) in PPA, their underlying origins are different and differentiable. Therefore, the cognitive approach to language assessment is useful for the differential diagnosis of PPA. Future studies should extend beyond the surface manifestations of language impairment in PPA to develop more comprehensive and distinctive diagnostic criteria.

In addition to its importance for the differential diagnosis of PPA, identifying the underlying cognitive deficit of the language impairment is crucial in order to plan effective therapeutic interventions based on restorative and compensatory approaches or teach communication strategies to patients and their relatives. For example, anomia is often targeted in PPA behavioral treatments. However, the underlying cause of anomia is functionally localized in the activation of phonological forms in the lexicon in lvPPA and nfvPPA, while it is caused by an impairment of semantic memory in svPPA [6]. The exact origin of anomia has an important role to play in how the intervention is planned. For example, when semantic memory is impaired, generalization is limited, as relearning primarily relies on episodic memory [62]. Moreover, treatment success in svPPA has been shown to be related to residual semantic knowledge and contextual information, which were more preserved for significant and familiar words [63]. Therefore, the selection of vocabulary based on personal interests is more crucially important in svPPA than in the two other PPA variants [64].

Identifying the underlying deficit is also important in compensatory approaches. For example, the choice of an app to compensate for language impairments depends directly on their functional origin. An app in which the content is organized by semantic categories (e.g., fruits/vegetables/meat) could be very effective in compensating for lexical-based anomia in lvPPA and nfvPPA. In contrast, teaching a patient with svPPA to search for information on the Internet to cope with comprehension problems (e.g., Wikipedia, Google Images) using keywords would be preferable due to the semantic origin of his/her difficulties [65].

Finally, the underlying deficit must also be considered when teaching communication strategies to patients and their relatives. For example, one popular strategy to compensate for word-finding difficulty is to encourage the patient to describe the object that he/she is unable to produce (e.g., Mug: What I use to drink coffee). However, while this could be a very efficient strategy for an impairment in the activation of phonological forms, it would be ineffective for an impairment that is localized in semantic memory because the spontaneous generation of a useful or reliable definition of the word would be compromised due to difficulty activating conceptual knowledge.

The aforementioned examples mainly concern word retrieval deficits. It is worth noting that the cognitive approach is similarly useful for treating, compensating, or teaching communication strategies for other deficits associated with PPA, such as agrammatism [66], apraxia of speech [67] and spelling deficits [68]. In summary, considering the underlying deficit in the clinical management of PPA allows for a tailored intervention that is likely to maximize benefits for patients and their relatives.

Author Contributions: All three authors contributed to the writing of the article: Conceptualization, J.M., A.L., M.L.; writing-original draft preparation: J.M., M.L.; writing-review and editing: J.M., A.L., M.L. All authors have read and agreed to the published version of the manuscript.

Funding: This research received no external funding.

Conflicts of Interest: The authors declare no conflict of interest.

References

1. Arvanitakis, Z.; Shah, R.C.; Bennett, D.A. Diagnosis and Management of Dementia: Review. *JAMA* **2019**, *322*, 1589–1599. [CrossRef] [PubMed]
2. Hugo, J.; Ganguli, M. Dementia and Cognitive Impairment: Epidemiology, Diagnosis, and Treatment. *Clin. Geriatr. Med.* **2014**, *30*, 421–442. [CrossRef] [PubMed]
3. Miller, B.L.; Boeve, B.F. *The Behavioral Neurology of Dementia*, 2nd ed.; Cambridge University Press: Cambridge, MA, USA, 2016; ISBN 9781107077201.
4. Macoir, J.; Turgeon, Y.; Laforce, R. *Language Processes in Delirium and Dementia*; Elsevier: Amsterdam, The Netherlands, 2015; ISBN 9780080970875.
5. Mesulam, M.-M.; Rogalski, E.J.; Wieneke, C.; Hurley, R.S.; Geula, C.; Bigio, E.H.; Thompson, C.K.; Weintraub, S. Primary progressive aphasia and the evolving neurology of the language network. *Nat. Rev. Neurol.* **2014**, *10*, 554–569. [CrossRef] [PubMed]
6. Gorno-Tempini, M.L.; Hillis, A.E.; Weintraub, S.; Kertesz, A.; Mendez, M.; Cappa, S.F.; Ogar, J.M.; Rohrer, J.; Black, S.; Boeve, B.F.; et al. Classification of primary progressive aphasia and its variants. *Neurology* **2011**, *76*, 1006–1014. [CrossRef]
7. Macoir, J.; Sylvestre, A.; Turgeon, Y. *Classical Tests for Speech and Language Disorders*; Elsevier: Amsterdam, The Netherlands, 2006; ISBN 9780080448541.
8. Levelt, W.J.M.; Roelofs, A.; Meyer, A.S. A theory of lexical access in speech production. *Behav. Brain Sci.* **1999**, *22*, 1–38. [CrossRef]
9. Kay, J.; Lesser, R.; Coltheart, M. *Psycholinguistic Assessments of Language Processing in Aphasia (PALPA)*; Lawrence Erlbaum Associates Publishers: Hove, UK, 1992.
10. Mar, R.A. The neuropsychology of narrative: Story comprehension, story production and their interrelation. *Neuropsychologia* **2004**, *42*, 1414–1434. [CrossRef]
11. Kaplan, E.F.; Goodglass, H.; Weintraub, S. *The Boston Naming Test*; Lea & Febiger: Philadelphia, PA, USA, 1983.
12. Macoir, J.; Chagnon, A.; Hudon, C.; Lavoie, M.; Wilson, A.M. TDQ-30—A New Color Picture-Naming Test for the Diagnostic of Mild Anomia: Validation and Normative Data in Quebec French Adults and Elderly. *Arch. Clin. Neuropsychol.* **2021**, *36*, 267–280. [CrossRef]
13. Woollams, A.M.; Cooper, E.; Hodges, J.R.; Patterson, K. Anomia: A doubly typical signature of semantic dementia. *Neuropsychologia* **2008**, *46*, 2503–2514. [CrossRef] [PubMed]
14. Adlam, A.L.; Patterson, K.; Rogers, T.T.; Nestor, P.J.; Salmond, C.H.; Acosta-Cabronero, J.; Hodges, J.R. Semantic Dementia and Fluent Primary Progressive Aphasia: Two Sides of the Same Coin? *Brain* **2006**, *129*, 3066–3080. [CrossRef]
15. Henry, M.L.; Gorno-Tempini, M.L. The logopenic variant of primary progressive aphasia. *Curr. Opin. Neurol.* **2010**, *23*, 633–637. [CrossRef]
16. Clark, D.G.; Charuvastra, A.; Miller, B.L.; Shapira, J.S.; Mendez, M.F. Fluent versus nonfluent primary progressive aphasia: A comparison of clinical and functional neuroimaging features. *Brain Lang.* **2005**, *94*, 54–60. [CrossRef]
17. Botha, H.; Josephs, K.A. Primary Progressive Aphasias and Apraxia of Speech. *Continuum* **2019**, *25*, 101–127. [CrossRef] [PubMed]
18. Botha, H.; Duffy, J.R.; Whitwell, J.L.; Strand, E.A.; Machulda, M.M.; Schwarz, C.; Reid, R.I.; Spychalla, A.J.; Senjem, M.L.; Jones, D.T.; et al. Classification and clinicoradiologic features of primary progressive aphasia (PPA) and apraxia of speech. *Cortex* **2015**, *69*, 220–236. [CrossRef] [PubMed]

19. Croot, K.; Ballard, K.; Leyton, C.E.; Hodges, J.R. Apraxia of Speech and Phonological Errors in the Diagnosis of Nonfluent/Agrammatic and Logopenic Variants of Primary Progressive Aphasia. *J. Speech Lang. Hear. Res.* **2012**, *55*, S1562–S1572. [CrossRef]
20. Croota, K.; Pattersona, K.; Hodges, J.R. Single Word Production in Nonfluent Progressive Aphasia. *Brain Lang.* **1998**, *61*, 226–273. [CrossRef] [PubMed]
21. Allen, R.J.; Baddeley, A.D. Working memory and sentence recall. In *Interactions between Short-Term and Long-Term Memory in the Verbal Domain*; Thorn, A., Page, M., Eds.; Psychology Press: New York, NY, USA, 2009; pp. 63–85. ISBN 978-1-84169-639-3.
22. Leyton, C.E.; Savage, S.; Irish, M.; Schubert, S.; Piguet, O.; Ballard, K.J.; Hodges, J.R. Verbal Repetition in Primary Progressive Aphasia and Alzheimer's Disease. *J. Alzheimer's Dis.* **2014**, *41*, 575–585. [CrossRef]
23. Meyer, A.M.; Snider, S.F.; Campbell, R.E.; Friedman, R.B. Phonological short-term memory in logopenic variant primary progressive aphasia and mild Alzheimer's disease. *Cortex* **2015**, *71*, 183–189. [CrossRef] [PubMed]
24. Gorno-Tempini, M.L.; Brambati, S.M.; Ginex, V.; Ogar, J.; Dronkers, N.F.; Marcone, A.; Perani, D.; Garibotto, V.; Cappa, S.F.; Miller, B.L. The logopenic/phonological variant of primary progressive aphasia. *Neurology* **2008**, *71*, 1227–1234. [CrossRef]
25. Leyton, C.E.; Hodges, J.R. Towards a Clearer Definition of Logopenic Progressive Aphasia. *Curr. Neurol. Neurosci. Rep.* **2013**, *13*, 1–7. [CrossRef]
26. Knott, R.; Patterson, K.; Hodges, J.R. Lexical and Semantic Binding Effects in Short-term Memory: Evidence from Semantic Dementia. *Cogn. Neuropsychol.* **1997**, *14*, 1165–1216. [CrossRef]
27. Ogar, J.M.; Dronkers, N.F.; Brambati, S.M.; Miller, B.L.; Gorno-Tempini, M.L. Progressive Nonfluent Aphasia and Its Characteristic Motor Speech Deficits. *Alzheimer Dis. Assoc. Disord.* **2007**, *21*, S23–S30. [CrossRef] [PubMed]
28. Bonner, M.F.; Ash, S.; Grossman, M. The New Classification of Primary Progressive Aphasia into Semantic, Logopenic, or Nonfluent/Agrammatic Variants. *Curr. Neurol. Neurosci. Rep.* **2010**, *10*, 484–490. [CrossRef] [PubMed]
29. Schumacher, R.; Halai, A.D.; Ralph, M.A.L. Assessing and mapping language, attention and executive multidimensional deficits in stroke aphasia. *Brain J. Neurol.* **2019**, *142*, 3202–3216. [CrossRef]
30. Hodges, J.R.; Patterson, K. Semantic dementia: A unique clinicopathological syndrome. *Lancet Neurol.* **2007**, *6*, 1004–1014. [CrossRef]
31. Wilson, S.M.; Henry, M.; Besbris, M.; Ogar, J.M.; Dronkers, N.F.; Jarrold, W.; Miller, B.L.; Gorno-Tempini, M.L. Connected speech production in three variants of primary progressive aphasia. *Brain* **2010**, *133*, 2069–2088. [CrossRef] [PubMed]
32. Thompson, C.K.; Cho, S.; Hsu, C.-J.; Wieneke, C.; Rademaker, A.; Weitner, B.B.; Mesulam, M.M.; Weintraub, S. Dissociations between fluency and agrammatism in primary progressive aphasia. *Aphasiology* **2012**, *26*, 20–43. [CrossRef]
33. Dunn, L.M.; Dunn, L.M. *Manual for the Peabody Picture Vocabulary Test-Revised*; American Guidance Service: Circle Pines, MN, USA, 1981.
34. Macoir, J.; Gauthier, C.; Jean, C.; Potvin, O. BECLA, a new assessment battery for acquired deficits of language: Normative data from Quebec-French healthy younger and older adults. *J. Neurol. Sci.* **2016**, *361*, 220–228. [CrossRef]
35. Howard, D.; Patterson, K. *The Pyramids and Palm Trees Test: A Test for Semantic Access from Words and Pictures*; Thames Valley Test Company: Bury St Edmunds, UK, 1992.
36. Monetta, L.; Légaré, A.; Macoir, J.; Wilson, M.A. Quebec Semantic Questionnaire (QueSQ). Development, Validation and Normalization. *Can. J. Aging* **2020**, *39*, 98–106. [CrossRef] [PubMed]
37. Rogers, T.T.; Patterson, K.; Jefferies, E.; Ralph, M.A.L. Disorders of representation and control in semantic cognition: Effects of familiarity, typicality, and specificity. *Neuropsychology* **2015**, *76*, 220–239. [CrossRef]
38. Riley, E.A.; Barbieri, E.; Weintraub, S.; Mesulam, M.M.; Thompson, C.K. Semantic Typicality Effects in Primary Progressive Aphasia. *Am. J. Alzheimer's Dis. Other Dement.* **2018**, *33*, 292–300. [CrossRef]
39. Thompson, C.K. *Northwestern Assessment of Verbs and Sentences (NAVS)*; Northwestern University: Evanston, IL, USA, 2012.
40. Caron, S.; Le May, M.-E.; Bergeron, A.; Bourgeois, M.E.; Fossard, M. *Batterie D'évaluation de La Compréhension Syntaxique (BCS)*; Institut de réadaptation en Déficience Physique de Québec (IRDPQ): Québec, QC, Canada, 2015.
41. Eikelboom, W.; Janssen, N.; Jiskoot, L.C.; Berg, E.V.D.; Roelofs, A.; Kessels, R.P. Episodic and working memory function in Primary Progressive Aphasia: A meta-analysis. *Neurosci. Biobehav. Rev.* **2018**, *92*, 243–254. [CrossRef]
42. Peelle, J.E.; Troiani, V.; Gee, J.; Moore, P.; McMillan, C.; Vesely, L.; Grossman, M. Sentence comprehension and voxel-based morphometry in progressive nonfluent aphasia, semantic dementia, and nonaphasic frontotemporal dementia. *J. Neurolinguist.* **2008**, *21*, 418–432. [CrossRef] [PubMed]
43. Coltheart, M.; Curtis, B.; Atkins, P.; Haller, M. Models of Reading Aloud: Dual-Route and Parallel-Distributed-Processing Approaches. *Psychol. Rev.* **1993**, *100*, 589–608. [CrossRef]
44. Coltheart, M. Cognitive neuropsychology and the study of reading. In *Attention and Performance*; Posner, M.I., Marin, O.S.M., Eds.; Lawrence Erlbaum Associates Inc.: Hillsdale, MI, USA, 1985; Volume 11, pp. 3–37.
45. Funnell, E. Phonological processes in reading: New evidence from acquired dyslexia. *Br. J. Psychol.* **1983**, *74*, 159–180. [CrossRef]
46. Coltheart, M.; Rastle, K.; Perry, C.; Langdon, R.; Ziegler, J.C. DRC: A dual route cascaded model of visual word recognition and reading aloud. *Psychol. Rev.* **2001**, *108*, 204–256. [CrossRef]
47. Wilson, S.M.; Brambati, S.M.; Henry, R.G.; Handwerker, D.; Agosta, F.; Miller, B.L.; Wilkins, D.P.; Ogar, J.M.; Gorno-Tempini, M.L. The neural basis of surface dyslexia in semantic dementia. *Brain* **2008**, *132*, 71–86. [CrossRef]

48. Funnell, E. Response Biases in Oral Reading: An Account of the Co-Occurrence of Surface Dyslexia and Semantic Dementia. *Q. J. Exp. Psychol.* **1996**, *49 A*, 417–446. [CrossRef]
49. Rohrer, J.; Rossor, M.; Warren, J.D. Syndromes of nonfluent primary progressive aphasia: A clinical and neurolinguistic analysis. *Neurology* **2010**, *75*, 603–610. [CrossRef]
50. Woollams, A.M.; Patterson, K. The consequences of progressive phonological impairment for reading aloud. *Neuropsychology* **2012**, *50*, 3469–3477. [CrossRef]
51. Brambati, S.; Ogar, J.; Neuhaus, J.; Miller, B.; Gorno-Tempini, M. Reading disorders in primary progressive aphasia: A behavioral and neuroimaging study. *Neuropsychology* **2009**, *47*, 1893–1900. [CrossRef] [PubMed]
52. Miceli, G.; Capasso, R. Spelling and dysgraphia. *Cogn. Neuropsychol.* **2006**, *23*, 110–134. [CrossRef]
53. Teichmann, M.; Sanches, C.; Moreau, J.; Ferrieux, S.; Nogues, M.; Dubois, B.; Cacouault, M.; Sharifzadeh, S. Does surface dyslexia/dysgraphia relate to semantic deficits in the semantic variant of primary progressive aphasia? *Neuropsychology* **2019**, *135*, 107241. [CrossRef] [PubMed]
54. Sepelyak, K.; Crinion, J.; Molitoris, J.; Epstein-Peterson, Z.; Bann, M.; Davis, C.; Newhart, M.; Heidler-Gary, J.; Tsapkini, K.; Hillis, A.E. Patterns of breakdown in spelling in primary progressive aphasia. *Cortex* **2011**, *47*, 342–352. [CrossRef]
55. Faria, A.V.; Crinion, J.; Tsapkini, K.; Newhart, M.; Davis, C.; Cooley, S.; Mori, S.; Hillis, A.E. Patterns of Dysgraphia in Primary Progressive Aphasia Compared to Post-Stroke Aphasia. *Behav. Neurol.* **2013**, *26*, 21–34. [CrossRef]
56. Graham, N.L. Dysgraphia in primary progressive aphasia: Characterisation of impairments and therapy options. *Aphasiology* **2014**, *28*, 1092–1111. [CrossRef]
57. Shim, H.; Hurley, R.S.; Rogalski, E.; Mesulam, M.-M. Anatomic, clinical, and neuropsychological correlates of spelling errors in primary progressive aphasia. *Neuropsychology* **2012**, *50*, 1929–1935. [CrossRef] [PubMed]
58. Tetzloff, K.A.; Utianski, R.L.; Duffy, J.R.; Clark, H.M.; Strand, E.A.; Josephs, K.A.; Whitwell, J.L. Quantitative Analysis of Agrammatism in Agrammatic Primary Progressive Aphasia and Dominant Apraxia of Speech. *J. Speech Lang. Hear. Res.* **2018**, *61*, 2337–2346. [CrossRef] [PubMed]
59. Ouellet, N.A.; Fossard, M.; Macoir, J. Consensual recommendations for the description of three variants of primary progressive aphasia: Limits and controversies regarding language impairments. *Gériatr. Psychol. Neuropsychiatrie Viellissement* **2015**, *13*, 441–451. [CrossRef]
60. Sajjadi, S.A.; Patterson, K.; Arnold, R.J.; Watson, P.C.; Nestor, P.J. Primary progressive aphasia: A tale of two syndromes and the rest. *Neurology* **2012**, *78*, 1670–1677. [CrossRef]
61. Hoffman, P.; Sajjadi, S.A.; Patterson, K.; Nestor, P.J. Data-driven classification of patients with primary progressive aphasia. *Brain Lang.* **2017**, *174*, 86–93. [CrossRef]
62. Cadório, I.; Lousada, M.; Martins, P.; Figueiredo, D. Generalization and maintenance of treatment gains in primary progressive aphasia (PPA): A systematic review. *Int. J. Lang. Commun. Disord.* **2017**, *52*, 543–560. [CrossRef] [PubMed]
63. Snowden, J.S.; Neary, D. Relearning of verbal labels in semantic dementia. *Neuropsychologia* **2002**, *40*, 1715–1728. [CrossRef]
64. Lavoie, M.; Bier, N.; LafoJrce, R.L.; Macoir, J. Improvement in functional vocabulary and generalization to conversation following a self-administered treatment using a smart tablet in primary progressive aphasia. *Neuropsychol. Rehabil.* **2020**, *30*, 1224–1254. [CrossRef] [PubMed]
65. Routhier, S.; Macoir, J.; Jacques, S.; Imbeault, H.; Pigot, H.; Giroux, S.; Cau, A.; Bier, N. From Smartphone to External Semantic Memory Device: The Use of New Technologies to Compensate for Semantic Deficits. *Non Pharmacol. Ther. Dement.* **2012**, *2*, 81–99.
66. Schneider, S.L.; Thompson, C.K.; Luring, B. Effects of verbal plus gestural matrix training on sentence production in a patient with primary progressive aphasia. *Aphasiology* **1996**, *10*, 297–317. [CrossRef]
67. Henry, M.L.; Meese, M.V.; Truong, S.; Babiak, M.C.; Miller, B.L.; Gorno-Tempini, M.L. Treatment for Apraxia of Speech in Nonfluent Variant Primary Progressive Aphasia. *Behav. Neurol.* **2013**, *26*, 77–88. [CrossRef] [PubMed]
68. Tsapkini, K.; Hillis, A.E. Spelling Intervention in Post-Stroke Aphasia and Primary Progressive Aphasia. *Behav. Neurol.* **2013**, *26*, 55–66. [CrossRef]

Article

Breakdowns in Informativeness of Naturalistic Speech Production in Primary Progressive Aphasia

Jeanne Gallée [1,*], Claire Cordella [2], Evelina Fedorenko [1,3], Daisy Hochberg [2], Alexandra Touroutoglou [2], Megan Quimby [2] and Bradford C. Dickerson [2,*]

1   Program in Speech and Hearing Bioscience and Technology, Harvard University, Cambridge, MA 02138, USA; evelina9@mit.edu
2   Frontotemporal Disorders Unit, Department of Neurology, Massachusetts General Hospital, Harvard Medical School, Boston, MA 02129, USA; ccordella@fas.harvard.edu (C.C.); dhochberg@mgh.harvard.edu (D.H.); atouroutoglou@mgh.harvard.edu (A.T.); mquimby@partners.org (M.Q.)
3   Department of Brain and Cognitive Sciences, Massachusetts Institute of Technology, Cambridge, MA 02139, USA
*   Correspondence: jgallee@g.harvard.edu (J.G.); brad.dickerson@mgh.harvard.edu (B.C.D.)

Abstract: "Functional communication" refers to an individual's ability to communicate effectively in his or her everyday environment, and thus is a paramount skill to monitor and target therapeutically in people with aphasia. However, traditional controlled-paradigm assessments commonly used in both research and clinical settings often fail to adequately capture this ability. In the current study, facets of functional communication were measured from picture-elicited speech samples from 70 individuals with mild primary progressive aphasia (PPA), including the three variants, and 31 age-matched controls. Building upon methods recently used by Berube et al. (2019), we measured the informativeness of speech by quantifying the content of each patient's description that was relevant to a picture relative to the total amount of speech they produced. Importantly, form-based errors, such as mispronunciations of words, unusual word choices, or grammatical mistakes are not penalized in this approach. We found that the relative informativeness, or efficiency, of speech was preserved in non-fluent variant PPA patients as compared with controls, whereas the logopenic and semantic variant PPA patients produced significantly less informative output. Furthermore, reduced informativeness in the semantic variant is attributable to a lower production of content units and a propensity for self-referential tangents, whereas for the logopenic variant, a lower production of content units and relatively "empty" speech and false starts contribute to this reduction. These findings demonstrate that functional communication impairment does not uniformly affect all the PPA variants and highlight the utility of naturalistic speech analysis for measuring the breakdown of functional communication in PPA.

Keywords: primary progressive aphasia; informativeness; speech production

Citation: Gallée, J.; Cordella, C.; Fedorenko, E.; Hochberg, D.; Touroutoglou, A.; Quimby, M.; Dickerson, B.C. Breakdowns in Informativeness of Naturalistic Speech Production in Primary Progressive Aphasia. *Brain Sci.* **2021**, *11*, 130. https://doi.org/10.3390/brainsci11020130

Academic Editor: Jordi A. Matias-Guiu
Received: 11 December 2020
Accepted: 15 January 2021
Published: 20 January 2021

1. Introduction

Primary progressive aphasia (PPA) is a clinical syndrome where aphasia is the initial predominant symptom due to neurodegenerative disease, most commonly frontotemporal lobar degeneration or Alzheimer's disease [1]. The characteristics of aphasia in PPA are heterogeneous, and many patients present with a profile of language impairments that can be classified into one of the following three subtypes: the semantic variant (svPPA), the logopenic variant (lvPPA), or the non-fluent/agrammatic variant (nfvPPA) [2–6]. As distinct as these subtypes may be, they all share a devastating prognosis, i.e., as a patient's aphasia progresses, his or her relationships will be adversely impacted by the breakdown of communication abilities [7–9].

A wide array of tests has been used to characterize language impairment in PPA [10,11] but concerns have been raised about highly constrained, decontextualized linguistic tasks being insufficient to describe or predict a person's ability to communicate in everyday life, which is often referred to as "functional communication" [12–15]. One way to measure functional communication in patients with aphasia is through relatively naturalistic picture description tasks or structured interviews [9,14,16]. Despite concerns about the reliability of these methods [16], abnormalities of a variety of elements of speech and language in PPA can be successfully captured through the analysis of connected speech samples from such tasks. For example, the production of verbs and complex grammatical sentence structures is reduced in nfvPPA relative to lvPPA and svPPA [17,18], mirroring reports from more structured experimental tasks. To determine the extent to which communication is functional, researchers have measured breakdowns in the informativeness of speech production in patients with chronic aphasia by identifying words or phrases that are relevant to a picture or question, while ignoring form-based errors, such as mispronunciations of words, unusual word choices, and grammatical mistakes [16,19]. However, impairment in the informativeness of speech in PPA has received little investigation [20].

Berube et al. (2019) recently evaluated the informativeness of speech output in both stroke aphasia patients and PPA patients using a contemporary version of the Cookie Theft picture description task. They measured the patients' production of words and phrases referring to concepts that were mentioned by control participants who described the same picture (i.e., "content units") [21,22], and found that individuals with PPA and those with stroke aphasia conveyed less information about the picture than controls. These results provided further evidence that connected speech elicited in such a paradigm can be useful for quantifying functional communication abilities. However, the sample of PPA patients in Berube et al.'s study was small and did not allow for a detailed examination of potential between-variant differences.

We sought to investigate the breakdown of functional communication in PPA using a similar approach, aiming to answer two questions. First, as compared with controls, do patients in the mild stages of each of the three PPA variants exhibit reduced ability to convey information in a picture description task? Secondly, if multiple variants exhibit an impairment, do different factors contribute to the reduction in speech informativeness? In particular, we aimed to replicate the Berube et al. (2019) [21] method and compare it with our own proposed method of examining content production in naturalistic speech. Prior studies on svPPA have reported lower speech rate [17,23–27] but similar total numbers of words produced [18], with the important observation of fewer nouns and frequent semantic errors [17,18,23,24], in the presence of relative preservation of syntactic abilities [24,27]. Semantic variant PPA patients tend to produce words with higher frequency and less specificity [8,17,18,23,28]. With respect to lvPPA, which is perhaps the least well-understood variant, prior studies have observed lower speech rate [17,18,24], fewer open-class words [17,24], frequent phonemic errors [17,18,24], as well as numerous false starts and filled pauses [17,18]. As in svPPA, syntactic abilities appear to be largely preserved [18,24]. Finally, prior studies have reported that the speech of nfvPPA patients is slower [17,18,29–31] and contains fewer words [17,18,23,24,28], contains errors in closed-class words [25,26,32], syntactic agreement errors [26,28], few or no complex syntactic structures [26,28], and also exhibits lower narrative coherence [24,26,27].

These findings, combined with clinical observation, led us to formulate the following hypotheses: With regard to the first research question, we expected to find differences among the three variants in the patients' ability to convey information relevant to the picture. The speech of svPPA patients was predicted to demonstrate reduced information content with each individual showing some degree of impairment; lvPPA patients' information content was predicted to be reduced as a group but also to exhibit variability within the group, with some patients showing preserved informativeness of speech; finally, the nfvPPA group was predicted to show preserved informativeness, with few, if any, individuals demonstrating impairment. With regard to the second research question, svPPA patients were predicted to produce an abnormally high number of self-referential tangents and empty utterances in addition to reduced information content, whereas lvPPA patients were predicted to produce a high number of empty utterances and false starts in addition to reduced information content.

2. Materials and Methods

2.1. Participants

Data analyzed in this study were obtained from 101 participants, including 70 patients with a diagnosis of PPA and 31 age-matched controls. The PPA patients participated in this study as part of the Massachusetts General Hospital Frontotemporal Disorders Unit longitudinal PPA cohort study. Participants in this cohort underwent a comprehensive clinical evaluation as previously described [7,10]. The evaluation included a structured interview by a neurologist or psychiatrist covering cognition, mood/behavior, sensorimotor function, and daily activities; a neurologic examination, including office-based cognitive testing (for cases in this report, performed by B.C.D.); a speech-language assessment by a speech-language pathologist (for cases in this report, performed by C.C., M.Q., or D.H.), including the Progressive Aphasia Severity Scale (PASS) to specifically assess language impairment relative to a patient's premorbid baseline [10]; and an MRI scan with T1- and T2-weighted sequences inspected visually by a neurologist. For all cases included here, visual inspection of the clinical MRI identified an atrophy pattern consistent with that typically seen in each PPA variant and ruled out other causes of focal brain damage [6].

For each participant, a clinician also performed a structured interview with an informant who knew the participant well (e.g., a spouse), augmented with standard questionnaires. For the participants in this report, the protocol included the National Alzheimer's Coordinating Center (NACC) Uniform Data Set measures (using version 2.0 for 65 of the assessments, version 3.0 for the remaining 5), including Clinical Dementia Rating (CDR) scale supplementary language box ratings [10,33]. As part of the standard battery, connected speech samples were elicited through the Western Aphasia Battery–Revised (WAB-R) "Picnic Scene" task [34].

Individuals selected for this study had been diagnosed with imaging-supported svPPA, lvPPA, or nfvPPA according to consensus diagnostic criteria [3,6]. All participants and their care partners denied a pre-existing psychiatric disorder, other neurological disorder, or developmental cognitive disorder. Given the focus on mild PPA in the current study, we excluded participants with CDR language box scores above 1 (0 = normal language, 0.5 = very mild language impairment, 1 = mild language impairment, 2 = moderate language impairment, and 3 = severe language impairment). The PPA patient sample included 19 svPPA patients (mean age 69.6, SD 8.35, 11 females), 26 lvPPA patients (mean age 70.2, SD 7.17, 12 females), and 25 nfvPPA patients (mean age 68.2, SD 8.28, 15 females) (see Table 1 for further demographic information).

Table 1. Summary demographic information and clinical characteristics for the participants included in this study (see Appendix A Table A1 for participant-specific characteristics).

	PPA (*N* = 70)			Healthy Controls
	svPPA (*N* = 19)	**lvPPA** (*N* = 26)	**nfvPPA** (*N* = 25)	**(HC)** (*N* = 31)
Age at testing, years (SD)	69.7 (8.35)	70.2 (7.17)	68.2 (8.28)	63.4 (8.20)
Female, *n* (%)	11.0 (57.8)	12.0 (46.2)	15.0 (60.0)	17.0 (54.8)
Education, years (SD)	16.2 (2.30)	15.7 (2.38)	15.5 (2.73)	14.9 (1.83)
CDR language (SD), *n* = 0.5	0.71 (0.25), 11	0.71 (0.25), 15	0.66 (0.24), 17	–
PASS sum of boxes (SD)	3.50 (1.30)	4.04 (1.25)	3.82 (1.26)	–

Data were also analyzed from thirty-one age-matched healthy controls, with no self-reported history of neurologic or psychiatric disorders, who participated in a longitudinal study conducted at the Speech and Feeding Disorders Laboratory at the MGH Institute of Health Professions (mean age 63.4, SD 8., 17 females). All participants in both samples were right-handed native English speakers. All participants (and their care partners for

PPA patients) gave written informed consent in accordance with guidelines established by the Mass General Brigham Healthcare System Institutional Review Boards which govern human subjects research at Massachusetts General Hospital (protocol no. 2015P001363, protocol no. 2012P000432, protocol no. 2016P001421, protocol no. 2019P003391, protocol no. 2016P001594, and protocol no. 2013P001746). No significant between-group differences (svPPA, lvPPA, nfvPPA, and HC) were observed in sex or education level. The control group was observed to be younger than the lvPPA ($p = 0.01$) and nfvPPA ($p = 0.004$) groups. In the comparison of the PPA variants, no significant between-group differences (svPPA, lvPPA, and nfvPPA) were observed for the CDR language or PASS sum of boxes scores.

2.2. Speech Sample Collection

As part of their standard assessment, all PPA participants completed speech elicitation tasks including the WAB-R [24] Picnic Scene task; the connected speech samples obtained from this task were analyzed for the present study. Participants were presented the picture and prompted to provide a description of what they saw. Audio recordings were collected using a handheld Zoom H4N Recorder (Hauppauge, NY, USA) or an Olympus VN-702PC Voice Recorder (Center Valley, PA, USA) in a quiet room. For participants included in this analysis, the average speech sample durations were 73.7 s for svPPA (SD 28.0 s), 81.7 s (s) for lvPPA (SD 28.9 s), 104s for nfvPPA (SD 53.4 s), and 48.9 s for HC (SD 11.3 s).

2.3. Speech Sample Transcription and Basic Coding

Following data collection, speech samples were uploaded to a local hard drive on an encrypted device and preprocessed in Audacity® Version 2.4.1 [35] to remove all clinician speech and background noise. Then, speech samples were transcribed and double-checked by two blinded (to diagnostic group) listeners. Inconsistencies in transcription were discussed and fixed upon consensus. Because the focus of the current study was on functional communication, phonemic errors (additions, omissions, and substitutions) where the intended word was clear were corrected during transcription (e.g., if the participant said, "they are sitting on a splanket" blanket was transcribed). Unrecognizable words were counted as empty words. Phonemic clusters followed by self-corrections or rerouting were counted as false starts. The average number of utterances (defined as any verbalization attempt, i.e., single words, including filler words, and false starts) was 135 for svPPA (SD 53.6), 135 for lvPPA (SD 60.3), 97.1 for nfvPPA (SD 34.1), and 150 for healthy controls (SD 65.9).

2.4. Content Unit (CUs) Coding

In order to analyze the informativeness of speech production, we followed the approach outlined by Berube et al. (2019) (see [36] for earlier work using this and related approaches). Using the healthy control samples, first, we created a corpus of 64 content units (CUs) (Table 2). A CU is defined as a unique concept. These concepts can correspond to an object/entity, an action, a property, or more abstract notions, such as spatial relations. A core motivation for the construct of CUs is to abstract over the potentially variable verbal descriptions that can be used for the same referent (e.g., "man", "dad", and "pops" all referring to a male individual). Following Berube et al. (2019) [21], the set of 64 CUs only included CUs that were each mentioned by at least three healthy controls. Then, we determined the number of CUs within each speech sample for all four groups. If a speech sample contained a CU more than once (even if the later occurrences used a different verbal description), only a single occurrence was counted. There were no penalties for the absence of any CUs.

Table 2. The complete set of 64 content units (CUs) identified in our healthy control sample (*n* = 31) based on Berube et al.'s (2019) methods [21]. The number of healthy control participants who produced a given CU in a speech sample is indicated in the "count" column. The unique referents for ambiguous CUs are represented in the "referent CU#" column, designating the agents, actions, and properties that at least three healthy control participants referred to by naming a particular CU.

Category	CU #	Referent CU#	Count	CU
Unambiguous entities/objects/events	1		31	Kite
	2		31	House, home, rental, cottage, cabin
	3		31	Wine, drink, liquid, drinks, beverage, soda, juice, champagne
	4		30	Dog, puppy
	5		29	Beach, sand
	6		28	Sailboat, boat
	7		28	Picnic, picnicking
	8		27	Woman, mother, mom, wife, lady
	9		26	Sandcastle(s), pile, castle
	10		26	Water, lake, ocean, river, sea
	11		26	Boy, son
	12		24	Girl, daughter
	13		21	Pier, dock
	14		18	Book
	15		18	Tree
	16		16	Car
	17		16	Driveway, path, road, street
	18		15	Garage, parked (car)
	19		15	Shoes, sandals, sneakers
	20		13	Fish (noun)
	21		13	Couple, two (people), parents
	22		13	Radio, boombox
	23		11	Flag, flagpole
	24		11	Glass, cup, glasses
	25		10	Blanket, carpet
	26		8	Shovel
	27		7	Pail, bucket
	28		7	Sky, clouds
	29		6	Fisherman
	30		4	Nasket
	31		3	(Man's) glasses
	32		3	Bottle, thermos
Unambiguous actions	33		31	Fishing, caught (a fish), catching (a fish)
	34		31	Flying (a kite), pulling (a kite)
	35		31	Building (a sandcastle), making (a sandcastle), playing (in the sand), built a sandcastle
	36		26	Reading
	37		21	Pouring
	38		18	Enjoying, relaxing, happy, having a good time, relaxed, (having) fun
	39		17	(Dog is) chasing, (dog is) following, chased
	40		8	Sailing, cruising
Unambiguous properties	41		17	Beautiful, idyllic, nice, lovely, pleasant, calm
	42		10	Big, large
Ambiguous-referent entities	43		31	Man, father, dad, gentleman, grandpa, husband, hubby, pops
	43a	42	18	
	43b	29	15	

Table 2. Cont.

Category	CU #	Referent CU#	Count	CU
	44	No Pattern	31	Grass, yard, shrubbery, park, enclave, grassy, place, bushes, environment, foliage, gable, hill, mountains, scenery, spot, trees, forest
	45		30	Someone, guy, fellow, somebody, person, jabroni, adult
	45a	11		
	45b	12		
	45c	42		
	45d	29		
	46		24	Family, people, everyone, families, occupants
	46a	11, 12, 8, 42		
	46b	11, 12, 8, 42, 29		
	46c	No Pattern		
	47		12	Inlet, lakeside, seashore, seaside, bay, oceanside, shore, wharf
	47a	58		
	47b	12		
	47c	29		
	48		8	Kid, kids, children
	48a	11		
	48b	12		
	49		5	Shorts
	49a	11		
	49b	8 + 42		
	49c	42		
	50		3	T-shirt
	50a	11		
	50b	8		
Ambiguous-referent actions	**51**		12	Running
	51a	11		
	51b	4		
	51c	11 + 4		
	52		6	Sitting
	52a	8		
	52b	42		
	52c	8 + 42		
	53		5	Wearing
	53a	11		
	53b	8 + 42		
	53c	42		
	53d	29		
	54	15	4	Bloom, blossomed, blooming
Ambiguous-referent properties	**55**		27	Little, young, younger
	55a	11		
	55b	12		
	55c	4		
	55d	2		
	55e	43		

Table 2. *Cont.*

Category	CU #	Referent CU#	Count	CU
Unambiguous but inferred (not physically present/abstract)	56		20	Summer, sun, sunny, summertime, weather, season, spring, warm
	57		16	Music, listening to (music), (playing) music
	58		16	Day, afternoon, vacation, retreat
	59		9	Scene, picture
	60		9	Activities, recreational, sport
	61		8	Outside
	62		4	Breeze, windy
	63		4	Sandwich(es), food
Ambiguous-referent spatial relations	64		31	Across, around, background, behind, beside, close, distance, distant, far, foreground, front, left, nearby, next to, right, side
	64a	2		
	64b	6		
	64c	11		
	64d	15		
	64e	23		
	64f	4		
	64g	21		
	64h	29		
	64i	43		
	64j	16		
	64k	22		
	64l	17		

In addition to Berube et al.'s (2019) [21] CU coding method, we developed an adjusted method to account for some ambiguity in the meaning of some CUs and compared the results of both methods. In particular, although the majority of CUs (42 of the 64, #1–42) unambiguously refer to a particular object/entity, action, or property of a specific object/entity in the picture (see Figure 1A for examples), the remaining 22 (#43–64) do not unambiguously refer to a particular aspect of the picture. Nine of these 22 CUs (#55–63) either refer to the scene overall or to concepts not physically present in the picture, inferred based on the schema of a picnic event (e.g., CU #63 "sandwich", which is not actually present) or abstract in nature (e.g., CU #57 "music"). The remaining 13 CUs (#43–54, 64) are ambiguous with respect to which aspect(s) of the picture they refer to: in 2 CUs (#44, 54), the speech output does not help disambiguate the identity/scope of the referent (e.g., CU #44 "grass"/"yard"/"shrubbery", and so forth), but in the remaining 12 CUs (#43, 45–53, 64), the speech output disambiguates toward one of two or more possible referents. For example, CU #45 where, e.g., "fellow" can refer to one of three entities (the man reading a book, the boy running with a kite, or the man fishing on the dock, see Figure 1B for examples), and based on the surrounding verbal context, it is possible to determine which of the referents the speaker is talking about. In an alternative CU coding scheme (CU-uniqueref, for "unique referent"), for this set of 12 CUs we broke each CU down into further CUs based on the referent in question. In this coding scheme, if the same word is used again but now refers to a different object/entity, it is coded as a new CU. This alternative coding method could reveal biases in terms of which objects/entities patients tend to refer to, which may be different from the controls' patterns and perhaps driven by the availability of different words.

Figure 1. Identifying the range of ambiguity of elements in the Western Aphasia Battery-Revised (WAB-R) "Picnic Scene" task (Kertesz, 2007) [34]. In the study, the original WAB-R Picnic Scene was used, but due to permissions issues, here, we use a freely available picture to illustrate non-exhaustive examples of the range of ambiguity of possible referents in a picture description task. (**A**) The examples highlighted here (CU #24 "glass," CU #25 "blanket", CU #30 "basket") are unambiguous entities similar to the ones found within the WAB-R Picnic Scene; (**B**) The examples highlighted here (CU #44 "shrubbery" (red) and CU #45 "fellow" (blue)) demonstrate the possible range of referents some CUs may represent. The image utilized for this example is freely available and reproduced with permission from Vecteezy (https://www.vecteezy.com/free-vector/picnic-basket).

For each of these two coding methods (the original Berube et al. (2019) [21] method and the CU-uniqueref method), we followed the original method by computing the total number of verbal descriptions that corresponded to CUs; thus, each participant was scored on the raw number of CUs. The counting of raw CUs in both the original method by Berube and colleagues (2019) [21] and our own CU-uniqueref method are non-normalized, and thus might be heavily influenced by sample length. This aspect of the CU coding process must be acknowledged as nfvPPA participants have been shown to produce less speech overall [17,18,23,24,28]. As such, between-group comparisons of the non-normalized measures of raw CU counts must be interpreted with caution and motivate a further, normalized analysis. For this reason, in contrast to the original Berube et al. (2019) [21] method, we also computed informativeness, i.e., raw CU count/all utterances, which places the raw CU count in the context of all the utterances produced by a given participant. Then, we used this normalized informativeness measure for subsequent interpretation of group differences.

2.5. Self-Referential, Empty, or Other Atypical Speech

To investigate the second research question regarding additional elements of speech samples that may contribute to reduced informativeness, such as self-referential or empty speech, we analyzed the speech samples for 5 additional types of output. A coder blinded to diagnostic group examined each sample for the presence of statements that (a) were self-referential (e.g., "I hate fish, I don't eat fish, my husband eats salmon all the time" or "We have one of these"), (b) referred to inability to decipher the picture or retrieve the right word (e.g., "I'm not quite sure what that person is" or "Can't think of that"), (c) were tangential to the contents of the picture (e.g., "Maybe an investor in a sailboat company" or "You don't take a picnic right next to your house usually"), (d) were "empty" speech (filler words or phrases, for example, e.g., "Uh no, yeah I guess", "um, uh"), or (e) were false starts (unsuccessful lexical retrieval attempts and meaningless phonemic cluster productions). As with our measure of informativeness, the atypical speech analyses were also normalized in that they were calculated as a proportion for every participant. Controlling for speech sample length was an essential component in order to reduce bias in our subgroup comparisons. Inter-rater reliability was assessed for 20 of the 101 of the transcripts (five from each group), where the presence of a rating and rating type were judged for every single word per transcript and was found to be 97%. Further examples of these ratings can be seen in Figure 2.

"You know playing uh kite I think I said that and then you got later h- he just wants his uh brew   h- he's is drinking, he is- he is in a book, I could never do that outside."

"I wonder what that is, wine or something, the kid is running with the dog, this guy is over there fishing … that's a house, and oh, kind of like my house, I think, I guess the water is right out here, house, car, tree, I have trees just like that."

"Ah the boy is running for the good kite, and uh, uh, grandfather is fishing, ah, is a sailboat ah I don't know whether they used them uh maybe an investor in a sailboat company ah there is a sail on a boat known as the four seventy ah ah I daughter is making a castle ah of sand."

"Flying, flying a-, is uh I can't remember what that called, uh there's a fisherman and a boat and some houses and car."

CODING
Self-Referential
Inability
Tangential
Empty Speech
False Start

Figure 2. Examples of speech excerpts that were coded as self-referential, reflecting inability to decipher the picture or retrieve the right word, tangential, empty, or reflective of a false start.

2.6. Statistical Analysis

We descriptively compared the number of raw CUs and informativeness (CUs/all utterances) across methods and between groups using both the original Berube et al. (2019) [21] and the CU-uniqueref methods. Then, we conducted one-way ANOVAs to examine the main effect of group on the number of CUs produced and the informativeness of speech samples, followed up by post hoc t-tests and Hedges' g effect sizes. Although length is a confound that must be considered in the interpretation of these ANOVAs, these analyses are still merited because there can be cases where even shorter sample lengths (as measured by total utterances) show relatively high numbers of CUs. Moreover, reduced sample length does not necessarily lead to reduced CU counts. While our measure of informativeness is normalized, we stress that our raw CU results are non-normalized and need to be interpreted with caution. To examine self-referential, empty, or other atypical speech, we calculated the proportion of words rated as belonging to one of five categories (self-referential, inability-related, tangential, empty speech, or false starts, see Section 2.5) relative to the total number of utterances. Then, we conducted one-way ANOVAs to examine the main effect of group on these 5 types of output. As in the prior analysis, post hoc *t*-tests and Hedge's *g* effect sizes were calculated for every comparison. We set alpha at 0.05 and corrected for multiple comparisons as described below for each analysis. Statistical analysis was performed using R (Version 3.5.3 (2019-03-11)) [37].

3. Results

3.1. CUs and Informativeness

The complete sets of CUs generated for the original Berube et al. (2019) [21] and our CU-uniqueref methods are shown in Table 2. In the original Berube et al. (2019) method, we were able to generate 64 CUs. In the CU-uniqueref method, our specification of referents added 35 additional possible CUs to be scored.

For 49 out of 101 transcripts (9 svPPA, 8 lvPPA, 8 nfvPPA, and 24 HC), the total number of CU scores increased using the CU-uniqueref method relative to the Berube et al. (2019) method [21]. For 87.8% of score changes, the increase was one or two points (see Appendix B Table A2). Similarly, the informativeness scores of 49 out of 101 transcripts (9 svPPA, 8 lvPPA, 8 nfvPPA, and 24 HC) increased with the CU-uniqueref method. For 18.0% of score increases, the change in informativeness was greater than 1.5%.

There was a significant effect of group on the non-normalized raw CU production for both the original (F(3, 97) = 36.2, $p < 0.001$) and CU-uniqueref (F(3, 97) = 35.6, $p < 0.001$) methods (see Table 3). Post hoc comparisons were conducted using pairwise t-tests using Bonferroni adjusted alpha levels of 0.008 (0.05/6) and reported alongside Hedges' *g* effect sizes. For both the original and CU-uniqueref methods, all three PPA variants demonstrated impaired production of unique CUs relative to HCs ($p < 0.001$ and $g > 1.28$). The nfvPPA participants produced fewer raw CUs than HCs, however, despite having the overall shorter speech samples relative to the other groups, raw CU count production was higher than that of the svPPA and lvPPA participants ($ps < 0.005$ and $gs > 0.96$, see Table 3). Across methods, there were no significant differences between lvPPA and svPPA.

Table 3. The average CU and informativeness scores derived from both the original Berube et al. (2019) [21] and CU-uniqueref methods for all participants.

Method	Group	CUs (SD)	CU Effect Size vs. HCs (Hedge's g)	Informativeness (SD)	Informativeness Effect Size vs. HCs (Hedge's g)
Original Berube et al. (2019) [21]	HC	28.5 (6.46)	-	20.9% (5.67)	-
	lvPPA	14.5 (5.81) ***♦	2.25	11.7% (5.18) ***♦	1.75
	nfvPPA	20.3 (6.07) ***	1.28	23.1% (9.45)	−0.29
	svPPA	13.0 (5.67) ***♦	2.47	11.1% (6.46) ***♦	1.69
CU-uniqueref	HC	30.0 (7.99)	-	21.9% (5.87)	-
	lvPPA	14.8 (6.07) ***♦	2.23	11.9% (5.26) ***♦	1.86
	nfvPPA	20.8 (6.19) ***	1.33	23.5% (9.31)	−0.22
	svPPA	13.7 (6.24) ***♦	2.34	11.7% (7.02) ***♦	1.65

*** $p < 0.001$ relative to HCs; ♦ $p < 0.005$ relative to nfvPPA.

Similarly, we found a significant effect of group on our normalized measure of informativeness (CUs/total utterances) for both the original (F(3, 97) = 20.6, $p < 0.001$) and CU-uniqueref (F(3, 97) = 21.1, $p < 0.001$) methods. As hypothesized, post hoc comparisons demonstrated that both svPPA and lvPPA exhibited impairment in the informativeness of speech relative to HCs ($p < 0.001$ and $g > 1.27$) across methods, whereas the informativeness of speech by the nfvPPA group was not impaired relative to HCs (see Figure 3). Across methods, the informativeness of nfvPPA output was greater than that of svPPA and lvPPA ($ps < 0.001$ and $gs > 1.40$), but there were no significant differences between svPPA and lvPPA. Contrary to one point in our first hypothesis, just as many individual lvPPA patients exhibited impaired informativeness as did svPPA patients (i.e., approximately eight or nine cases in each group fell within the 95% confidence interval for the HC distribution).

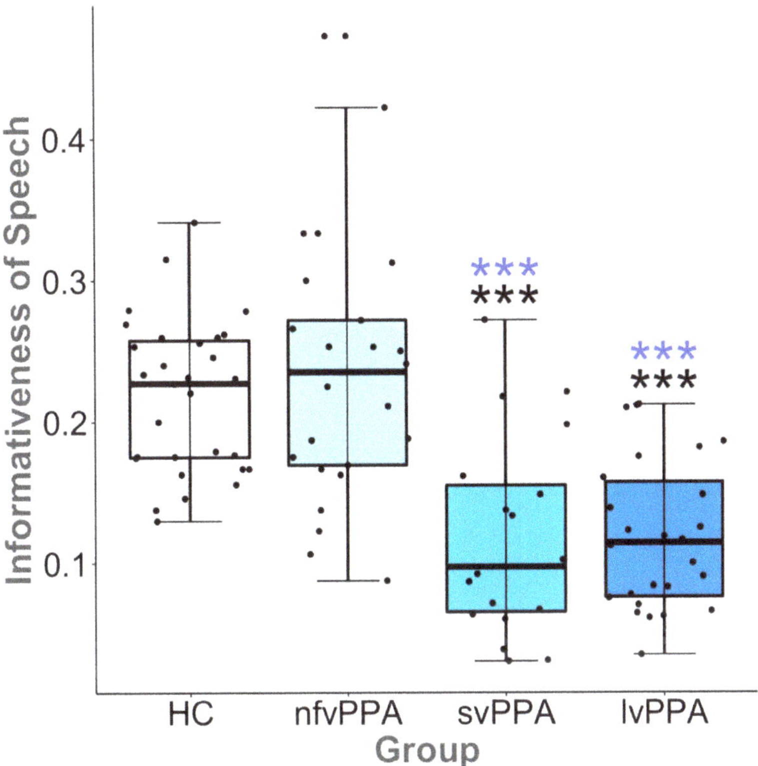

Figure 3. Illustrated here with the CU-uniqueref results, both the logopenic variant primary progressive aphasia (lvPPA) and semantic variant (svPPA) exhibited reduced informativeness of speech relative to controls, whereas the non-fluent/agrammatic variant (nfvPPA) group did not. The informativeness of speech output by the nfvPPA group was greater than that of lvPPA and svPPA. The error bars represent the 95% confidence interval of the median for each group. Significant differences are represented by asterisks (HC, black; nfvPPA, blue; *** $p < 0.001$).

3.2. Self-Referential, Empty, or Other Atypical Speech

There was a significant main effect of group for the proportion of statements that were self-referential ($F(3, 97) = 10.7$, $p < 0.001$) or the statements in which the patient described their inability to do the task ($F(3, 97) = 14.5$, $p < 0.001$), as well as empty speech ($F(3, 97) = 29.2$, $p < 0.001$) and false starts ($F(3, 97) = 17.9$, $p < 0.001$) (see Figure 4 and Appendix C Table A3). A group effect for tangential statements was not present ($F(3, 97) = 2.42$, $p = 0.071$). Post hoc comparisons demonstrated that, as hypothesized, the speech samples of svPPA patients contained a greater proportion of self-referential statements than HC, lvPPA, and nfvPPA ($ps < 0.001$ and $gs > 0.94$). There were no significant differences among the HC, lvPPA, and nfvPPA groups ($ps > 0.150$). For statements about inability, both svPPA and lvPPA made more comments about their own difficulties with the task than nfvPPA ($p < 0.001$ and $g > 1.09$) and HC ($ps < 0.001$ and $gs > 1.21$), whereas nfvPPA and HC did not differ ($p = 0.790$). Whereas the overall effect of group only trended towards significance, svPPA produced numerically more tangential statements than HC ($p = 0.058$), however, there were no differences among the lvPPA, nfvPPA, and HC groups ($ps > 0.536$).

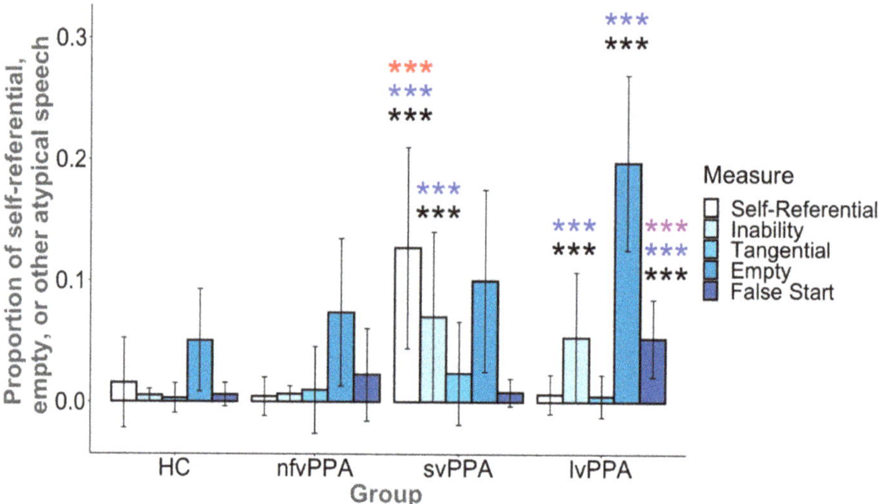

Figure 4. Relative to controls, the svPPA group exhibited more self-referential speech output and statements about inability, whereas the lvPPA group also exhibited more statements about inability, empty speech, and false starts. The nfvPPA group did not differ from HCs on any measure. The svPPA group exhibited more self-referential speech than lvPPA, whereas the lvPPA group exhibited more empty speech and false starts than svPPA. There were no significant group differences for tangential speech. The error bars represent one standard deviation from group means. Asterisks represent a significant difference relative to another group (HC, black; svPPA, purple; lvPPA, red; nfvPPA, blue; *** $p < 0.001$).

With regard to empty speech, as hypothesized, lvPPA produced significantly more empty output than HC ($p < 0.001$ and $g > 1.29$), svPPA and nfvPPA ($ps < 0.001$ and $g > 1.01$), whereas svPPA ($p = 0.046$) and nfvPPA did not differ from HC ($p = 0.115$). Finally, as hypothesized, lvPPA produced significantly more false starts than svPPA, nfvPPA, and HC ($ps < 0.001$ and $gs > 0.84$), however, there were no significant differences between svPPA and HC ($p = 0.503$), nfvPPA and HC ($p = 0.043$), nor svPPA and nfvPPA ($p = 0.082$).

4. Discussion

In this work, we asked two questions about functional communication in primary progressive aphasia (PPA). First, we asked whether patients in the mild stages of each of the three variants of PPA exhibit reduced ability to convey information in a picture description task. Second, we examined whether the increased production of types of speech output not directly relevant to the task may contribute to the reduction in speech informativeness. In particular, we asked whether patients with one of the three variants of PPA differ with respect to the production of self-referential, empty, or other atypical speech during this task. Such characteristics are often observed by clinicians and investigators, but they have not received sufficient attention in the prior literature.

To tackle the first question, we built on recent work by Berube et al. (2019) [21], who collected speech samples from individuals with aphasia using a picture description task and coded these for content units or CUs, i.e., concepts that are present in the output of healthy controls. Using this approach, Berube et al. (2019) [21] reported fewer CUs in the speech output of individuals with chronic stroke aphasia and of individuals with PPA. We proposed an adjustment to the coding method (which we termed CU-uniqueref) to disambiguate cases where the same CU could refer to one of multiple referents in the picture. Then, we compared the results of both of these methods. Furthermore, an additional unique contribution of our work was our normalized measure of informativeness, where we took participant-specific sample length into account. Given that raw CU counts do not account for semantically empty or irrelevant speech output, this measure more precisely targets functional communication ability as it reflects content efficiency. The reason this is important is that, for example, if a patient says five words or phrases that correspond to target concepts and nothing else, a listener could likely understand the patient's point even if the patient's speech is relatively sparse as compared with normal speech. In contrast, if another patient says five words or phrases that correspond to target concepts but also produces twenty words or phrases that include false starts, statements about how they cannot find the right word, and tangents, a listener would likely have greater difficulty understanding that person's point. Raw CU count would be the same between the two, but the informativeness would be much lower in the second case.

We replicated Berube et al.'s finding that PPA patients produce fewer CUs than controls [21]. Critically, as predicted, we observed important differences among the three PPA variants in their ability to convey information relevant to the picture. In particular, svPPA and lvPPA patients' speech demonstrated reduced informativeness, although the former group exhibited greater variability. The nfvPPA patients did not differ from controls in informativeness. These results suggest that grammatical impairment (a core feature of nfvPPA [6]) does not lead to a reduction in informativeness, whereas anomia (a core feature of both lvPPA and svPPA) and semantic memory impairment (a core feature of svPPA) likely do. Importantly, raw CU counts alone show reductions in all three variants relative to controls, consistent with prior work [17–20]; nonetheless, the magnitude of raw CUs produced by nfvPPA is still greater than for svPPA and lvPPA. Importantly, raw CU counts must be interpreted with caution as they do not reflect sample length. The risk of bias in the interpretation of raw CU counts motivated our measure of informativeness. Informativeness, the proportion of CUs to total utterances, demonstrates that the relative number of CUs that nfvPPA patients communicate is similar to that of healthy age-matched adults (i.e., nfvPPA patients do not produce more non-content related speech than controls, in contrast to the other two variants). While the content of nfvPPA speech samples were not diluted by atypical speech patterns (i.e., empty speech or false starts), the omission of closed-class words may have also contributed to preserved informativeness. This result speaks to the overall efficiency of nfvPPA speech relative to the other PPA variants. While previous work has reported upon the relative paucity of output in nfvPPA relative to the other variants [24,26,28], our results demonstrate that the amount of content relative to total output (i.e., informativeness) is preserved. This finding aligns with a recent systematic review [20], which concluded that word meanings and semantic structure appear to be largely preserved in this PPA variant. In our view, these findings demonstrate the value of examining both the raw volume of information, as well as the proportion of information in PPA. In line with prior reports of reduced informativeness [24,27,28,38], we found a reduction in informativeness in svPPA and lvPPA relative to controls and nfvPPA. These results are consistent with reports of reduced content word production in both of these variants [17–20].

To address the second question about additional factors that contribute to reduced informativeness of speech in svPPA and lvPPA, we quantified the presence of five types of task-irrelevant speech output, i.e., self-referential speech, speech referring to one's inability to perform the task, tangential statements, empty speech, and false starts. As predicted, svPPA patients produced a large number of self-referential utterances about their own lives related to the scene depicted in the picture but tangential to the task, in addition to statements about their inability to decipher the picture, retrieve a word, or accurately describe a referent in the picture. Similar to svPPA, lvPPA patients produced many statements related to their inability to perform the task, but in contrast to svPPA, they did not produce self-referential statements, and instead had a large amount of empty (uninformative or indecipherable) speech (see Section 2.5 and Figure 2 for examples), as well as numerous false starts. The latter were plausibly related to failed lexical retrieval attempts. Thus, the reasons for reduced informativeness in these two variants are at least partially dissociable. As for nfvPPA, where the level of informativeness was similar to that of controls, the production of task-irrelevant speech was also similar to that of controls, both qualitatively and quantitatively.

A limitation of this work is that our naturalistic speech productions were prompted through a visual aid, rather than an open-ended prompt. As such, our findings speak to the informativeness of naturalistic speech within the constraints of a specific task. As such, we were unable to examine either the cohesion or pragmatics of discourse [20,37]. Another limitation is that the length of the speech samples varied by variant and by participant, where the shortest were produced by the nfvPPA group. However, as our primary measures of the proportion of content (informativeness) and atypical speech to total output were normalized, we consider this to be a minor issue in the interpretation of our results. Further consideration must be given to our methods of coding atypical speech. While our inter-rater reliability was quite high, only 20% of the transcriptions were cross-checked. This brings us to our final limitation, i.e., our methodology required extensive hand-coding and could not be automated. Thus, the replication or upscaling of our procedure with a larger sample size would be time-intensive for both the initial coding and reliability checks. However, given the irregularities of speech output in PPA, our hand-coding allowed for sensitivities to task-relevant speech, such as circumlocutions, and empty speech at the phrase-level, features that are unique to the individual.

In conclusion, the current results demonstrate that functional communication assessed in a task that closely approximates everyday interactions is not ubiquitously impaired across the PPA variants in the mild stages of the disease. Whereas both svPPA and lvPPA produce fewer CUs than controls, they each produce a larger amount of less meaningful speech, the types of which are partially dissociable, leading to an overall reduction in the informativeness of communication. In contrast, nfvPPA patients produce fewer CUs than controls but the speech they produce is informative. These findings highlight the value of assessing functional communication using paradigms that elicit naturalistic speech, and the utility of scoring the reductions in task-relevant speech output and also the increases in task-irrelevant speech output. Future directions of this work include a longitudinal analysis of informativeness in naturalistic speech production to monitor changes in the different variants as the disease progresses, and the potential application of this approach to evaluate outcomes of speech-language therapy.

5. Conclusions

Naturalistic speech samples can be used to identify differences between PPA variants and to shed light on the nature of language impairments. In the current study, we found that the informativeness of speech varies across groups, and critically, the nfvPPA group performed similarly to controls. Similarly, atypical patterns of speech vary across the PPA variants, where the nfvPPA group performs similarly to controls, whereas there are differences between lvPPA and svPPA.

Author Contributions: Conceptualization, J.G., C.C., M.Q., A.T., and B.C.D.; Methodology, J.G., E.F., M.Q., C.C., D.H., and B.C.D.; Software, J.G.; Validation, C.C.; Formal analysis, J.G.; Investigation, J.G., C.C., M.Q., D.H., and B.C.D.; Resources, B.C.D.; Data curation, J.G.; Writing—original draft preparation, J.G.; Writing—review & editing, J.G., E.F., C.C., M.Q., A.T., and B.C.D.; Visualization, J.G.; Supervision, B.C.D.; Project administration, B.C.D.; Funding acquisition, B.C.D. All authors have read and agreed to the published version of the manuscript.

Funding: This work was supported by the National Institute of Deafness and Communication Disorders grants T32 DC-00038 (J.G.), R01 DC014296 (B.C.D.), K23 DC016912 (A.T.), F31 DC015703-02 (C.C.), and R01 DC013547-01 (Jordan R. Green). J.G. was additionally supported by the William Orr Dingwall Dissertation Fellowship in Cognitive, Clinical, and Neural Foundations awarded by the Dingwall Foundation.

Institutional Review Board Statement: The study was approved by the Mass General Brigham Healthcare System Institutional Review Boards which govern human subjects research at Massachusetts General Hospital (protocol no. 2015P001363 (10/23/15), protocol no. 2012P000432

(05/31/12), protocol no. 2016P001421 (11/10/16), protocol no. 2019P003391 (11/26/19), protocol no. 2016P001594 (08/16/16), and protocol no. 2013P001746 (11/07/13).

Informed Consent Statement: Informed consent was obtained from all subjects involved in the study.

Data Availability Statement: The data presented in this study are available on request from the corresponding author. The data are not publicly available as they contain information that could compromise the privacy of research participants.

Acknowledgments: We would like to thank Jordan R. Green, Director of the Speech and Feeding Disorders Laboratory at the MGH Institute of Health Professions, for sharing his naturalistic speech samples from the age-matched controls. Furthermore, we gratefully acknowledge the participation and contributions of our patients, without whom this work would not have been possible.

Conflicts of Interest: B.C.D.: Research support from NIH, Alzheimer's Drug Discovery Foundation; consulting for Acadia, Alector, Arkuda, Axovant, Biogen, Eisai, Life Molecular Sciences, Lilly, Merck, Novartis, Wave LifeSciences; editorial duties with payment for Elsevier (Neuroimage, Clinical and Cortex); royalties from Oxford University Press and Cambridge University Press. All other authors declare no conflict of interest.

Appendix A

Table A1. Participant-specific demographic information.

Group	Participant	Sex	Age	Highest Educational Degree	CDR: Language	PASS Sum of Boxes
Semantic variant PPA	svPPA 1	F	68	Master's	1	6
	svPPA 2	F	62	Bachelor's	1	3
	svPPA 3	F	53	Bachelor's	0.5	3
	svPPA 4	F	64	High School	0.5	3.5
	svPPA 5	M	54	Master's	0.5	1.5
	svPPA 6	M	59	Bachelor's	0.5	3
	svPPA 7	M	83	Bachelor's	0.5	5
	svPPA 8	F	63	Bachelor's	1	4.5
	svPPA 9	M	80	Bachelor's	0.5	2
	svPPA 10	F	63	High School	0.5	4
	svPPA 11	F	70	Master's	0.5	4.5
	svPPA 12	M	65	Bachelor's	1	5
	svPPA 13	M	81	Bachelor's	0.5	3
	svPPA 14	M	64	Doctorate	1	5
	svPPA 15	F	54	Doctorate	1	3
	svPPA 16	F	74	Bachelor's	0.5	2.5
	svPPA 17	F	72	High School	0.5	4
	svPPA 18	F	65	Bachelor's	1	3
	svPPA 19	M	68	Master's	1	1
Logopenic variant PPA	lvPPA 1	M	68	Bachelor's	1	5
	lvPPA 2	F	68	High School	0.5	3.5
	lvPPA 3	M	79	High School	0.5	3
	lvPPA 4	M	71	Master's	0.5	0.5
	lvPPA 5	M	71	Master's	0.5	3.5
	lvPPA 6	M	79	Master's	1	3.5
	lvPPA 7	F	59	Bachelor's	1	6
	lvPPA 8	F	70	Associate's	1	5
	lvPPA 9	F	59	Master's	1	6
	lvPPA 10	F	64	Bachelor's	0.5	5
	lvPPA 11	M	71	Master's	1	3.5
	lvPPA 12	M	72	High School	1	5
	lvPPA 13	F	53	Bachelor's	0.5	4.5
	lvPPA 14	F	68	High School	1	3
	lvPPA 15	F	75	High School	0.5	5

Table A1. *Cont.*

Group	Participant	Sex	Age	Highest Educational Degree	CDR: Language	PASS Sum of Boxes
	lvPPA 16	M	79	Bachelor's	0.5	4
	lvPPA 17	F	70	Bachelor's	0.5	3.5
	lvPPA 18	F	69	Bachelor's	0.5	3
	lvPPA 19	M	75	Bachelor's	1	5
	lvPPA 20	M	76	Master's	0.5	3.5
	lvPPA 21	F	78	Master's	0.5	5
	lvPPA 22	F	55	High School	0.5	5
	lvPPA 23	M	69	Doctorate	0.5	3
	lvPPA 24	M	72	Master's	1	4
	lvPPA 25	M	73	Bachelor's	1	5
	lvPPA 26	M	69	Bachelor's	0.5	2
Non-fluent variant PPA	nfvPPA 1	F	72	Bachelor's	0.5	3.5
	nfvPPA 2	M	64	Bachelor's	0.5	5.5
	nfvPPA 3	F	69	Master's	1	3
	nfvPPA 4	F	74	Bachelor's	0.5	5
	nfvPPA 5	M	60	Bachelor's	1	5
	nfvPPA 6	M	63	Bachelor's	0.5	4
	nfvPPA 7	M	70	Doctorate	0.5	2
	nfvPPA 8	F	69	Master's	1	3
	nfvPPA 9	F	63	High School	0.5	4
	nfvPPA 10	M	68	Bachelor's	1	4
	nfvPPA 11	F	75	Doctorate	0.5	2
	nfvPPA 12	M	67	Bachelor's	0.5	3.5
	nfvPPA 13	F	65	Bachelor's	0.5	5.5
	nfvPPA 14	F	79	High School	0.5	5.5
	nfvPPA 15	M	55	Master's	0.5	3
	nfvPPA 16	M	80	Master's	0.5	3
	nfvPPA 17	F	71	Master's	0.5	1
	nfvPPA 18	F	78	High School	0.5	3.5
	nfvPPA 19	M	69	Bachelor's	1	5
	nfvPPA 20	F	75	High School	0.5	3.5
	nfvPPA 21	F	79	High School	0.5	5
	nfvPPA 22	F	82	High School	1	6
	nfvPPA 23	M	77	High School	0.5	3
	nfvPPA 24	F	74	High School	1	4
	nfvPPA 25	F	62	Master's	1	3
Healthy Controls	HC 1	F	61	Bachelor's		
	HC 2	F	61	High School		
	HC 3	M	53	Bachelor's		
	HC 4	F	50	High School		
	HC 5	M	52	Bachelor's		
	HC 6	M	51	Bachelor's		
	HC 7	M	68	Bachelor's		
	HC 8	F	71	Bachelor's		
	HC 9	F	71	Bachelor's		
	HC 10	M	62	High School		
	HC 11	F	71	Bachelor's		
	HC 12	F	68	Bachelor's		
	HC 13	M	69	Bachelor's		
	HC 14	M	53	Bachelor's		
	HC 15	F	55	Bachelor's		
	HC 16	M	68	Bachelor's		
	HC 17	F	66	High School		
	HC 18	M	65	Bachelor's		
	HC 19	F	62	Bachelor's		

Table A1. *Cont.*

Group	Participant	Sex	Age	Highest Educational Degree	CDR: Language	PASS Sum of Boxes
	HC 20	F	66	Bachelor's		
	HC 21	M	59	Bachelor's		
	HC 22	F	57	High School		
	HC 23	M	54	Bachelor's		
	HC 24	F	60	Bachelor's		
	HC 25	F	72	High School		
	HC 26	M	73	Bachelor's		
	HC 27	F	55	High School		
	HC 28	F	65	Bachelor's		
	HC 29	F	63	Bachelor's		
	HC 30	M	76	High School		
	HC 31	M	83	Bachelor's		

Appendix B

Table A2. The number of participants whose score increased by 1-4 points with the CU-uniqueref method.

Group	0 pt.	1 pt.	2 pt.	3 pt.	4 pt.	No. Participants Changed
HC	7	9	11	2	2	24
lvPPA	18	7	0	1	0	8
nfvPPA	17	4	4	0	0	8
svPPA	10	6	2	1	0	9

Appendix C

Table A3. The proportion of self-referential and other off-topic or empty speech.

Group	Self-Referential Statements	SD	Ability Statements	SD	Tangential Statements	SD	Empty Speech	SD	False Starts	SD
HC	0.016	0.037	0.005	0.013	0.003	0.012	0.051	0.042	0.006	0.009
lvPPA	0.006	0.016	0.053	0.056	0.005	0.017	0.198	0.072	0.053	0.032
nfvPPA	0.004	0.016	0.007	0.019	0.009	0.036	0.074	0.061	0.023	0.038
svPPA	0.127	0.183	0.070	0.067	0.024	0.043	0.100	0.075	0.008	0.011

References

1. Mesulam, M.-M. Primary Progressive Aphasia—A Language-Based Dementia. *N. Engl. J. Med.* **2003**, *349*, 1535–1542. [CrossRef] [PubMed]
2. Neary, D.; Snowden, J.S.; Gustafson, L.; Passant, U.; Stuss, D.; Black, S.; Freedman, M.; Kertesz, A.; Robert, P.H.; Albert, M.; et al. Frontotemporal lobar degeneration: A consensus on clinical diagnostic criteria. *Neurology* **1998**, *51*, 1546–1554. [CrossRef] [PubMed]
3. Mesulam, M.-M. Primary progressive aphasia. *Ann. Neurol.* **2001**, *49*, 425–432. [CrossRef]
4. Gorno-Tempini, M.L.; Dronkers, N.F.; Rankin, K.P.; Ms, J.M.O.; Ba, L.P.; Rosen, H.J.; Johnson, J.K.; Weiner, M.W.; Miller, B.L. Cognition and anatomy in three variants of primary progressive aphasia. *Ann. Neurol.* **2004**, *55*, 335–346. [CrossRef] [PubMed]
5. Gorno-Tempini, M.L.; Brambati, S.M.; Ginex, V.; Ogar, J.; Dronkers, N.F.; Marcone, A.; Perani, D.; Garibotto, V.; Cappa, S.F.; Miller, B.L. The logopenic/phonological variant of primary progressive aphasia. *Neurology* **2008**, *71*, 1227–1234. [CrossRef] [PubMed]
6. Gorno-Tempini, M.L.; Hillis, A.E.; Weintraub, S.; Kertesz, A.; Mendez, M.; Cappa, S.F.; Ogar, J.M.; Rohrer, J.D.; Black, S.; Boeve, B.F.; et al. Classification of primary progressive aphasia and its variants. *Neurology* **2011**, *76*, 1006–1014. [CrossRef] [PubMed]
7. Sapolsky, D.; Domoto-Reilly, K.; Negreira, A.; Brickhouse, M.; McGinnis, S.; Dickerson, B.C. Monitoring progression of primary progressive aphasia: Current approaches and future directions. *Neurodegener. Dis. Manag.* **2011**, *1*, 43–55. [CrossRef]
8. Grossman, M. Linguistic Aspects of Primary Progressive Aphasia. *Annu. Rev. Linguist.* **2018**, *4*, 377–403. [CrossRef]
9. Beales, A.; Whitworth, A.; Cartwright, J.; Panegyres, P.K.; Kane, R.T. Determining stability in connected speech in primary progressive aphasia and Alzheimer's disease. *Int. J. Speech Lang. Pathol.* **2018**, *20*, 361–370. [CrossRef]

10. Sapolsky, D.; Domoto-Reilly, K.; Dickerson, B.C. Use of the Progressive Aphasia Severity Scale (PASS) in monitoring speech and language status in PPA. *Aphasiology* **2014**, *28*, 993–1003. [CrossRef]
11. Grasso, S.M.; Henry, M.L. Assessment of Individuals with Primary Progressive Aphasia. *Semin. Speech Lang.* **2018**, *39*, 231–241. [CrossRef] [PubMed]
12. Armstrong, E.; Ferguson, A. Language, meaning, context, and functional communication. *Aphasiology* **2010**, *24*, 480–496. [CrossRef]
13. Linnik, A.; Bastiaanse, R.; Höhle, B. Discourse production in aphasia: A current review of theoretical and methodological challenges. *Aphasiology* **2016**, *30*, 765–800. [CrossRef]
14. Doedens, W.J.; Meteyard, L. Measures of functional, real-world communication for aphasia: A critical review. *Aphasiology* **2019**, *34*, 492–514. [CrossRef]
15. Volkmer, A.; Spector, A.; Meitanis, V.; Warren, J.D.; Beeke, S. Effects of functional communication interventions for people with primary progressive aphasia and their caregivers: A systematic review. *Aging Ment. Health* **2020**, *24*, 1381–1393. [CrossRef]
16. Pritchard, M.; Hilari, K.; Cocks, N.; Dipper, L. Reviewing the quality of discourse information measures in aphasia. *Int. J. Lang. Commun. Disord.* **2017**, *52*, 689–732. [CrossRef]
17. Wilson, S.M.; Henry, M.L.; Besbris, M.; Ogar, J.M.; Dronkers, N.F.; Jarrold, W.; Miller, B.L.; Gorno-Tempini, M.L. Connected speech production in three variants of primary progressive aphasia. *Brain* **2010**, *133*, 2069–2088. [CrossRef]
18. Ash, S.; Evans, E.; O'Shea, J.; Powers, J.; Boller, A.; Weinberg, D.; Haley, J.; McMillan, C.; Irwin, D.J.; Rascovsky, K.; et al. Differentiating primary progressive aphasias in a brief sample of connected speech. *Neurology* **2013**, *81*, 329–336. [CrossRef]
19. Bryant, L.; Ferguson, A.; Spencer, E. Linguistic analysis of discourse in aphasia: A review of the literature. *Clin. Linguist. Phon.* **2016**, *30*, 489–518. [CrossRef]
20. Boschi, V.; Catricalà, E.; Consonni, M.; Chesi, C.; Moro, A.; Cappa, S.F. Connected Speech in Neurodegenerative Language Disorders: A Review. *Front. Psychol.* **2017**, *8*, 269. [CrossRef]
21. Berube, S.; Nonnemacher, J.; Demsky, C.; Glenn, S.; Saxena, S.; Wright, A.; Tippett, D.C.; Hillis, A. Stealing Cookies in the Twenty-First Century: Measures of Spoken Narrative in Healthy Versus Speakers With Aphasia. *Am. J. Speech Lang. Pathol.* **2019**, *28*, 321–329. [CrossRef]
22. Yorkston, K.M.; Beukelman, D.R. An Analysis of Connected Speech Samples of Aphasic and Normal Speakers. *J. Speech Hear. Disord.* **1980**, *45*, 27–36. [CrossRef] [PubMed]
23. Fraser, K.C.; Meltzer, J.A.; Graham, N.L.; Leonard, C.; Hirst, G.; Black, S.E.; Rochon, E. Automated classification of primary progressive aphasia subtypes from narrative speech transcripts. *Cortex* **2014**, *55*, 43–60. [CrossRef]
24. Ash, S.; Grossman, M. Why study connected speech production? In *Cognitive Neuroscience of Natural Language Use*; Willems, R.M., Ed.; Cambridge University Press: Cambridge, UK, 2015; pp. 29–58.
25. Meteyard, L.; Patterson, K.E. The relation between content and structure in language production: An analysis of speech errors in semantic dementia. *Brain Lang.* **2009**, *110*, 121–134. [CrossRef] [PubMed]
26. Sajjadi, S.A.; Patterson, K.; Tomek, M.; Nestor, P.J. Abnormalities of connected speech in the non-semantic variants of primary progressive aphasia. *Aphasiology* **2012**, *26*, 1219–1237. [CrossRef]
27. Ash, S.; Moore, P.; Antani, S.; McCawley, G.; Work, M.; Grossman, M. Trying to tell a tale: Discourse impairments in progressive aphasia and frontotemporal dementia. *Neurology* **2006**, *66*, 1405–1413. [CrossRef] [PubMed]
28. Graham, N.L.; Patterson, K.; Hodges, J.R. When More Yields Less: Speaking and Writing Deficits in Nonfluent Progressive Aphasia. *Neurocase* **2004**, *10*, 141–155. [CrossRef]
29. Rogalski, E.; Cobia, D.; Harrison, T.M.; Wieneke, C.; Thompson, C.K.; Weintraub, S.; Mesulam, M.-M. Anatomy of Language Impairments in Primary Progressive Aphasia. *J. Neurosci.* **2011**, *31*, 3344–3350. [CrossRef]
30. Cordella, C.; Dickerson, B.C.; Quimby, M.; Yunusova, Y.; Green, J.R. Slowed articulation rate is a sensitive diagnostic marker for identifying non-fluent primary progressive aphasia. *Aphasiology* **2017**, *31*, 241–260. [CrossRef]
31. Cordella, C.; Quimby, M.; Touroutoglou, A.; Brickhouse, M.; Dickerson, B.C.; Green, J.R. Quantification of motor speech impairment and its anatomic basis in primary progressive aphasia. *Neurology* **2019**, *92*, e1992–e2004. [CrossRef]
32. Knibb, J.A.; Woollams, A.M.; Hodges, J.R.; Patterson, K. Making sense of progressive non-fluent aphasia: An analysis of conversational speech. *Brain* **2009**, *132*, 2734–2746. [CrossRef] [PubMed]
33. Knopman, D.S.; Weintraub, S.; Pankratz, V.S. Language and behavior domains enhance the value of the clinical dementia rating scale. *Alzheimer's Dement.* **2011**, *7*, 293–299. [CrossRef] [PubMed]
34. Kertesz, A. *WAB-R: Western Aphasia Battery-Revised*; PsychCorp: Toronto, ON, Canada, 2007.
35. Audacity Team. Audacity(R): Free Audio Editor and Recorder, Version 2.4.1. 2014. Available online: http://audacity.sourceforge.net/ (accessed on 21 April 2020).
36. Agis, D.; Goggins, M.B.; Oishi, K.; Oishi, K.; Davis, C.; Wright, A.; Kim, E.H.; Sebastian, R.; Tippett, D.C.; Faria, A.; et al. Picturing the Size and Site of Stroke With an Expanded National Institutes of Health Stroke Scale. *Stroke* **2016**, *47*, 1459–1465. [CrossRef]
37. R Core Team. *R: A Language and Environment for Statistical Computing*; R Foundation for Statistical Computing: Vienna, Austria, 2014; Available online: http://www.R-project.org/ (accessed on 21 April 2020).
38. Sajjadi, S.A.; Patterson, K.; Tomek, M.; Nestor, P.J. Abnormalities of connected speech in semantic dementia vs Alzheimer's disease. *Aphasiology* **2012**, *26*, 847–866. [CrossRef]

MDPI

St. Alban-Anlage 66

4052 Basel

Switzerland

Tel. +41 61 683 77 34

Fax +41 61 302 89 18

www.mdpi.com

Brain Sciences Editorial Office

E-mail: brainsci@mdpi.com

www.mdpi.com/journal/brainsci

9 783036 544083